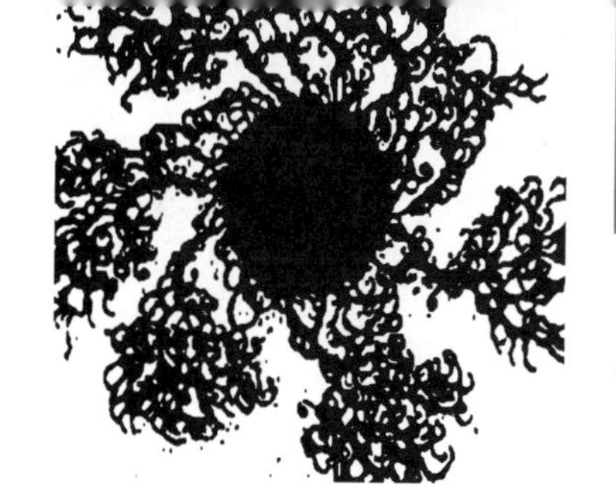

FIFTH EDITION

MICROBIOLOGY for the Health Sciences

Gwendolyn R. W. Burton, Ph.D.
Professor Emeritus
Department of Science
Environmental Science and Technology Department
Front Range Community College
Westminster, Colorado

Paul G. Engelkirk, Ph.D.
Faculty
Department of Science
Central Texas College
Killeen, Texas

Acquisitions Editor: Andrew Allen
Production Editor: Molly E. Dickmeyer
Production Services: Caslon, Inc.
Compositor: The Composing Room of Michigan, Inc.
Printer/Binder: R. R. Donnelley & Sons Company
Color Insert and Cover Printer: Lehigh Press
Cover Designer: Louis Fuiano

Fifth Edition

Library of Congress Cataloging in Publication Data

Burton, Gwendolyn R. W. (Gwendolyn R. Wilson)
Microbiology for the health sciences / Gwendolyn R. W. Burton,
Paul G. Engelkirk. — 5th ed.
p. cm.
Includes bibliographical references and index.
ISBN 0-397-55187-8
1. Microbiology. 2. Medical microbiology. I. Engelkirk, Paul G.
QR41.2.B88 1995
616'.01—dc20 95-38570
CIP

The material contained in this volume was submitted as previously unpublished material, except in the instances in which credit has been given to the source from which some of the illustrative material was derived.

Any procedure or practice described in this book should be applied by the health-care practitioner under appropriate supervision in accordance with professional standards of care used with regard to the unique circumstances that apply in each practice situation. Care has been taken to confirm the accuracy of information presented and to describe generally accepted practices. However, the authors, editors, and publisher cannot accept any responsibility for errors or omissions of for any consequences from application of the information in this book and make no warranty, express or implied, with respect to the contents of the book.

The authors and publisher have exerted every effort to ensure that drug selection and dosage set forth in this text are in accordance with current recommendations and practice at the time of publication. However, in view of ongoing research, changes in government regulations, and the constant flow of information relating to drug therapy and drug reactions, the reader is urged to check the package insert for each drug for any change in indications and dosage and for added warnings and precautions. This is particularly important when the recommended agent is a new or infrequently employed drug.

9 8 7 6 5 4 3 2 1

The cover image illustrates a colony of *Bacillus subtilis* and is based on a photograph courtesy of Dr. Eshel Ben-Jacob, University of Tel Aviv, Tel Aviv, Israel. Previously published in Ben-Jacob E. Bacterial Chatter: how patterns reveal clues about bacteria's chemical communication. Science News 1995;147(9):136.

The color plates were previously published in Koneman EW, Allen SD, Janda WM, Schreckenberger PC, Winn WC. Color atlas and textbook of diagnostic microbiology, 4th ed. Philadelphia, JB Lippincott, 1992.

For our spouses, Lynn and Janet, and our extended families

ABOUT THE AUTHORS

Dr. Gwendolyn R. Wilson Burton is retired Chairperson and Professor Emeritus of the Science and the Environmental Science and Technology Departments at Front Range Community College, Westminster, Colorado. She has studied at Colorado State University, the University of Oklahoma, and the University of Denver. She is a member of the American Society for Microbiology and a Past President of the Rocky Mountain Branch of that organization. Honors presented to Dr. Burton include the Academic Excellence Science Award from the American Association of Community and Junior Colleges, Outstanding Educators of America, and Distinguished Leadership Award for her many accomplishments in the educational fields of hazardous materials and microbiology.

Dr. Paul G. Engelkirk is a faculty member in the Science Department at Central Texas College in Killeen, Texas, where he teaches introductory microbiology to approximately 300 students per year. Before joining Central Texas College, he was an Associate Professor at the University of Texas Health Science Center in Houston, Texas, where he taught diagnostic microbiology to medical technology students for 7 years. Before that, Dr. Engelkirk spent 22 years in the U.S. Army Medical Department as a clinical microbiologist and clinical laboratory officer; he retired with the rank of Lieutenant Colonel. Dr. Engelkirk is the author or co-author of 3 textbooks, 10 additional book chapters, 2 self-study courses, and more than a dozen scientific articles. Together with his wife, Dr. Janet Duben-Engelkirk, he writes and publishes two quarterly newsletters, *Anaerobe Abstracts* and *Clinical Parasitology Abstracts.* Dr. Engelkirk has been a registered medical technologist for more than 30 years, a clinical microbiologist for more than 25 years, and is a Past President of the Rocky Mountain Branch of the American Society for Microbiology.

PREFACE

Microbiology, the study of microorganisms, is a fascinating topic to those of us who feel its importance in our daily lives. Others find it necessary to learn microbiologic concepts and vocabulary to function well in their chosen vocations. For example, those who plan to work in any area of health care, such as prevention of disease or treatment of disease, must be aware of the principles of sterilization, infectious disease causation, and infectious disease prevention.

Microbiology for the Health Sciences will aid those who want to learn the basic microbiologic concepts that apply to the health-care field. It is also intended for students who have little or no science background and for mature students returning to school after an absence of several years.

There is a need for a fundamental microbiology text that presents major concepts clearly and concisely for persons entering the health-care occupations. This book is appropriate for use in a one-term, allied health microbiology class or as one unit in a basic science class for health-care–oriented students.

Microbiology is an enormous and complex subject with many interrelated facets and hundreds of scientific terms. The authors have attempted a very fundamental approach to the subject matter by presenting, at the beginning of each chapter, the basic information necessary to understand the more complex concepts with which the chapters conclude. Specialized vocabulary has been kept to a minimum. New terms are listed at the beginning of each chapter for easy reference and are highlighted and defined in the text. A complete glossary, with pronunciation keys, can be found at the end of the book. Objectives and a brief outline are presented at the beginning of each chapter to enable the student to survey the topics covered. Discussion questions and self tests are included for review at the end of each chapter.

In this fifth edition, some chapters have been rearranged for better continuity. The whole book has been updated and partially rewritten for clarity. Color figures have been added, and each chapter contains an Insight Box. These provide a more detailed look at particular aspects of the topics being discussed in the chapters. The student workbook has been incorporated into each chapter to encourage students to review the material and to take the self test while learning the information. In this way, the student may have greater insight into the important facts that will be stressed on examinations.

Although this book is intended primarily for nonscience majors, it is not an easy text—microbiology is not an easy topic. As the student will discover, the concise nature of this text has made each sentence significant. Thus, the reader will be intellectually challenged to learn each new concept as it is presented. It is our hope that students will enjoy their study of microbiology and be motivated to further explore this fascinating field, especially as it relates to their occupations.

We are deeply indebted to those colleagues, friends, family members, and students who provided illustrations used in the text (especially Elmer Koneman, MD) or who served as sources of advice and encouragement, especially Pat Hidy and our spouses and children. We particularly wish to acknowledge Andrew Allen, sponsoring editor, and Laura Dover, editorial assistant at Lippincott-Raven Publishers Company, for their editorial assistance in the preparation of this manuscript as well as Caslon, Inc. for designing and coordinating the production of this book.

Gwendolyn R. W. Burton, Ph.D.
Paul G. Engelkirk, Ph.D.

CONTENTS

APPENDICES

MICROBIOLOGY
for the Health Sciences

CHAPTER 1

Introduction to Microbiology

OBJECTIVES

After studying this chapter, you should be able to

- Define microbiology
- List some important functions of microbes in the environment
- Explain the relevance of microbiology to the health professions
- List some areas of microbiological study
- Outline some contributions of Leeuwenhoek, Pasteur, and Koch to microbiology
- Explain the biological theory of fermentation
- Explain the germ theory of disease
- Learn Koch's postulates and cite some circumstances in which they may not apply
- Describe the differences between light microscopes and electron microscopes and the applications of both
- List the metric units used in microscopic measurements and indicate their relative sizes

CHAPTER OUTLINE

Microbiology, the Science
The Scope of Microbiology
- *General Microbiology*
- *Medical Microbiology*
- *Veterinary Microbiology*
- *Agricultural Microbiology*
- *Sanitary Microbiology*
- *Industrial Microbiology*
- *Microbial Physiology and Genetics*
- *Environmental Microbiology*

Milestones of Microbiology

The Tools of Microbiology
Microscopes
- *Light Microscope*
- *Electron Microscope*

Units of Measure

NEW TERMS

Abiogenesis
Algae (sing. *alga*)
Angstrom
Antiseptic surgery
Archaebacteria
Aseptic techniques
Bacteria (sing. *bacterium*)
Bacteriologist
Bacteriology
Biogenesis
Compound microscope
Contagious
Cyanobacteria
Electron microscope
Endospore
Eucaryotes
Fungi (sing. *fungus*)
Indigenous microflora
Infectious disease
Koch's postulates
Light microscope
Microbiologist
Microbiology
Micrometer
Microorganisms (microbes)
Microscope
Mycologist
Mycology
Nanometer
Nonpathogen
Opportunists
Parasite
Pasteurization
Pathogen
Petri dish
Phycologist
Phycology
Prion
Procaryotes
Protozoa (sing. *protozoan*)
Protozoologist
Protozoology
Resolving power
Saprophyte
Simple microscope
Sterile techniques
Tyndallization
Variolation
Viroid
Virologist
Virology
Viruses (sing. *virus*)

MICROBIOLOGY, THE SCIENCE

What is *microbiology? Micro* means very small, anything so small that it must be viewed with a microscope; *bio* refers to living organisms; and *ology* means "the study of." Therefore, microbiology is the study of very small living organisms. These *microorganisms* include *bacteria, algae, protozoa, fungi,* and *viruses* (Fig. 1–1). They are often called *microbes,* single-celled organisms, and germs.

You cannot see microorganisms without the aid of a *microscope,* and you may not be aware of the effect they have on your daily life. Occasionally, you may become conscious of their effect on your body when a cut or burn becomes infected or when you have a sore throat. Have you ever been very sick after a picnic and wondered which of the foods you ate contained harmful germs?

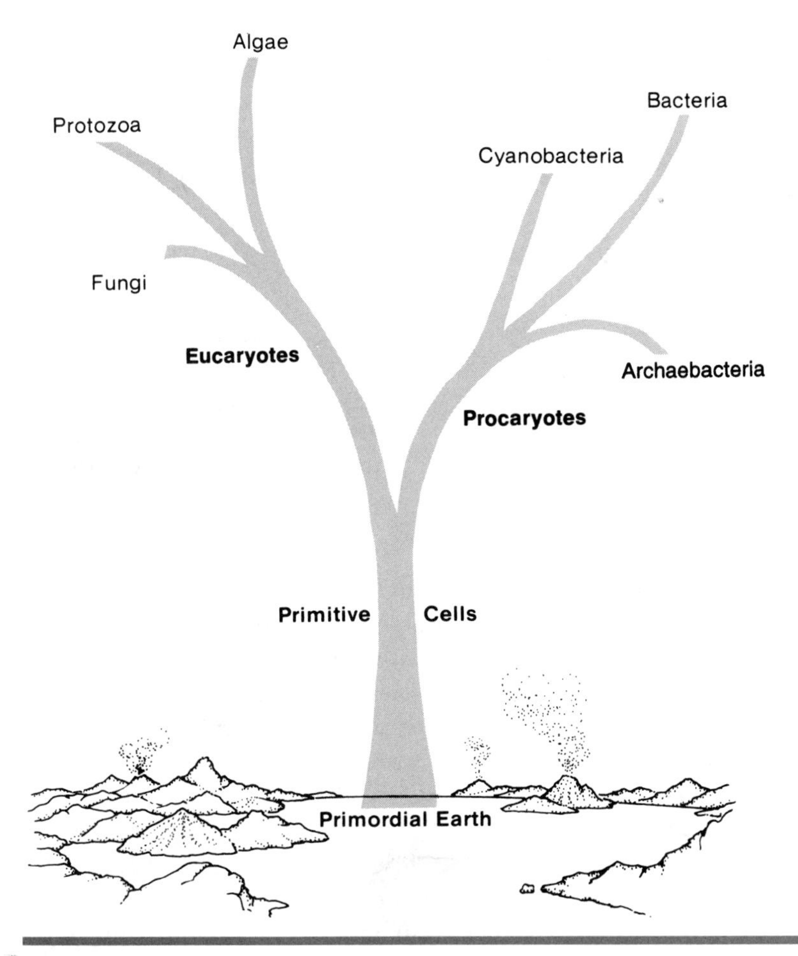

FIGURE 1 – 1
Family tree of microorganisms. The primitive cells are divided into *eucaryotes* (organisms with a true nucleus, such as algae, fungi, and protozoa) and *procaryotes* (organisms without a true nucleus, such as *archaebacteria,* bacteria, and *cyanobacteria*). Although viruses are microbes, they are not considered to be cells (they are acellular) and are, therefore, not included on this family tree. (see Chapter 2).

Most of us are aware of *pathogens,* or disease-causing microorganisms, only when we are affected by them. Actually, only a small percentage of microbes are pathogenic, that is, capable of causing disease (Fig. 1 – 2). The others (*nonpathogens*) are considered beneficial or harmless. Some cause disease only if they accidentally invade the "wrong place" at the "right time," such as when the host's resistance is low and growth conditions are right. These microbes are considered opportunistic. Usually, these *opportunistic pathogens* (*opportunists*) are microbes that live on and in the healthy human body (*e.g.,* on the skin, in

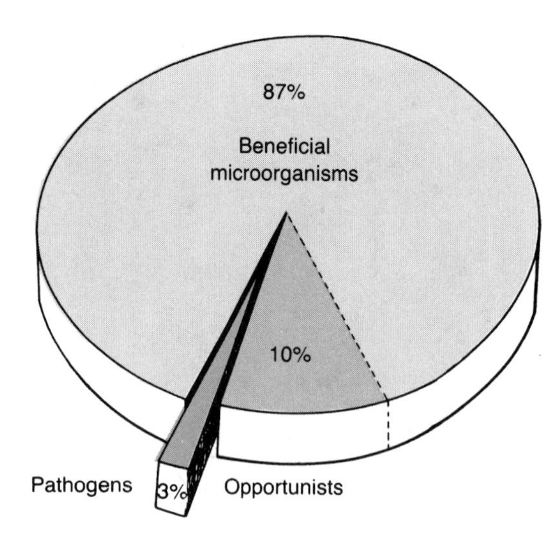

FIGURE 1 – 2
Pathogens comprise about 3% of all microorganisms. Another 10% are opportunistic pathogens that may cause disease if they land in an appropriate place.

the mouth, and in the intestine). They are called the *indigenous microflora* or *indigenous microbiota.* They ordinarily cause humans no harm, but because many of them are potentially pathogenic, they can cause disease if they gain entrance to wounds, the bloodstream, or organs such as the urinary bladder. Diseases caused by microorganisms are called *infectious diseases.*

Microorganisms can be found nearly everywhere and with relatively few exceptions, they contribute to the welfare of humans. The indigenous microflora actually inhibit the growth of pathogens in those areas of the body where they live by occupying the space, using the food supply, and secreting materials (waste products, toxins, antibiotics) that may prevent or reduce the growth of pathogens. Other nonpathogens make it possible to produce yogurt, cheese, raised bread, beer, wine, and many other foods and beverages.

Many bacteria and fungi are *saprophytes,* which aid in fertilization by returning inorganic nutrients to the soil. Saprophytes break down dead and dying organic materials (plants and animals) into nitrates, phosphates, carbon dioxide, water, and other chemicals necessary for plant growth (Fig. 1 – 3). These saprophytes also destroy paper, feces, and other biodegradable substances, although they cannot break down most plastics and glass. Nitrogen-fixing bacteria, which live with certain plants called legumes (peas, peanuts, alfalfa, clover), are able to return nitrogen from the air to the soil in the form of nitrates for use by other plants (Fig. 1 – 4). Knowledge of these important microbes is important to farmers who practice crop rotation to replenish their fields and to gardeners who use compost as a natural fertilizer. In both cases, dead organic material is broken down into inorganic nutrients (nitrates and phosphates) by microorganisms.

The purification of waste water is partially accomplished by bacteria in the holding tanks of sewage disposal plants, where feces, garbage, and other or-

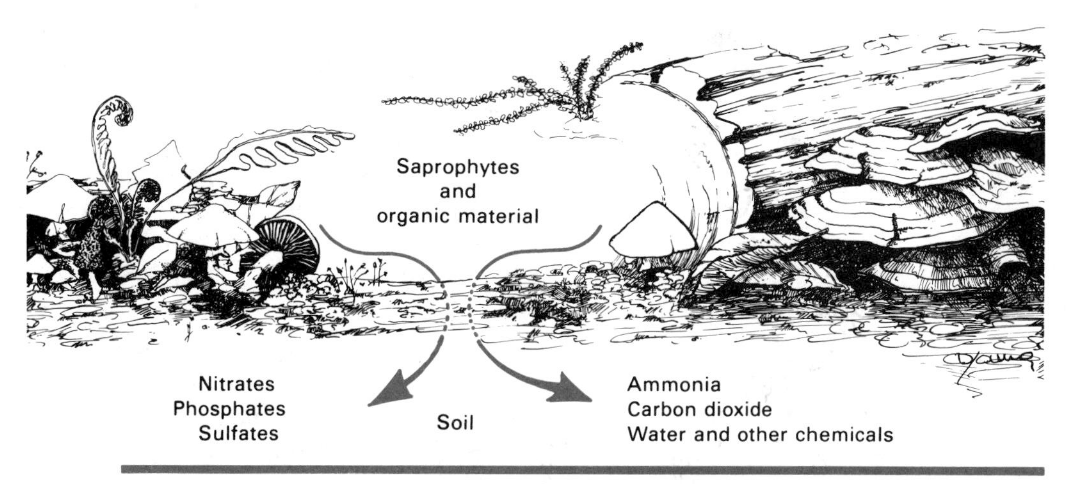

FIGURE 1 – 3
Saprophytes break down dead and decaying organic material into inorganic nutrients in the soil.

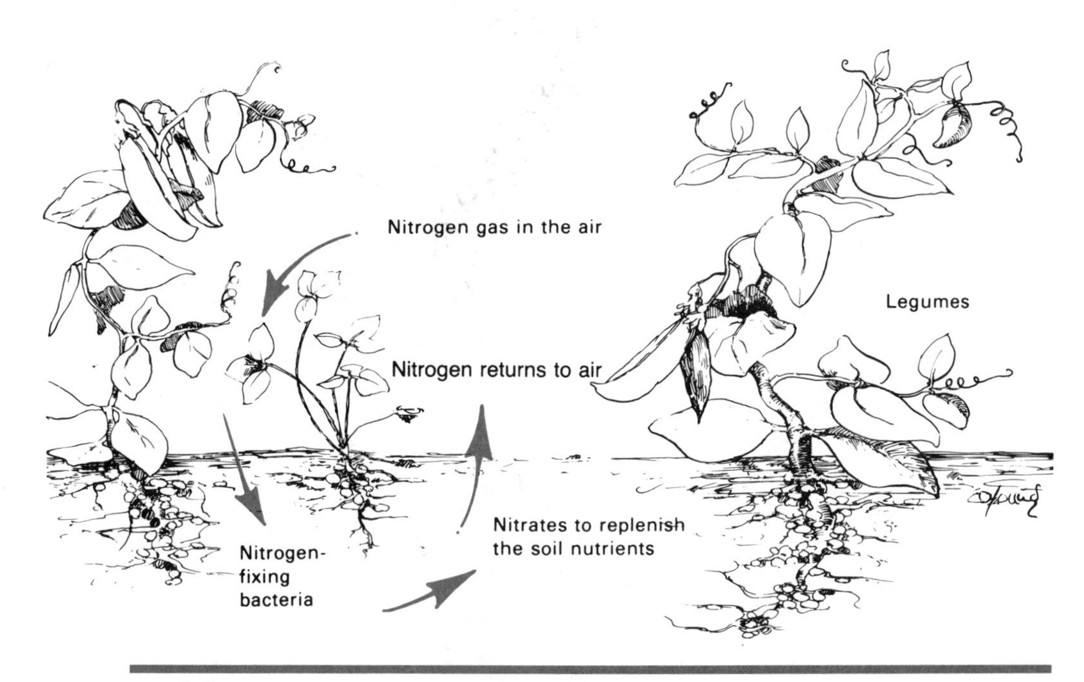

FIGURE 1 – 4
Nitrogen fixation. The nitrogen-fixing bacteria that live on or near the roots of legumes convert free nitrogen from the air into nitrates to replenish the soil nutrients.

ganic materials are collected and reduced to harmless waste. Some microorganisms, such as the iron- and sulfur-utilizing bacteria, even break down metals and minerals. The beneficial activities of microbes affect every part of our environment—in the land, water, and air.

Those who work in the health professions must be particularly aware of pathogens, their sources, and how they may be transmitted from one person to another. Physicians' assistants, dental assistants, nurses, laboratory technicians, respiratory therapists, orderlies, nurses' aides, and all others associated with patient care must take precautions to prevent the spread of pathogens. Harmful microorganisms may be transferred from health workers to patients; from patient to patient; from contaminated mechanical devices, instruments, and syringes to patients; from contaminated bedding, clothes, dishes, and food to patients; and from patients to health workers and other susceptible persons.

The Scope of Microbiology

There are many fields of study within microbiology. One may specialize in the study of many different types of microorganisms. For example, a *bacteriologist* concentrates on *bacteriology,* the study of structure, functions, and activities of bacteria. *Phycology* is the study of the various types of algae by scientists called *phycologists.* Those who specialize in the study of fungi, or *mycology,* are called *mycologists. Protozoologists* explore the area of *protozoology,* the study of protozoa and their activities. *Virology* encompasses the study of viruses and their effects on living cells of all types. *Virologists* and cell biologists may become genetic engineers who manipulate the genetic material (deoxyribonucleic acid or DNA) from one cell to another. Virologists may also study *prions* and *viroids,* infectious agents that are even smaller than viruses.

Within the general field of microbiology there are many specialized areas in which a knowledge of all types of microorganisms and of their applications is important. These areas include medical, veterinary, and agricultural microbiology as well as applications in sewage disposal, industrial production, space research, microbial ecology, biodegradation, and genetic engineering.

General Microbiology The study and classification of microorganisms and how they function is known as general microbiology. It encompasses all areas of microbiology.

Medical Microbiology The field of medical microbiology involves the study of pathogens, the diseases they cause, and the body's defenses against disease. This field is concerned with epidemiology, transmission of pathogens, disease-prevention measures, aseptic techniques, treatment of infectious diseases, immunology, and the production of vaccines to protect against infectious diseases. The complete or almost complete eradication of diseases like smallpox and diphtheria, the safety of modern surgery, and the treatment of victims of acquired immunodeficiency syndrome (AIDS) are due to the many technological advances in this field.

Veterinary Microbiology The spread and control of infectious diseases among animals is the concern of veterinary *microbiologists*. The production of food from livestock, the raising of other agriculturally important animals, the care of pets, and the transmission of diseases from animals to humans are areas of major importance in this field.

Agricultural Microbiology Included in the field of agricultural microbiology are studies of both the beneficial and harmful roles of microbes in soil formation and fertility; in carbon, nitrogen, phosphorus, and sulfur cycles; in diseases of plants; in the digestive processes of cows and other ruminants; and in the production of crops and foods. A food microbiologist is concerned with the production, processing, storage, cooking, and serving of food as well as the prevention of food spoilage, food poisoning, and food toxicity. A dairy microbiologist oversees the grading, pasteurizing, and processing of milk and cheeses to prevent contamination, spoilage, and transmission of diseases from environmental, animal, and human sources.

Sanitary Microbiology The field of sanitary microbiology includes the processing and disposal of garbage and sewage wastes as well as the purification and processing of water supplies to ensure that no pathogens are carried to the consumer by drinking water. Sanitary microbiologists also inspect food processing installations and eating establishments to ensure that proper food handling procedures are being enforced.

Industrial Microbiology Many businesses and industries depend on the proper growth and maintenance of certain microbes to produce beer, wine, alcohol, and organic materials such as enzymes, vitamins, and antibiotics. Industrial microbiologists monitor and maintain the essential microorganisms for these commercial enterprises. Applied microbiologists conduct research aimed at producing new products and more effective antibiotics. The scope of microbiology, indeed, has broad, far-reaching effects on humans, pathogens, and their relationships to the environment.

Microbial Physiology and Genetics Research in microbial physiology has contributed greatly to a clearer understanding of the function of microorganisms, the structure of DNA, and the science of genetics (the study of heredity) in general. Genetic manipulation is much easier and faster with viruses and bacteria than with more complex cells; thus, everyday organisms such as the intestinal bacterium, *Escherichia coli*, are invaluable tools in this field.

Environmental Microbiology The field of environmental microbiology, or microbial ecology, has become important because of increased concern about the environment. This field encompasses the areas of soil, air, water, sewage, food, and dairy microbiology as well as the cycling of elements by microbial, environmental, and geochemical processes. In addition, the biodegradation of toxic chemicals by various microorganisms is being used as a new method for cleaning up hazardous materials found in soil and water.

Milestones of Microbiology

Parasites are microorganisms, such as protozoa, helminths (parasitic worms), and arthropods, that live on or in other living organisms (called hosts). Parasites are usually pathogenic. They have existed almost as long as living organisms have existed, long before humans inhabited the earth. We know that human pathogens have existed for ages because damage caused by them has been observed in the bones of mummies and early humans, indicating that diseases such as osteomyelitis (inflammation of bones and bone marrow) and syphilis were present.

In the ancient civilizations of Egypt and China, people kept clean by washing with water in an effort to prevent disease. They knew that some diseases were easily transmitted from one person to another, and they learned to isolate the sick to prevent the spread of these diseases, which we recognize as being *contagious.* The Egyptians were aware of the effectiveness of biological warfare. They often used the blood and bodies of their diseased dead to contaminate water supplies of their enemies and to spread diseases among them.

The Book of Leviticus in the Bible was probably the first recording of laws concerning public health. The Hebrew people were told to practice personal hygiene by washing and keeping clean. They were also instructed to bury their waste material away from their campsites, to isolate those who were sick, and to burn soiled dressings. They were prohibited from eating animals that had died of natural causes. The procedure for killing an animal was clearly described, and the edible parts were designated.

Most of the knowledge about public sanitation and transmission of disease was lost in Europe during the Middle Ages, when there was a general stagnation of culture and learning for almost 1,000 years. However, during the Renaissance, widespread epidemics of smallpox, syphilis, rabies, and other diseases prompted physicians and alchemists to search for explanations as to how contraction and transfer of diseases occurred. Most people believed that diseases were caused by curses of the gods and, as a result, many bizarre treatments (bleeding, drilling holes in the head, attaching leeches) were used to drive away the devils or evil spirits and relieve the symptoms.

An Italian physician, Girolamo Fracastorius, having observed the syphilis epidemic of the 1500s, proposed in 1546 that the agents of communicable diseases were living germs that could be transmitted by direct contact with humans and animals and indirectly by objects. Proof for vague theories such as this was long delayed because the agents of disease could not be observed and experimental evidence was lacking.

Until the development of magnifying lenses and microscopes that could sufficiently magnify microorganisms to allow them to be visualized, the discovery of disease-causing agents was impossible. Actually, it is not known who built the first microscope, but *compound microscopes,* which use two or more lenses to increase magnification, were developed by Johannes Janssen (1590), Galileo Galilei (1609), and Robert Hooke (1660). When Antony van Leeuwenhoek first described bacteria in 1667, he used a small, *simple microscope* with one lens the size of a large pinhead and observed material that he placed on the point of a pin (see Insight Box). Leeuwenhoek is called the "Father of Microbiology" because he described the three shapes of bacteria as well as protozoa, sperm, and blood cells as a result of his microscopic observations of pepper water, tooth scrapings, gutter water, semen, blood, urine, and feces. His letters to the Royal Society of London convinced the scientists of the late 17th century of the existence of microorganisms, which Leeuwenhoek called "animalcules." He did not speculate on the origin of these microbes nor did he associate them with the cause of disease. Such relationships were not established until the work of Louis Pasteur and Robert Koch in the late 19th century. Leeuwenhoek's fine art of grinding a single lens that would magnify an object to 300 times its size was lost at his death because he had not taught this skill to anyone.

Detailed descriptions of microorganisms did not occur until the development of better microscopes in the 19th and 20th centuries. Modern microscopy is discussed at the end of this chapter.

Although Leeuwenhoek was probably not concerned about the origin of microorganisms, many other scientists were searching for an explanation of the appearance of living creatures in decaying meat, stagnating ponds, fermenting grain, and infected wounds. On the basis of observation, many of the "scientists" of that time believed that life could develop spontaneously from decomposing nonliving material. This is called the theory of spontaneous generation or *abiogenesis.* For more than two centuries, from 1650 to 1850, this theory was debated and tested. Following the work of many others, Louis Pasteur and John Tyndall finally disproved the theory of spontaneous generation and proved that life must arise from preexisting life; this is called the theory of *biogenesis,* first proposed by Rudolf Virchow in 1858.

Experiments devised by Pasteur and other scientists in the mid-1800s resulted in several major advances in the field of microbiology: (1) the concept

INSIGHT
Leeuwenhoek, Father of Bacteriology and Protozoology

The discovery of various bacteria and protozoa by Antony van Leeuwenhoek, using his small, simple microscopes over 300 years ago, has long been an area for study and discussion. How did he build the microscopes and use them so effectively to observe the "wee beasties," as he called them? Some believe he may have used polished clear glass beads of various sizes for lenses and various intensities of outdoor light to view the microbes. By changing the direction and intensity of the light source, he could have developed the darkfield capability to enable him to observe microbial movement. See Insight Figure 1 – 1 for an idea of what Leeuwenhoek's microscopes looked like and what he saw.

Perhaps it was his curiosity about the reasons for pepper's potent taste that led Leeuwenhoek to the discovery of bacteria. He steeped peppercorns for 3 weeks to soften them, examined the water, and observed the "incredibly small organisms" we now know as bacteria. Though this has been acknowledged as the first recorded observation of bacteria, definitive evidence for the discovery of bacteria was not provided until he wrote a letter to the Royal Society of London on September 17, 1686. In this letter he described his regimen for keeping his teeth clean, detailing his examination of the white matter (plaque) that grew between his teeth. When he added this white matter to saliva (which he thought was free of microorganisms) and examined it microscopically, he described the "many very small living animals that moved very prettily."

Leeuwenhoek, determined to keep his microscopic methods to himself, shared his techniques with no one. Thus, to this day, we don't know how he was able to grind such marvelous little lenses or to observe such tiny organisms.

that life must arise from preexisting life; (2) the techniques of sterilization and pasteurization; (3) the understanding of the biological process of fermentation; (4) the development of the germ theory of disease; and (5) the development of vaccines from killed anthrax bacteria and from attenuated, or weakened, rabies viruses.

The sterilization techniques used by Pasteur showed that boiled broth remains sterile until it is contaminated by particles in the air. While repeating Pasteur's experiment, Tyndall found bacterial fragments, called *endospores,* that were not destroyed by the first boiling and that germinated into reproducing (vegetative) bacteria. This discovery led to the development of the fractional sterilization process, often called *tyndallization,* in which endospores of bacteria are destroyed by boiling and cooling three times, allowing the spores to germinate between boilings.

When Pasteur was investigating reasons for the spoilage of beer and wine, he developed the "biological theory of fermentation," which states that a specific microbe produces a specific change in the substance on which it grows or that a specific microorganism produces a specific fermentation product. Just as yeasts ferment sugar in grape juice to produce ethyl alcohol (ethanol) in wine, some contaminating bacteria, such as *Acetobacter,* may change the alcohol into acetic acid (vinegar); this, of course, ruins the taste of the wine. To eliminate

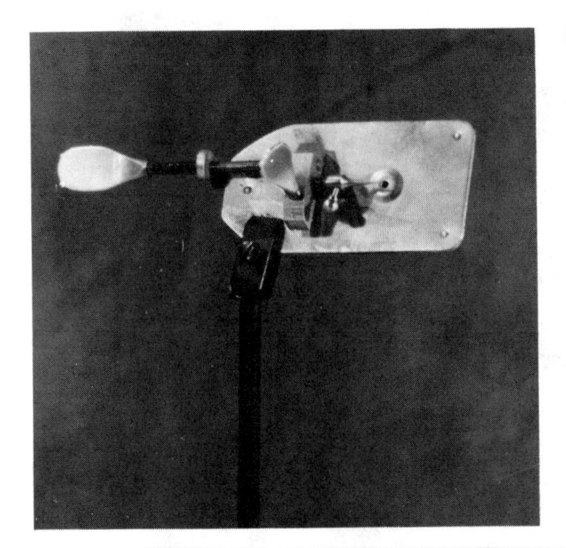

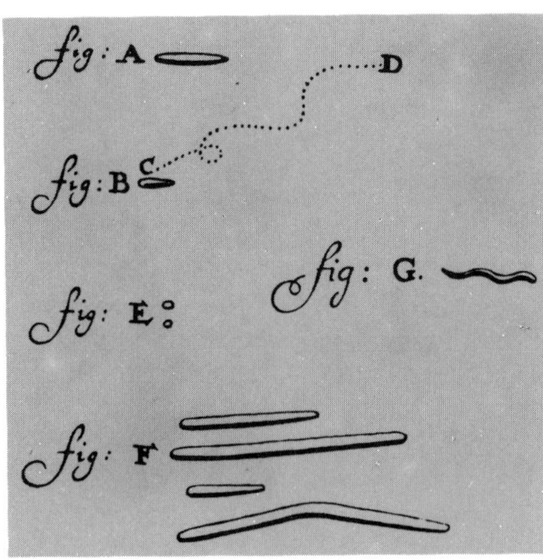

INSIGHT FIGURE 1 – 1
(Left) Leeuwenhoek used a microscope with a single biconvex lens to view bacteria suspended in a drop of liquid placed on a moveable pin. (Right) Although his microscope was capable of only 200- to 300-fold magnification, Leeuwenhoek was able to create these remarkable drawings of different bacteria types. (Volk WA, et al.: Essentials of Medical Microbiology, 4th ed. Philadelphia, JB Lippincott, 1991)

harmful contaminating bacteria from beer and wines, Pasteur heated them to 50° to 60° C (122° to 140° F). This process, now called *pasteurization,* has been adapted to destroy the pathogens in milk by heating it to 63° C (145.4° F) for 30 minutes or to 72° C (161.6° F) for 15 seconds.

By extending the biological theory of fermentation to animals and humans, Pasteur (together with others such as Robert Koch) developed the "germ theory of disease," which states that a specific disease is caused by a specific type of microorganism (Fig. 1 – 5). After isolating the causative pathogens of chicken cholera and rabies, he prepared vaccines against these diseases. The attenuated or weakened pathogen, which was no longer pathogenic but which made the injected animal immune to the disease, was used. For example, Pasteur found the rabies virus in the brain and spinal cord of rabid dogs. He discovered that by transferring the virus from rabbit to rabbit many times, the virus became so weakened it no longer caused rabies in rabbits or dogs. It was this attenuated virus that he used to immunize animals and humans against rabies.

The beginnings of immunology are actually found in ancient China, where a type of vaccination against smallpox was practiced in which healthy people inhaled a powder made from the scabs of healing pustules of smallpox. Although many Chinese developed the disease as a result of this custom, others

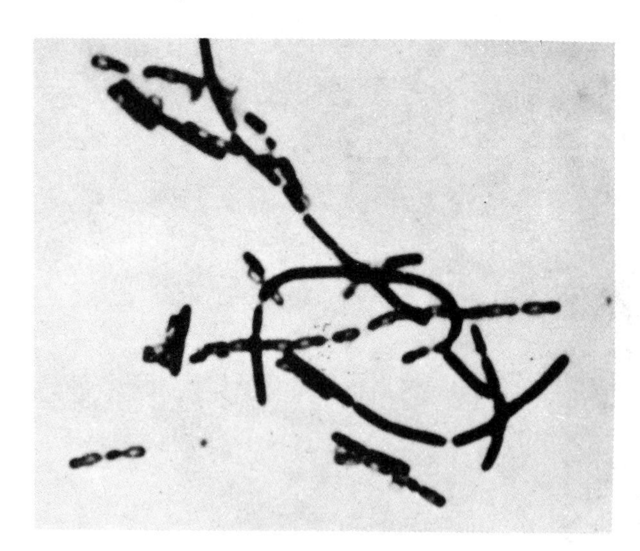

FIGURE 1 – 5
Bacillus anthracis, from a 48-hour culture, showing endospores. This bacterium causes anthrax and was used by Pasteur and Koch in many experiments on disease causation (original magnification ×1,200). (Burrows W: Textbook of Microbiology. Philadelphia, WB Saunders, 1963)

became immune. This method is called *variolation,* because the dying smallpox virus (variola) was used. In the late 1700s, Edward Jenner used cowpox virus (vaccinia) to vaccinate people against smallpox after he observed that milkmaids who caught cowpox, a mild disease transmitted to them from cows, were protected against smallpox, a far more serious disease.

Although Pasteur used the technique of isolating a microorganism and growing it in a pure culture in nutrient media in the laboratory, Robert Koch, a German physician, is usually given credit for most pure-culture research methods in microbiology. He and his assistant, Julius Petri, developed the *petri dish,* which is still in use today for microbial growth on solid media. At the suggestion of an associate's wife, Frau Hesse, they used agar—an extract from a red marine alga used at that time to make jelly—to solidify the growth medium so that distinct colonies of bacteria could be observed. A bacterial colony contains millions of bacteria.

In addition, Robert Koch, in 1876, established an experimental procedure to prove the germ theory of disease, which states that "a specific disease is caused by a specific pathogen." This scientific procedure is known as *Koch's postulates* (Fig. 1 – 6).

Koch's Postulates

1. The causative agent must be present in every case of the disease and must not be present in healthy animals.
2. The pathogen must be isolated from the diseased host animal and must be grown in pure culture.

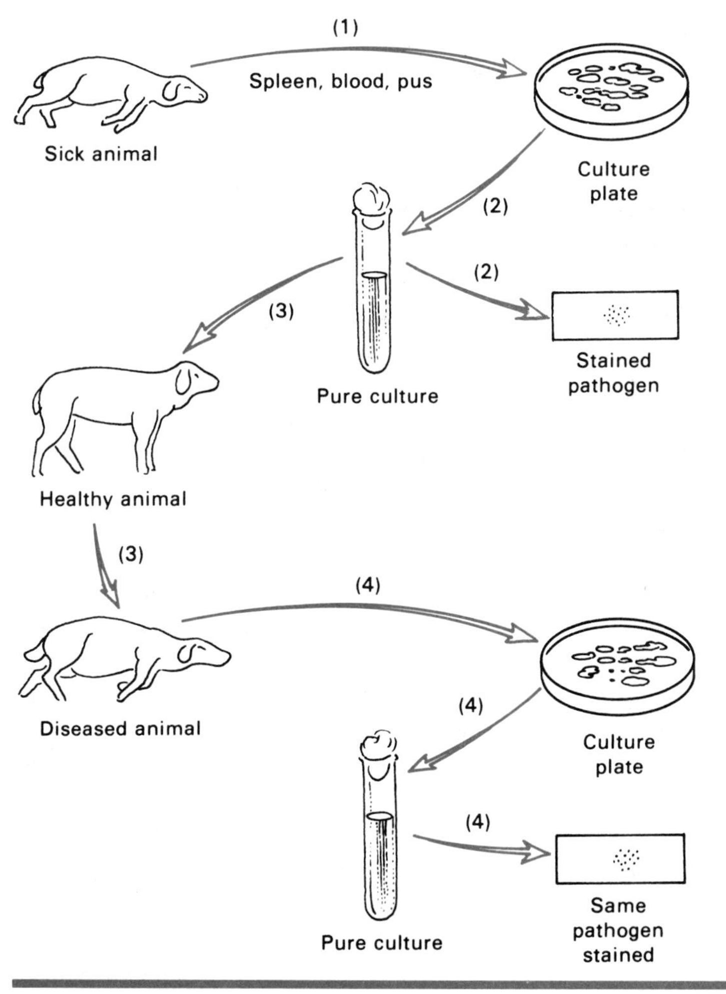

FIGURE 1 – 6
Koch's postulates: proof of the germ theory of disease.

3. The same disease must be produced when microbes from the pure culture are inoculated into healthy susceptible animals.
4. The same pathogen must be recoverable once again from this artificially infected host animal, and it must be able to be grown again in pure culture.

Koch's postulates not only proved the germ theory of disease but also gave a tremendous boost to the development of microbiology by stressing laboratory culture and identification of microorganisms. However, some circumstances exist in which these postulates may not be easy to apply, as outlined below:

Exceptions to Koch's Postulates

1. Many healthy people carry pathogens but do not exhibit symptoms of the disease. These "carriers" may transmit the pathogens to others who then may become diseased. This is usually the case with epidemics of certain hospital-acquired (nosocomial) infections, gonorrhea, typhoid fever, diphtheria, pneumonia, and AIDS.
2. Some microbes are very difficult or impossible to grow *in vitro* (in the laboratory) in artificial media; these include viruses, rickettsias, chlamydias, and the bacteria that cause leprosy and syphilis. Thus, pure cultures of such pathogens are difficult to obtain. However, many of these fastidious pathogens (those having complex nutritional requirements) can be grown in cultures of living human or animal cells of various types, in embryonated chicken eggs, or in certain animals. The leprosy pathogen thrives in armadillos; the spirochetes of syphilis grow well in the testes of rabbits and chimpanzees; and human immunodeficiency virus (HIV), also known as "the AIDS virus," proliferates in human lymphocyte cultures.
3. To induce a disease from a pure culture, the experimental animal must be susceptible to that pathogen. Many animals, such as rats, are very resistant to microbial infections. Many pathogens are species-specific, which means that they grow in only one species of animal. For example, the pathogen that causes cholera in humans does not cause hog cholera, and *vice versa.* Because human volunteers are difficult to find and ethical considerations limit their use, the researcher may only be able to observe the changes caused by the pathogen in human cells that can be grown in the laboratory.
4. Certain diseases develop only when an opportunistic pathogen invades a weakened host. These secondary invaders or opportunists cause disease in a person who is ill or recovering from another disease. Examples are pneumonia and ear infections, which may follow influenza. If researchers were looking for the influenza virus, they might be misled by isolating the bacteria that caused the pneumonia.

It is also important to remember that not all diseases are caused by microorganisms. Many diseases, such as rickets and scurvy, result from dietary deficiencies. Some diseases are inherited or are caused by an abnormality in the chromosomes, as in sickle cell anemia. Others, such as diabetes, result from malfunction of a body organ or system; still others, such as cancer of the lungs and skin, are influenced by environmental factors. However, all infectious diseases are caused by microorganisms.

The period of rapid development of microbiological techniques in the late 1800s is known as the "Golden Age of Microbiology." Efficient surgical techniques based on Pasteur's fermentation theories and sterilization techniques were developed. During this period, a physician named Ignaz Semmelweis showed that puerperal sepsis (an infection that follows childbirth) was caused by infectious agents present on the hands of doctors and midwives. In 1847, he demonstrated that washing and disinfecting the hands with a solution of chlorinated lime greatly reduced the number of these infections. However, his work was not widely accepted. It was nearly 20 years later (1865) when Joseph Lister showed there were fewer complications from infections following surgery and childbirth if the surgical instruments were boiled and if the hands of the surgeon and the wound were disinfected with carbolic acid (phenol) (Table 1 – 1).

This technique of using sterilization and disinfectants to prevent microorganisms from entering a surgical wound became known as *antiseptic surgery*.

TABLE 1 – 1
Major Contributors to the Development of Microbiology as a Science before 1900

Contributor	Contribution	Date(s)
Antony van Leeuwenhoek	First to observe live microorganisms, using a simple microscope	1685
Francisco Redi	Demonstrated that animals do not arise spontaneously from dead organic matter	1660
Abbe Spallanzani	One of the first to demonstrate that heated broth, in the absence of air, did not support spontaneous generation	1770
Schröder and von Dusch	Demonstrated that broth heated in the presence of filtered air did not support spontaneous generation	1854
John Tyndall	Demonstrated that open tubes of broth remained free of bacteria if air was free of dust. Developed tyndallization to destroy spores	1860
Louis Pasteur	Disproved the theory of spontaneous generation (1861). Contributed to the understanding of fermentation (1858). Developed technique for selective destruction of microorganisms (pasteurization) (1866). Study of bacterial contamination of wine (1866) and diseases of silkworms (1868). Attenuated vaccines for anthrax (1881) and chicken cholera. Immunization against rabies (1885)	1855–1890s
Joseph Lister	Contributed to concept of aseptic technique	1865–1870
Robert Koch	Developed postulates for proving the cause of infectious disease (1884) and pure culture concept. Observed anthrax bacilli (1876). Developed solid culture media (1882). Discovered organisms causing tuberculosis (1882)	1870s to 1890s
Paul Ehrlich	Formulated humoral theory of resistance (see Chap. 9) Developed new staining techniques. Developed first chemotherapeutic agent	1890s to 1900
Elie Metchnikoff	Formulated cellular theory of resistance	1890s
Emil von Behring	Developed method for producing immunity by using antitoxin against diphtheria	1890s

The application of antiseptic principles to surgery paved the way for many advances in surgical techniques, including *sterile techniques* and *aseptic techniques* (techniques that exclude pathogens), which are practiced throughout the modern world in operating rooms and research laboratories.

Modern nursing techniques logically followed a better understanding of the concepts of disease causation, transmission, and sterilization. Florence Nightingale, an English nurse of the 19th century, developed modern principles of nursing, methods of training nurses, and procedures for organizing hospitals to reduce the spread of disease.

THE TOOLS OF MICROBIOLOGY

Microscopes

Because microorganisms cannot be seen without the aid of powerful magnifying lenses or a microscope, the expansion of the field of microbiology beyond the advances made during the 19th century depended on the development of better microscopes to properly observe these organisms. Although extremely small infectious agents, such as rabies and smallpox viruses, were known to exist, they could not be seen until the electron microscope was developed.

LIGHT MICROSCOPE

A single-lens magnifying glass usually magnifies the image of an object from about 3 to 20 times the object's actual size. The *light microscope* used in laboratories today is a compound brightfield microscope with two lens systems and a visible light source (hence the term "brightfield") that passes through the specimen and lenses to the observer's eye (Fig. 1 – 7). The eyepiece contains an ocular lens. The second lens system is in the objective, which is positioned near the object to be viewed.

The two-lens system of the typical compound microscope can magnify 40 to 1,000 times. The magnification is usually indicated by a numeral preceded by an "×," such as "×1,000," in which "×" means "times." The total magnification of a compound microscope is obtained by multiplying the magnifying power of the ocular lens (usually ×10) by the magnifying power of the objective lens (usually ×4, 10, 40, or 100). Thus, with the low-power (×10) objective in place, the total magnification is 10 multiplied by 10, or ×100. Usually this objective is used to locate the microorganism to be studied. With the high-power or "high-dry" ×40 lens, the total magnification is 10 times 40, or ×400; this lens is used to study algae, protozoa, and other large microorganisms. With the oil-immersion (×100) objective, a total magnification of ×1,000 is obtained, which is useful for observing the general characteristics of bacteria.

The oil-immersion objective must be used with a drop of immersion oil between the specimen and the objective lens; the oil reduces the scattering of

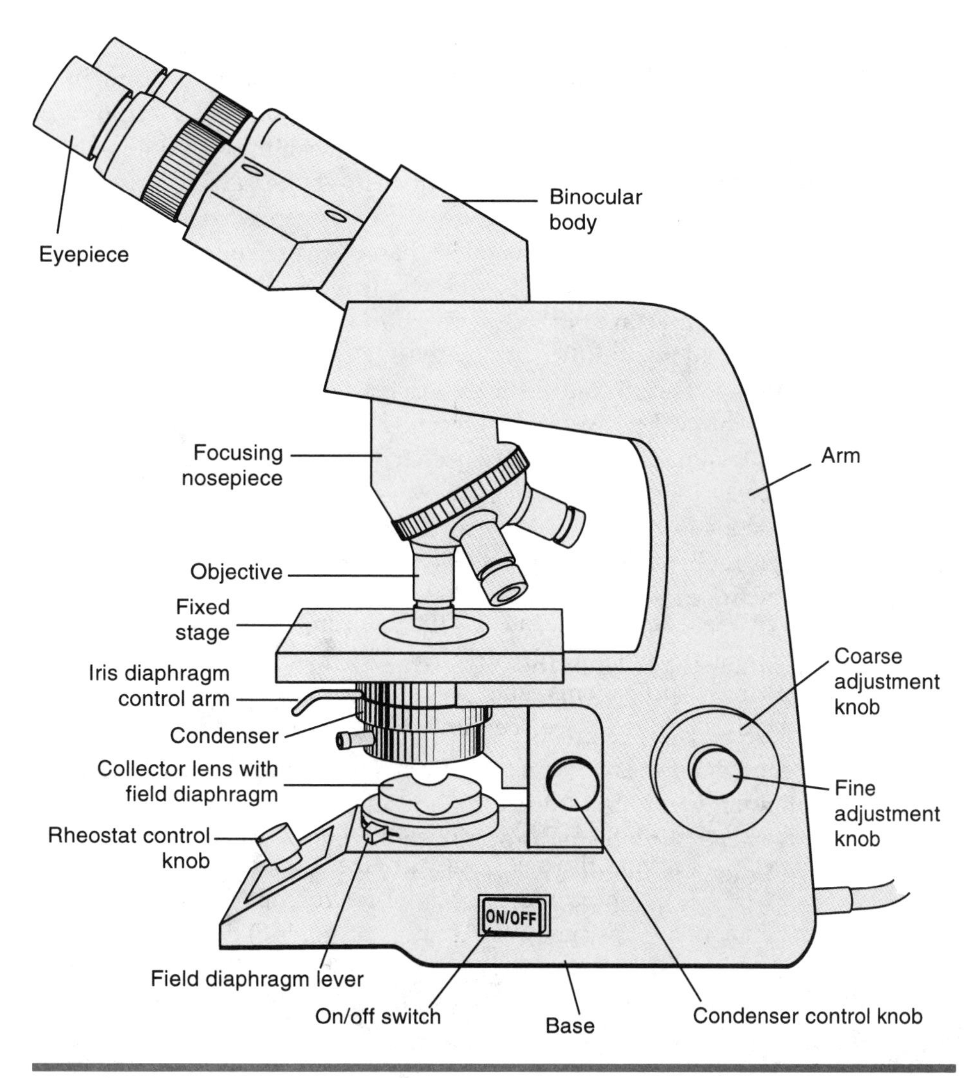

FIGURE 1 – 7
A modern light microscope.

light. For clear observation of the specimen, the light must be properly adjusted and focused. The condenser, located beneath the fixed stage, focuses light onto the specimen, adjusts the amount of light, and shapes the cone of light entering the objective. Generally, the higher the magnification, the more light that is needed.

Magnification alone is of little value unless the enlarged image possesses increased detail and clarity. The image clarity depends on the microscope's *resolving power* (or *resolution*), which is the ability of the lens to distinguish two adjacent points or objects at a particular distance apart. The resolving power depends on the wavelength of the light source and the numerical aperture of the microscope. The greater the numerical aperture, the greater the resolving power will be. The resolving power of the unaided human eye is approximately 0.2 millimeters (mm). The resolving power of a compound microscope is approximately 0.2 micrometers (μm) when the oil-immersion lens is used at a maximum numerical aperture (focus). This means that two bacteria could be distinguished as separate entities if they were separated by 0.2 μm or more; it also means that objects smaller than 0.2 μm cannot be seen.

Other light microscopes are built and adjusted for darkfield techniques, phase-contrast microscopy, and fluorescence. In darkfield microscopy, the light is directed toward the specimen from the side so that the only light to reach the objective is reflected from the bacteria or object being studied. Thus, the microbe appears as a bright object against a dark background (a "dark field"). This technique is frequently used to study very thin bacteria, like the spirochete *Treponema pallidum,* which causes syphilis (Fig. 1 – 8). The phase-contrast microscope may be used to observe living microbes without staining because the light refracted by living cells is different from the surrounding medium; thus, they are more easily seen. The microscope used for fluorescence microscopy has an ultraviolet (UV) light source that illuminates the object but does not pass into the objective of the microscope. When UV light strikes certain dyes and pigments, they emit a certain type of light. For example, different types of chlorophyll emit green, yellow, or orange light that can be seen in the microscope against a dark background. The UV light microscope is often used in immunology laboratories to show that antibodies stained with a fluorescent dye have combined with specific antigens on bacteria. The fluorescent antibody technique is frequently used as a diagnostic test in medical bacteriology (see Fig. 9 – 15 in Chapter 9).

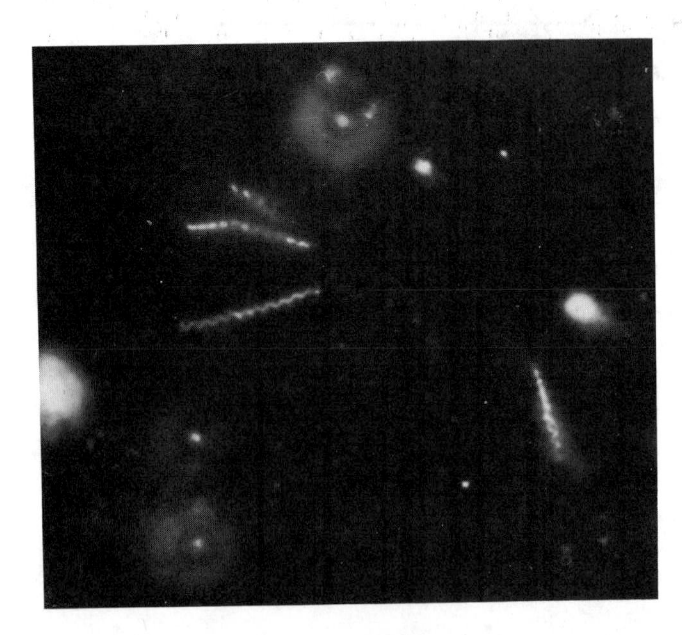

FIGURE 1 – 8
Treponema pallidum, the etiologic agent of syphilis, as seen by darkfield microscopy. (Pelczar MJ, Reid RD: Microbiology. Copyright © 1958, McGraw Hill. Used with permission of McGraw-Hill Book Company)

ELECTRON MICROSCOPE

The *electron microscope* uses an electron beam as a source of illumination instead of visible light; it also uses magnets instead of lenses to focus the beam. In transmission electron microscopy, electrons pass through the dry specimen, which is mounted in resin, and the image is seen on a fluorescent screen. The image may then be photographed and enlarged to magnify the object up to approximately 1 million times, about 1 thousand times higher magnification than with light microscopes. Thus, very tiny microbes (*e.g.*, viruses) may be observed. By examining thin sections of cells with the transmission microscope, their internal structure can be studied. A modification is the scanning electron microscope, which is very useful for observing surfaces and the three-dimensional structure of an object. Figures 1 – 9, 1 – 10, and 1 – 11 show the differences in magnification and detail between electron micrographs and light photomicrographs. Table 1 – 2 lists the characteristics of various types of microscopes.

Units of Measure

In microbiology, several common metric units are used to describe the size of microorganisms. The meter, the basic unit of the metric system, is equivalent to approximately 39 inches. The meter may be divided into 10 decimeters, 100 centimeters, 1,000 millimeters, 1 million micrometers, or 1 billion nanometers. Relationships among these units are shown in Figure 1 – 12.

It should be noted that the old term micron (μ) has been replaced by the term *micrometer* (μm); the term millimicron (mμ) has been replaced by the term *nanometer* (nm); and an *angstrom unit* (Å) is 0.1 nanometer (0.1 nm), according to the International System adopted by scientists worldwide. On this scale, red blood cells are about 7 μm in diameter. A typical bacterium is about 1×3 μm, although they can be as small as 0.2 μm or can be very long filaments. Most

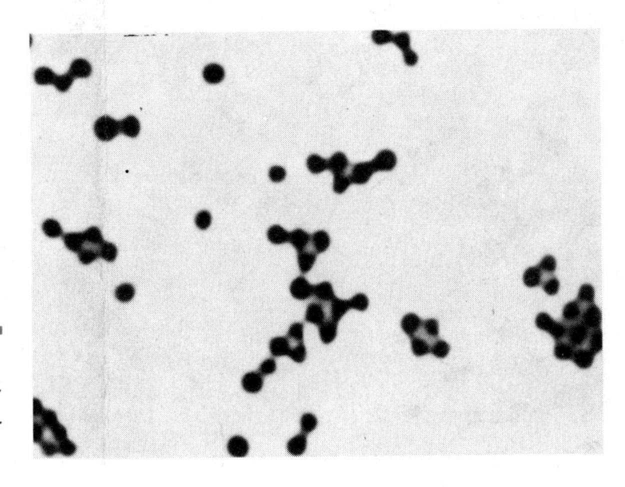

FIGURE 1 – 9
Staphylococcus aureus, as seen by light microscopy (magnification ×1,000). (Photograph courtesy of W. L. Wong)

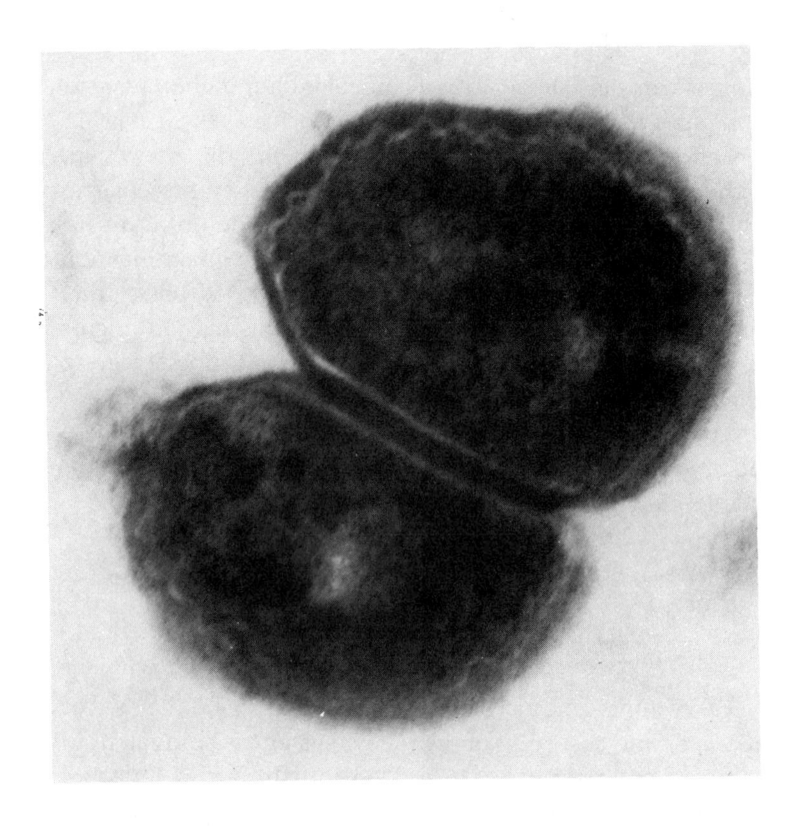

FIGURE 1 – 10
Staphylococcus aureus, as seen by transmission electron microscopy (magnification ×40,000). (Photograph courtesy of Ray Rupel)

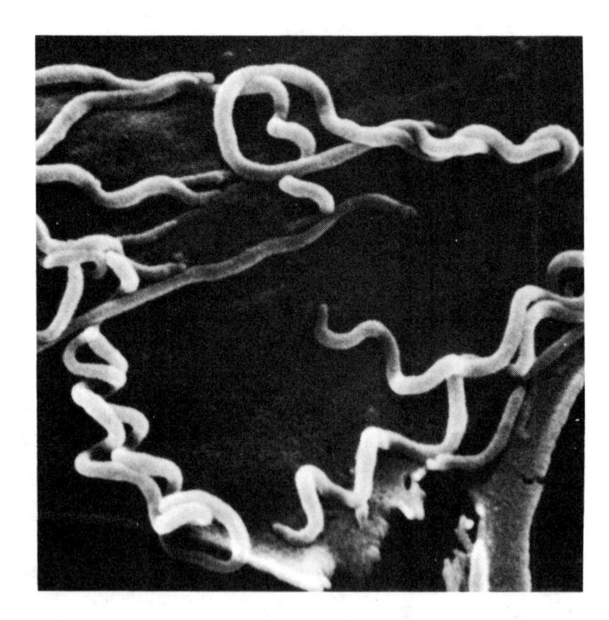

FIGURE 1 – 11
The three-dimensional qualities of scanning electron microscopy clearly reveal the corkscrew shape of cells of the syphilis-causing spirochete, *Treponema pallidum*, attached here to rabbit testicular cells grown in culture (original magnification ×8,000). (Volk WA, et al.: Essentials of Medical Microbiology, 4th ed. Philadelphia, JB Lippincott, 1991)

TABLE 1 – 2
Characteristics of the Various Types of Microscopes

Type	Resolving Power	Useful Magnification	Characteristics
Brightfield	0.2000 μm	1,000	Used to observe morphology of microorganisms, such as bacteria, protozoa, fungi, and algae in living (unstained) and nonliving (stained) state. Cannot resolve organisms less than 0.2 μm, such as spirochetes and viruses.
Darkfield	0.2000 μm	1,000	Background is dark, and unstained organisms can be seen. Useful for examining spirochetes. Slightly more difficult to operate than brightfield.
Phase contrast	0.2000 μm	1,000	Can observe dense structures in living procaryotic and eucaryotic microorganisms.
Fluorescence	0.2000 μm	1,000	Fluorescent dye attached to organism. Primarily a diagnostic technique (immunofluorescence) to detect microorganisms in cells, tissue, and clinical specimens. Training required in specimen preparation and microscope operation.
Transmission electron microscope (TEM)	0.0002 μm (0.2 nm)	200,000	Specimen can be viewed on screen. Excellent resolution. Allows examination of cellular ultrastructure and viruses. Specimen is nonliving. Image is two dimensional.
Scanning electron microscope (SEM)	0.0200 μm (20 nm)	10,000	Specimen can be viewed on screen. Three-dimensional view of specimen. Useful in examining surface structure of cells and viruses. Specimen is nonliving. Resolution limited compared with TEM.

(Adapted from Boyd, 1988)

viruses range in size from about 10 to 300 nm. Some very large protozoa reach a length of 2,000 μm or 2 mm.

The sizes of microorganisms are measured using an ocular micrometer, a tiny ruler within the eyepiece (ocular) of the light microscope. Before it can be used to measure objects, however, the ocular micrometer must first be calibrated using a microscope stage measuring device called a stage micrometer. Calibration must be performed for each of the objective lenses to determine the distance between the marks on the ocular micrometer. The ocular micrometer

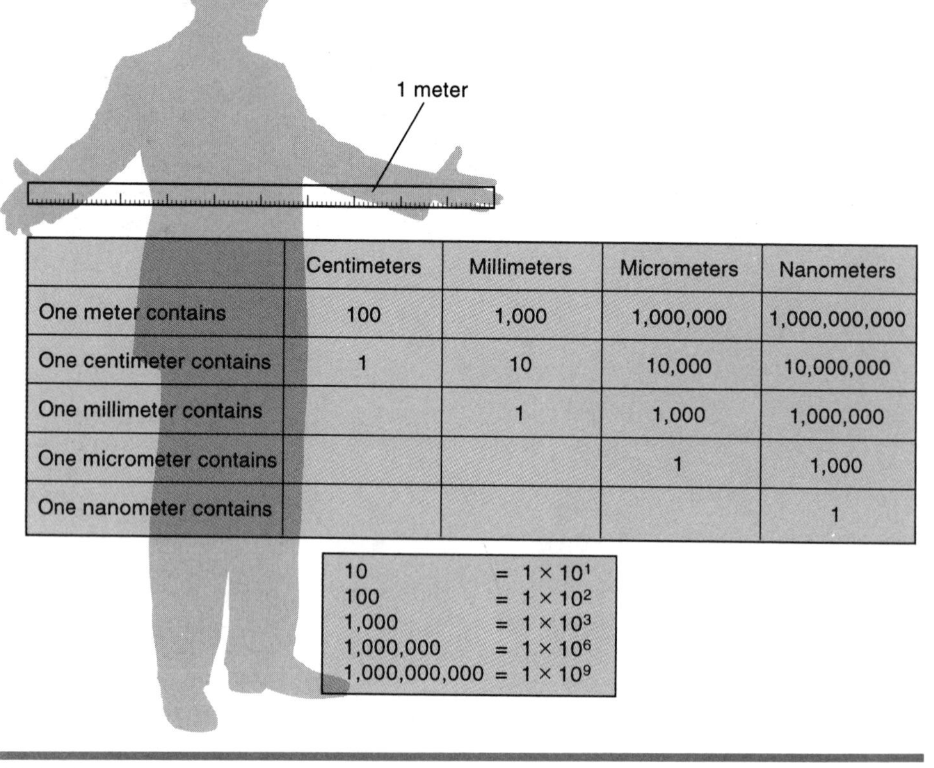

	Centimeters	Millimeters	Micrometers	Nanometers
One meter contains	100	1,000	1,000,000	1,000,000,000
One centimeter contains	1	10	10,000	10,000,000
One millimeter contains		1	1,000	1,000,000
One micrometer contains			1	1,000
One nanometer contains				1

10	$= 1 \times 10^1$
100	$= 1 \times 10^2$
1,000	$= 1 \times 10^3$
1,000,000	$= 1 \times 10^6$
1,000,000,000	$= 1 \times 10^9$

FIGURE 1 – 12
Representations of metric units of measure and numbers.

can then be used to measure lengths and widths of microbes and other objects on the specimen slide.

Each type of microscope has its limits of visibility. The light microscope can be used for observation of cells larger than 0.2 μm, whereas the electron microscope can discern objects as small as 0.2 nm (0.0002 μm) in diameter.

SUMMARY

In this chapter, the science of microbiology was introduced. You should now be aware of the presence and importance of the many different types of microorganisms that exist in the world around you. Individual fields of study within the science of microbiology were explained. As you read about the milestones in the historical development of microbiology, you should realize that this is a relatively new field of study that could only be developed with refined microscopes and modern techniques. These microscopic and biochemical discoveries enabled microbiologists to explore and understand more about the characteristics of these organisms that cannot be seen with the unaided eye—the microorganisms.

PROBLEMS AND QUESTIONS

1. What does the study of microbiology include? What types of microorganisms?
2. Why is the study of microbiology important to people working in health occupations?
3. Why is Leeuwenhoek called the "Father of Bacteriology?"
4. Why was the theory of spontaneous generation debated for 200 years?
5. What contributions did Pasteur make to microbiology?
6. What contributions to microbiology were made by Semmelweis, Lister, Jenner, and Nightingale?
7. What are Koch's postulates used to prove?
8. List Koch's postulates and the circumstances under which they might not be easily applied.
9. What types of microscopes are used to observe bacteria and viruses?
10. What units of measure are used to describe the size of bacteria and viruses?

Self Test

After you have read Chapter 1, examined the objectives, reviewed the chapter outline, studied the new terms, and answered the questions above, complete the following self test.

Matching Exercises

Complete each statement from the list of words provided within each section.

HISTORIC MILESTONES OF MICROBIOLOGY

Leeuwenhoek, Tyndall, Lister, Fracastorius, Nightingale, Hesse, Janssen, Pasteur, Petri, Jenner, Koch, Semmelweis

1. Modern nursing techniques that reduce the spread of disease were developed by ________________.
2. The man who proved the germ theory of disease using laboratory procedures and a list of postulates was ________________.
3. The chemist who stated the biological theory of fermentation and the germ theory of disease and who disproved the theory of spontaneous generation was ________________.
4. The physician who demonstrated the effectiveness of hand washing in re-

ducing infections following childbirth was ________________.

5. The "Father of Microbiology," who first described living microorganisms, was ________________.
6. The physician who described the transmission of diseases in 1546 was ________________.
7. An early compound microscope was built by ________________.
8. The process of boiling and cooling repeatedly to destroy spores, which bears his name, was discovered by ________________.
9. The person who developed vaccines for chicken cholera and rabies was ________________.
10. Antiseptic surgical technique was proved effective by ________________.
11. A small, flat glass dish for the growth of microorganisms was developed by ________________.
12. The wife of a researcher in Koch's laboratory who suggested the use of agar growth medium was ________________.
13. The physician who demonstrated that cowpox virus could be used to vaccinate against smallpox was ________________.
14. The technique for pure culture research methods is attributed to the laboratory group led by ________________.

SOME TYPES OF MICROBES

pathogenic
opportunists
indigenous microflora
saprophytes
nitrogen-fixing microbes
nonpathogenic

1. The microorganisms usually found on or within a person are called the ________________.
2. Microbes that are usually harmless but that may cause disease when a person's normal resistance is low are called ________________.
3. The soil microbes that return nitrogen from the air to the soil are ________________.
4. Bacteria and fungi that break down decaying organic materials to return plant nutrients to the soil are called ________________.
5. Microorganisms that cause diseases are known as ________________ microorganisms.
6. Legumes are plants such as clover, alfalfa, and peanuts that have microbes living on or near their roots. Some of these microbes replenish the soil because they are ________________.

THEORIES AND TERMS

abiogenesis
sterile technique
biogenesis
contaminant
fermentation
germ theory of disease
biological theory of fermentation
medium
micrometer
pasteurization
pathogen
pure culture
saprophytes
spontaneous generation
sterilization
vaccination

1. The procedure for killing all microorganisms is ________.
2. The process used to destroy the harmful microbes in milk and beer is ________.
3. The surgical technique in which all microorganisms are excluded from the surgical field is ________.
4. The theory stating that a specific disease is caused by a specific pathogen is the ________.
5. The theory stating that a specific microorganism produces a specific change in the material on which it grows is the ________.
6. The theory explaining that life must arise from preexisting life is the theory of ________.
7. The technique of isolating and growing one species of microorganism on a growth medium results in a ________.
8. For centuries it was believed that living animals could arise spontaneously from nonliving material. This is the theory of ________.
9. The artificial process by which people can be made immune to certain diseases is ________.
10. The unit of measurement equal to 1/1,000 of a millimeter is a ________.
11. A ________ is a disease-causing microbe.
12. The breakdown of carbohydrates to produce alcohol is called ________.
13. A culture containing only one species of organism is a ________.
14. A substance used to provide nutrients for the growth of microorganisms is a ________.
15. Organisms that live on dead organic matter are called ________.
16. An unwanted organism in an otherwise pure culture is called a ________.
17. The concept of a living organism originating from dead organic matter without prior existence of organisms of its own type is ________.
18. The ________ is an explanation of the cause of disease based on the existence of pathogenic microorganisms.

True or False (T or F)

___ 1. Most microorganisms are harmful to humans.
___ 2. Most viruses can be seen with a light microscope.
___ 3. Pasteurization kills all organisms in milk.
___ 4. Pasteur believed in abiogenesis.
___ 5. All bacteria live on animals.
___ 6. Sanitary microbiologists are concerned with the microorganisms found in sewage, garbage, water, and food.
___ 7. Tyndallization is a process of repeated boiling and cooling to destroy spores.
___ 8. Koch's postulates can be applied when studying animals with viral infections.
___ 9. Pneumonia and ear infections are usually caused by opportunistic bacteria that invade a weakened host.
___ 10. Bacteria can best be described when using the low-power objective of a light microscope.

Multiple Choice

1. To see viruses one must use
   a. a light microscope.
   b. a phase-contrast microscope.
   c. an electron microscope.
   d. Viruses are too small to be seen.
2. Magnification of ×1000 would be achieved by
   a. an ocular lens of ×10 and an objective lens of ×100.
   b. an ocular lens of ×15 and an objective lens of ×40.
   c. an ocular lens of ×100 and an objective lens of ×100.
3. Which of the following is an exception to Koch's postulates?
   a. The causative agent is present in every case of the disease.
   b. The causative agent may be present in healthy animals that are carriers.
   c. The pathogen inoculated into a healthy animal produces the disease.
4. The study of algae is
   a. virology.
   b. phycology.
   c. mycology.
   d. algology.
5. Pasteur is credited with all of the following *except*
   a. the development of most pure culture techniques.
   b. germ theory of disease.
   c. pasteurization.
   d. the development of vaccines.
6. Attenuated viruses are used
   a. to grow pure cultures.
   b. for vaccines against viral diseases.
   c. to prove Koch's postulates.
   d. during tyndallization.
7. The spirochete *Treponema pallidum* is best seen using
   a. a phase-contrast microscope.
   b. a fluorescence microscope.
   c. a darkfield microscope.
   d. a light microscope.
8. Which of the following are types of light microscopes: (1) ultraviolet; (2) fluorescence; (3) electron; (4) darkfield; (5) phase contrast?
   a. 1,2,4, and 5 only.
   b. 2,3, and 4 only.
   c. 3,4, and 5 only.
   d. 1 and 2 only.
   e. all of the above (1 through 5).

CHAPTER 2

Types of Microorganisms

OBJECTIVES

After studying this chapter, you should be able to

- State the cell theory
- Give a function for each part of the eucaryotic animal cell
- Cite a function for each part of the bacterial cell
- Explain the differences among plant, animal, and bacterial cells
- List the characteristics used to classify bacteria
- State the differences among rickettsias, chlamydias, and mycoplasmas
- Name several important bacterial diseases
- List the classes of protozoa and characteristics for classifying them
- List five pathogenic protozoa
- State some important characteristics of fungi
- List five diseases caused by fungi
- Discuss the important characteristics of procaryotic and eucaryotic algae

- Discuss the important characteristics that make algae different from protozoa and fungi
- Describe the characteristics used to classify viruses
- Compare some of the differences between viruses and bacteria
- List several important viral diseases

CHAPTER OUTLINE

NEW TERMS

Acid-fast stain
Aerotolerant anaerobe
Ameba (pl. *amebae*)
Amphitrichous bacteria
Anaerobe
Autolysis
Axial filament
Bacillus (pl. *bacilli*)
Bacteriophage
Binary fission
Capnophile
Capsid
Capsomeres
Capsule
Cell
Cell membrane
Cell wall
Cellulose
Centrioles
Chitin
Chloroplast
Chromatin
Chromosome
Cilium (pl. *cilia*)
Ciliophora
Ciliates (sing. *ciliate*)
Coccobacilli
Coccus (pl. *cocci*)
Conidium (pl. *conidia*)
Conjugation

Cytology
Cytoplasm
Deoxyribonucleic acid (DNA)
Diplobacilli
Diplococci
Endoplasmic reticulum (ER)
Facultative anaerobes
Fimbriae (sing. *fimbria*)
Flagella (sing. *flagellum*)
Flagellate
Flagellin
Genes
Genus (pl. *genera*)
Glycocalyx
Golgi complex
Gram stain
Hyphae (sing. *hypha*)
Inclusion bodies
L-forms
Lophotrichous bacteria
Lysogenic bacteria
Lysogenic conversion
Lysogenic cycle
Lysogeny
Lysosomes
Lysozyme
Lytic cycle
Mastigophora
Mesosome
Metabolism
Microaerophiles
Microtubules
Mitochondria (sing. *mitochrondrion*)
Mitosis
Monotrichous bacteria
Motile
Mycelium (pl. *mycelia*)
Mycosis (pl. *mycoses*)
Negative stain
Nuclear membrane
Nucleolus
Nucleoplasm
Nucleus (pl. *nuclei*)
Obligate aerobe
Obligate anaerobe
Organelles
Peptidoglycan
Peritrichous bacteria
Phagocyte
Phagocytosis
Photosynthesis
Pili (sing. *pilus*)
Pinocytosis
Pleomorphism
Polyribsosomes
Prophage
Protoplasm
Pseudopodium (pl. *pseudopodia*)
Ribonucleic acid (RNA)
Ribosomes
Rough endoplasmic reticulum
Sarcodina
Sarcomastigophora
Selective permeability
Sex pilus
Slime layer
Smooth endoplasmic reticulum
Species (pl. *species*)
Specific epithet
Spirochetes
Sporozoea
Sporulation
Staphylococci
Streptobacilli
Streptococci
Taxonomy
Teichoic acids
Temperate bacteriophage
Tetrads
Vacuoles
Vector
Virion
Virulent bacteriophage

CELLS: EUCARYOTES AND PROCARYOTES

In 1665, while peering through his crude microscope, Robert Hooke observed the small empty chambers in the structure of cork. He named them *cells* because they reminded him of the bare rooms in a monastery. More than a century later, when biologists had access to more advanced microscopes, they found that cells are not empty, but rather contain a sticky (viscous) fluid. They learned that various chemicals within the cell enable it to live and reproduce. They called this material *protoplasm,* meaning "the substance of life." Now we know that a cell is composed of many different substances and contains tiny particles called *organelles* that have important functions. The living material of the cell is still occasionally called protoplasm. It consists of two parts, the *cytoplasm* outside the nucleus and the *nucleoplasm* inside the nucleus.

Two German biologists, Matthias Schleiden and Theodore Schwann, proposed the cell theory in 1838. They theorized that all living things are composed of cells. Hooke had first used the term "cells" in the mid-17th century. Rudolf Virchow completed the cell theory with the idea that cells must arise from preexisting cells. In other words, life must arise from life.

In biology, the cell is defined as the fundamental living unit of any organism because, like the organism, the cell exhibits the basic characteristics of life. A cell obtains food from the environment to produce energy and nutrients for metabolism (Fig. 2 – 1). *Metabolism* is an inclusive term to describe all the chemical reactions by which food is transformed for use by the cells (see Chapter 4 for a detailed discussion of metabolism.) Through its metabolism, a cell can grow and reproduce. It can respond to changes in its environment such as light, heat, cold, and the presence of chemicals. It can mutate (change) as a result of accidental changes in its genetic material—the *DNA (deoxyribonucleic acid)* that makes up the genes of its chromosomes—and thus become better or less suited to its environment. As a result of these genetic changes, the mutant organism may be better adapted for survival and development into a new species of organism.

Much evidence exists to indicate that almost 4 billion years ago the first bit of life to appear on earth was a very primitive cell similar to the simple bacteria of

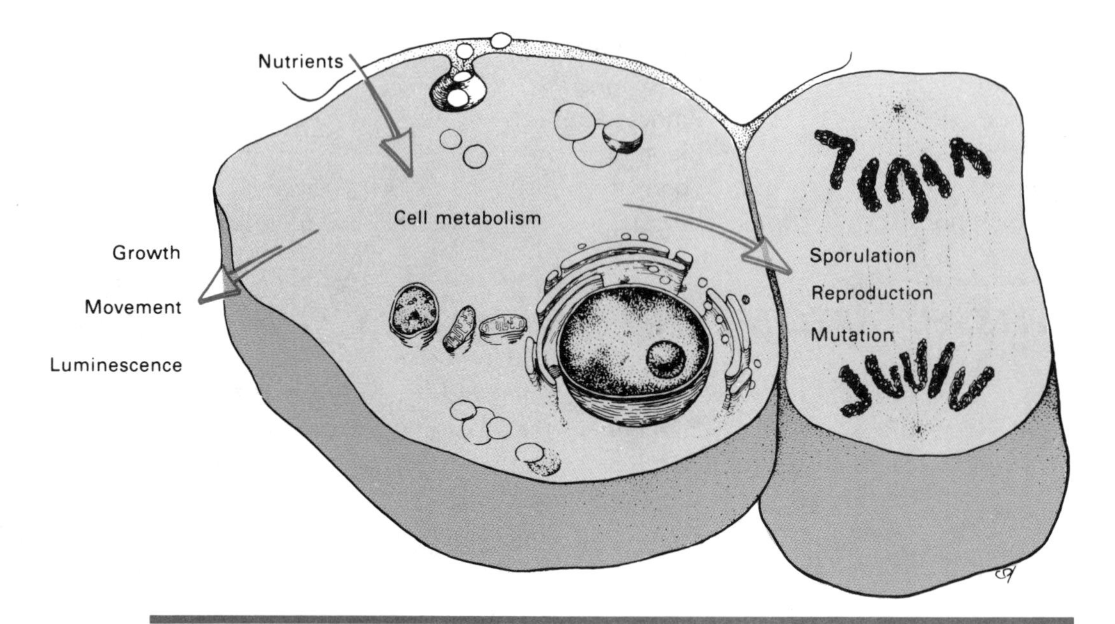

FIGURE 2 – 1
Cell metabolism. The metabolic cycles within the cells enable them to use nutrients for growth, movement, luminescence, sporulation, reproduction, and mutation.

today. Bacterial cells exhibit all the characteristics of life, although they do not have the complex system of membranes and organelles found in the more-advanced single-celled organisms. These less-complex cells, which include bacteria and cyanobacteria, are called procaryotes or procaryotic cells (Fig. 2 – 2). The more complex cells, containing a true nucleus and many membrane-bound organelles, are called eucaryotes or eucaryotic cells; these include such organisms as protozoa; fungi; green, brown, and red algae; and all plant and animal cells, including those that make up the human body.

Viruses appear to be the result of regressive or reverse evolution because they are composed of only a few genes protected by a protein coat and sometimes a few enzymes. Viruses depend on the energy and metabolic machinery of the host cell to reproduce. Therefore, because they are not truly viable cells, they are usually placed in a completely separate category and are not classified with the simple procaryotic cells. Viruses are said to be acellular.

For those in the health professions, it is important to understand the structure of different types of cells not only to classify the microorganisms, but also to understand differences in their structure and metabolism. These factors must be known before we can determine or explain how the chemicals of modern chemotherapy can destroy pathogens but not healthy human cells.

Cytology, the study of the structure and function of cells, has developed during the past 30 years with the aid of the electron microscope and sophisticated biochemical research. Books have been written about the details of these tiny functional factories, but only a brief discussion of their structure and activities is presented here.

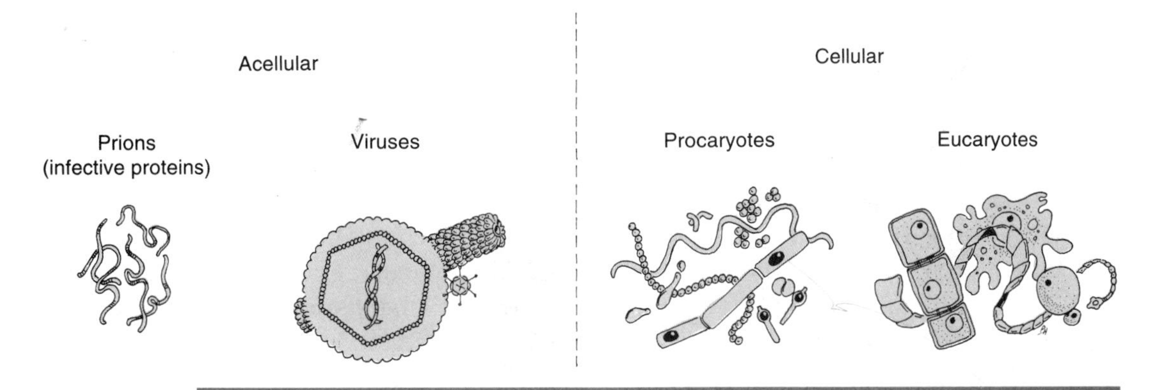

FIGURE 2 – 2
Acellular and cellular microbes. Acellular microbes include viruses, prions, and viroids (not shown). Cellular microbes include the less complex procaryotes (bacteria) and the more complex eucaryotes (algae, fungi, protozoa, and multicellular helminths).

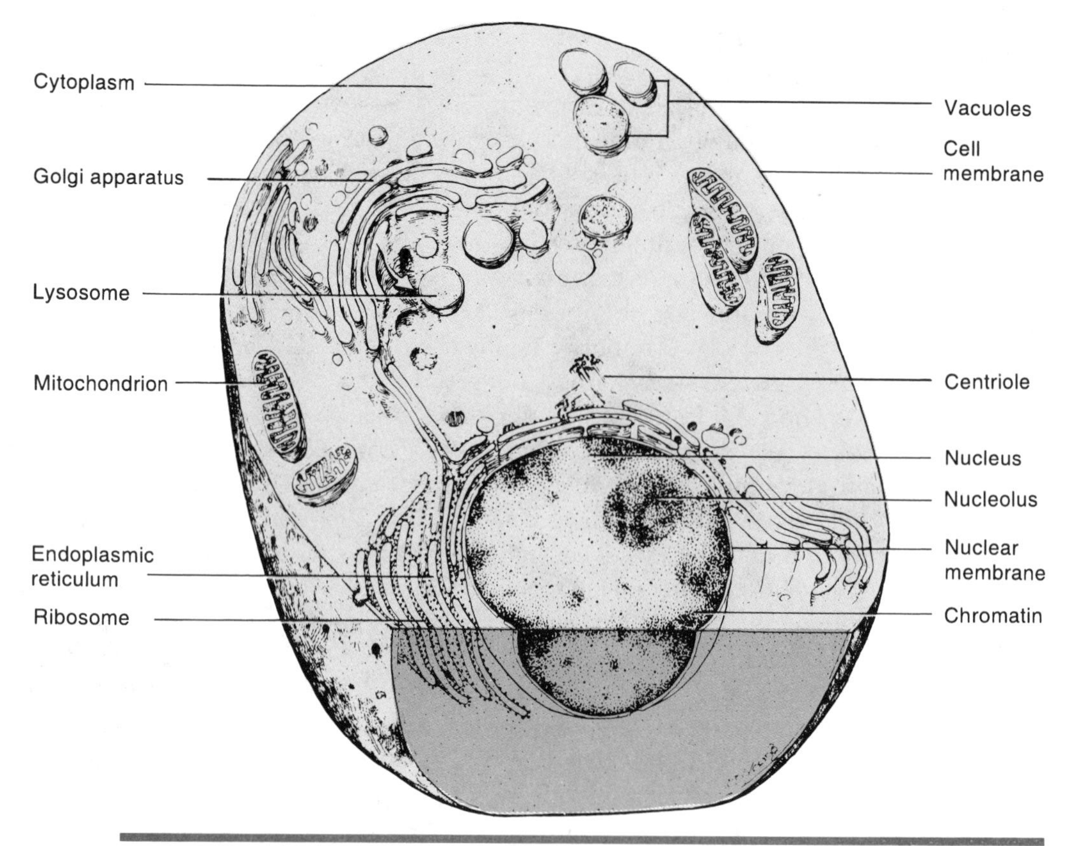

FIGURE 2 – 3
A typical eucaryotic animal cell.

Eucaryotic Cell Structure

Eucaryotes ("eu" = true; "caryo" refers to a nut or nucleus) are so named because they have a true nucleus, in that their DNA is enclosed by a nuclear membrane. Figure 2 – 3 illustrates a typical eucaryotic animal cell. This illustration is a composite of most of the structures that might be found in the various types of human body cells. The electron photomicrograph in Figure 2 – 4 is of an actual yeast cell. A discussion of the functional parts of eucaryotic cells can be better understood by keeping the illustrated structures in mind.

Cell Membrane The cell is enclosed and held intact by the *cell membrane,* which is also often called the plasma membrane, cytoplasmic membrane, or cellular membrane. Structurally, it is a mosaic composed of large molecules of proteins and phospholipids (certain types of fats). These large molecules regulate the passage of nutrients, waste products, and secretions across the cellular

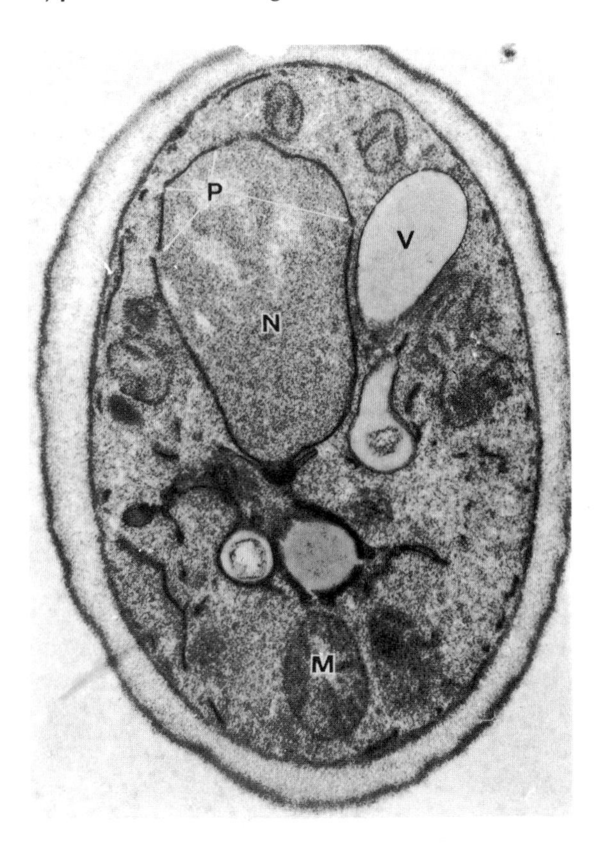

FIGURE 2 – 4
Cross section through a yeast cell showing the nucleus (*N*) with pores (*P*), mitochondrion (*M*), and vacuole (*V*). The cytoplasm is surrounded by the cell membrane. The thick, outer portion is the cell wall. (Lechavalier HA, Pramer D: The Microbes. Philadelphia, JB Lippincott, 1970)

membrane. Because the cell membrane has the property of *selective permeability,* only certain substances may enter and leave the cell. The cell membrane is similar in structure and function to all the other membranes that are a part of the organelles of eucaryotic cells.

Nucleus The organelle within the cell that unifies, controls, and integrates the functions of the entire cell is the *nucleus,* which is enclosed in the *nuclear membrane.* The nucleus contains one or more *chromosomes.* The number and composition of chromosomes and the number of genes on each chromosome are characteristic of the organism's species. Different species have different numbers and sizes of chromosomes. Human cells, for example, have 46 chromosomes (23 pairs), each consisting of thousands of *genes.* A gene is the unit that codes for, or determines, a particular trait or characteristic of an individual organism. When genes are broken apart chemically, they are found to be coiled strands of DNA and proteins. DNA contains the genetic information for the production of essential proteins that enable the cell to function properly. To understand more about how the base coding of the DNA can chemically control the entire organism, refer to Chapter 3.

Close observation of the nucleus of nondividing cells reveals the *chromatin,* which are loosely wound strands of chromosomes that are suspended in the nucleoplasm. Nucleoplasm is the nutrient gelatinous matrix or base material of the nucleus. These chromatin strands condense into tightly coiled chromosomes just before the cell divides. In the very dense dark area of the nucleus, called the *nucleolus,* ribosomal *ribonucleic acid* (rRNA) is manufactured before the molecules move to the cytoplasmic portion of the cell.

Cytoplasm The part of the cell where most of the work is performed is the cytoplasm, which is controlled by information carried in the DNA of the nucleus. Cytoplasm is the cellular material (protoplasm) outside the nucleus, enclosed by the cell membrane. It is composed of a semifluid, gelatinous, nutrient matrix and the cytoplasmic organelles, including the *endoplasmic reticulum, ribosomes, Golgi complex, mitochondria, centrioles, microtubules, lysosomes,* and other *vacuoles.* Each of these organelles has a highly specific function, and all of the functions are interrelated to maintain the cell and allow it to properly perform its activities.

The endoplasmic reticulum (ER) is a system of membranes that are interconnected and arranged to form a network of tubules connecting the outside of the cell to the nucleus. The ER transports nutrients to the nucleus and also provides some structural support for the cell. Much of the ER has a rough appearance and is designated as *rough endoplasmic reticulum* (RER). This rough appearance is due to the many ribosomes attached to the outer surface of the membranes. Ribosomes consist mainly of rRNA and protein and play an important role in the synthesis (manufacture) of essential proteins for use in the cell and elsewhere in the organism. Endoplasmic reticulum to which ribosomes are not attached is called *smooth ER* (SER).

The Golgi complex, also known as the Golgi apparatus or Golgi body, usually connects or communicates with the ER. This group of membranes completes the synthesis of secretory products and packages them into small sacs called vesicles for storage or export outside the cell.

Lysosomes are small sacs that originate from the Golgi complex. They contain *lysozyme* and other digestive enzymes that break down foreign material taken into the cell by *phagocytosis* (the engulfing of large particles by certain types of white blood cells called *phagocytes*). These enzymes also aid in breaking down worn out parts of the cell and may destroy the entire cell by a process called *autolysis* if the cell is damaged or deteriorating.

The energy necessary for cellular function is provided by the formation of high-energy phosphate molecules such as adenosine triphosphate (ATP). The mitochondria are the "power plants" or "energy factories" of the cell where most of the energy-carrying ATP molecules are formed by cellular respiration. During this process, energy is released from glucose molecules and other food nutrients to drive other cellular functions (see Chapter 4). The number of mitochondria in a cell varies greatly depending on the activities required of that cell.

Two cylindrical organelles called centrioles lie perpendicular to each other near the nucleus. The centrioles are involved in the formation of spindle fibers for eucaryotic cell division. This process, which results in two daughter cells with the same number of chromosomes as the parent cell, is called *mitosis,* or "the dance of the chromosomes." Eucaryotic cilia and flagella also appear to arise from centriole material because their complex internal protein fibril configuration is very similar.

Other structures that may be present in the cell include microfilaments, microtubules, granules, and vacuoles containing food, secretory products, and pigments. *Chloroplasts* are always present in plant cells. They contain the pigment chlorophyll, which is required for photosynthesis. *Photosynthesis* is the process by which light energy is used to convert carbon dioxide and water into carbohydrates and oxygen.

Cell Wall 'he eucaryotic *cell wall* is an external structure found on plant cells, algae, and fungi. It usually consists mainly of *cellulose* but may also contain pectin, lignin, chitin, and some mineral salts (usually found in algae). The eucaryotic cell wall is much simpler in structure than the procaryotic bacterial cell wall. Cell walls provide rigidity and protection. Although the cell walls of plants and algae contain cellulose, those of fungi do not. The cell walls of fungi contain a substance called *chitin,* which is not found in the cell walls of other microorganisms.

Flagella and Cilia Some eucaryotic cells (*e.g.,* spermatozoa and certain types ot protozoa and algae) possess relatively long, thin structures called *flagella.* Such cells are said to be flagellated. The whipping motion of the flagella enables flagellated cells to "swim" through liquid environments. Thus, flagella are said to be organelles of locomotion (cell movement). Flagellated cells may possess one *flagellum* or two or more flagella. Some protozoa and some algae are flagellated. *Cilia* (singular, *cilium*) are also organelles of locomotion, but they tend to be shorter (hair-like) and more numerous than flagella. Cilia can be found on some species of protozoa and on certain types of cells in the human body (*e.g.,* the ciliated epithelial cells that line the respiratory tract). Unlike flagella, cilia tend to beat with a coordinated, rhythmic movement. Except for size differences, eucaryotic flagella and cilia are virtually identical in structure; however, they differ from procaryotic flagella, which are far less complex.

Procaryotic Cell Structure

Structurally, procaryotes are very simple cells when compared with the eucaryotic system of membranes, yet they are able to carry on the necessary processes of life. In these cells, division is by *binary fission*—the simple division of the cell into two parts following formation of a separating membrane and cell wall. All bacteria are procaryotes.

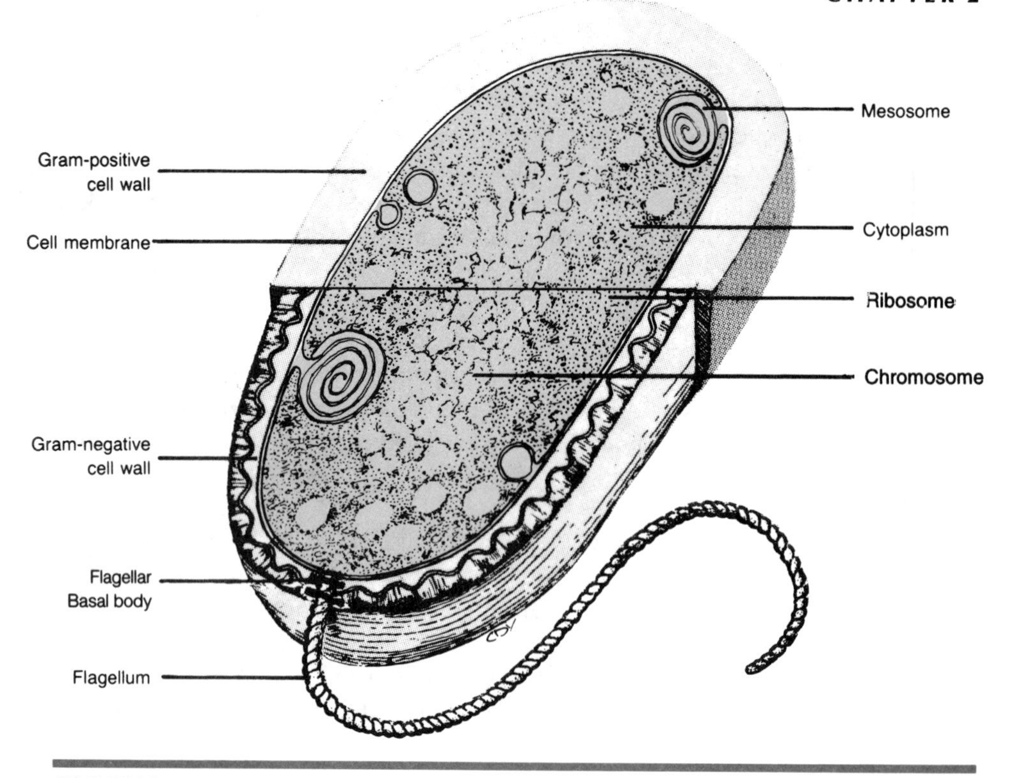

FIGURE 2 – 5
A composite procaryotic bacterial cell showing the difference between gram-negative and gram-positive cell walls.

Refer to the composite drawing (Fig. 2 – 5) of a "typical" gram-positive or gram-negative (explanation follows) bacterial cell and to the electron micrographs in Figures 2 – 6 and 2 – 7 as we discuss this cell from internal to external structures. Within the cytoplasm, the chromosome, *mesosomes,* ribosomes, *polyribosomes,* and other cytoplasmic particles can be seen. Unlike eucaryotic cells, the cytoplasm of procaryotic cells is not filled with internal membranes. The cytoplasm is surrounded by a cell membrane, a cell wall (usually), and sometimes a *capsule* or *slime layer.* These latter three structures make up the bacterial cell envelope. Depending on the particular genus and species of bacterium, flagella or *pili* (description follows) or both may be observed outside the cell envelope, and spores may sometimes be seen within the cell.

Chromosome The procaryotic chromosome is not surrounded by a nuclear membrane, does not have a definite shape, and has little or no protein material associated with it. It usually consists of a single, circular DNA molecule and serves as the control center of the bacterial cell, containing the genetic informa-

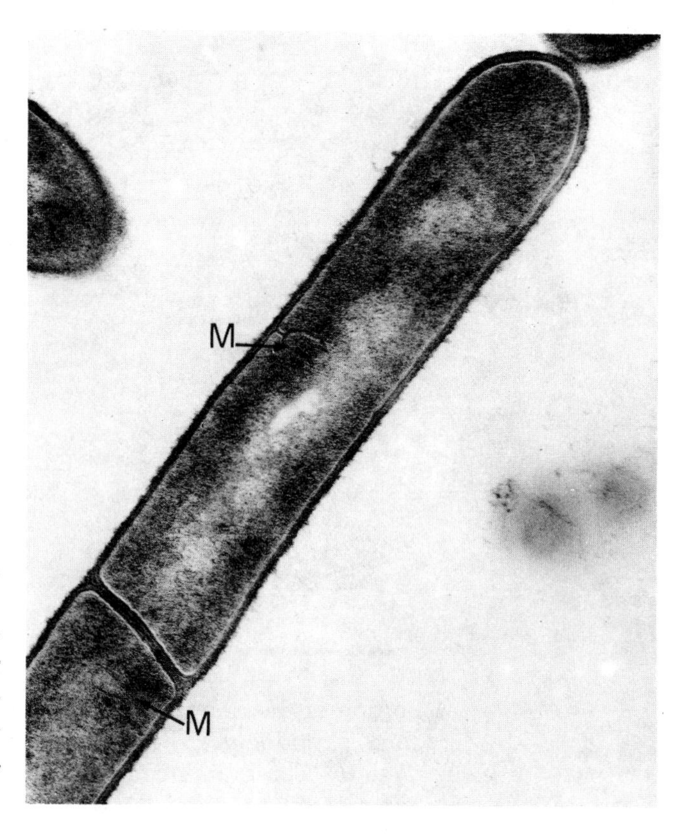

FIGURE 2 – 6
Recently divided cells of *Bacillus licheniformis* showing the light nuclear material and mesosomes (*M*). (Lechavalier HA, Pramer D: The Microbes. Philadelphia, JB Lippincott, 1970)

tion needed for producing several thousand enzymes and other proteins. It is capable of duplicating itself, guiding cell division, and directing cellular activities. A typical bacterial chromosome contains approximately 10,000 genes.

Cytoplasm The semiliquid cytoplasm, which surrounds the chromosome, is contained within the plasma membrane. Cytoplasm consists of water, enzymes, oxygen (in some cases), waste products, essential nutrients, proteins, carbohydrates, and lipids—a complex mixture of all the materials required by the cell for its metabolic functions.

Cytoplasmic Particles Within the bacterial cytoplasm, many submicroscopic particles have been observed. Most of these are ribosomes, often occurring in clusters called polyribosomes ("poly" meaning "many"). Procaryotic ribosomes are smaller than eucaryotic ribosomes, but their function is the same—they are the sites of protein synthesis.

Cytoplasmic granules occur in certain species of bacteria. These may be stained, by use of a suitable stain, and then identified microscopically. The granules may consist of starch, lipids, sulfur, iron, or other stored substances.

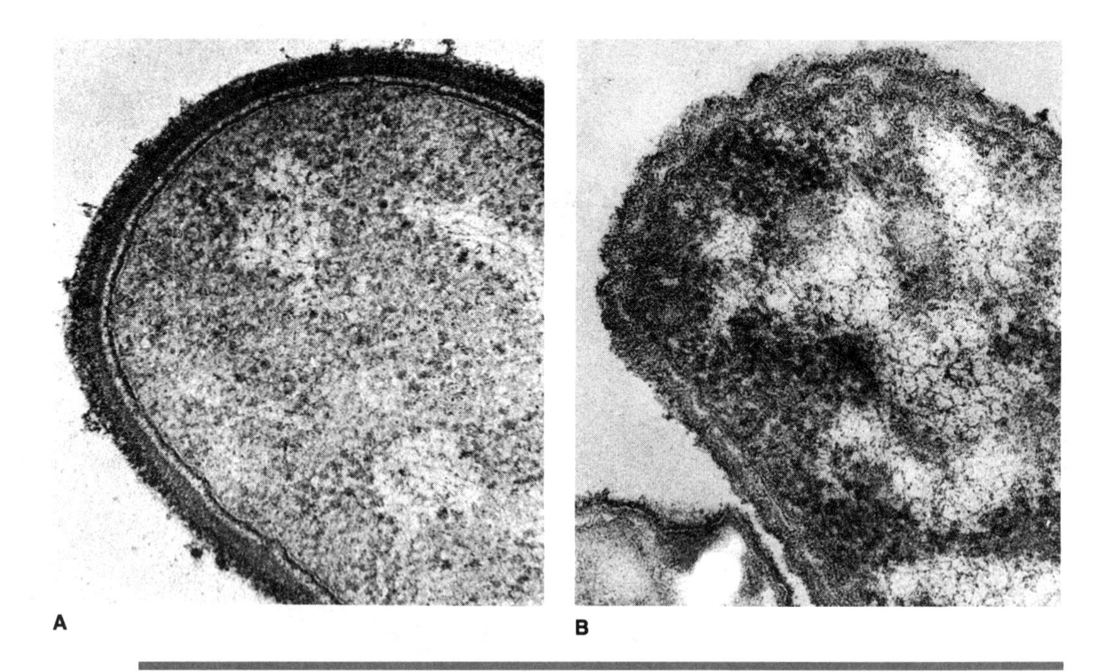

FIGURE 2 – 7
(*A*) A portion of the gram-positive bacterium *Bacillus fastidiosus;* note the cell wall's thick peptidoglycan layer beneath which can be seen the cytoplasmic membrane. (*B*) The gram-negative bacterium *Enterobacter aerogenes;* both the cytoplasmic membrane and the outer membrane are visible along some sections of the cell wall. (Volk WA, et al.: Essentials of Medical Microbiology, 4th ed. Philadelphia, JB Lippincott, 1991)

Cell Membrane Enclosing the cytoplasm is the cytoplasmic, plasma, or cell membrane. This membrane is similar to the eucaryotic cell membrane. Chemically, the plasma membrane consists of proteins and phospholipids, which are discussed further in Chapter 3. Being selectively permeable, the membrane controls which substances may enter or leave the cell. It is flexible and so thin that it cannot be seen with a light microscope. However, it is frequently observed in electron micrographs of bacteria.

Many metabolic reactions take place on the cell membrane. Mesosomes are inward foldings of these membranes, and some scientists believe that this is where cellular respiration takes place in bacteria. This process is similar to that occurring in the mitochondria of animal cells, in which food nutrients are broken down to produce energy in the form of ATP molecules to be used in the cell's metabolic activities.

In photosynthetic bacteria and cyanobacteria (certain bacteria that use light as an energy source) some internal membranes, which are derived from the cell membrane, contain chlorophyll and other pigments that serve to trap light energy for photosynthesis. However, bacteria do not have internal membrane systems similar to the endoplasmic reticulum and Golgi complex of eucaryotic cells.

Bacterial Cell Wall The rigid exterior cell wall that defines the shape of bacterial cells is chemically complex. Thus, it is quite different from the simple cellulose plant cell wall, although it performs the same functions. The main constituent of most bacterial cell walls is a complex macromolecular polymer known as *peptidoglycan* (or murein), consisting of many polysaccharide chains linked together by small peptide (protein) chains. The thickness of this wall and its exact composition vary with the species of bacteria. Certain bacteria, grouped as gram-positive (to be explained later) cells, have many layers of peptidoglycan combined with *teichoic acid* components. Gram-negative (also explained later) bacteria have a much thinner layer of peptidoglycan, but this layer is covered with a complex layer of lipid macromolecules, usually referred to as the outer membrane, as shown in Figure 2 – 7. These macromolecules are discussed in Chapter 3.

Capsules Some bacteria have a layer of material (*glycocalyx*) outside the cell wall. This layer is called a capsule if it is highly organized and firmly attached to the cell wall) or a slime layer if it is not highly organized and not firmly attached. Glycocalyx is a thick layer of slimy, gelatinous material produced by the plasma membrane and secreted outside of the cell wall. Capsules usually consist of complex sugars, or polysaccharides, which may combine with lipids and proteins, depending on the species of microorganism. Knowledge of the chemical composition of capsules is useful in differentiating between different types of bacteria within a particular species; for example, different strains of *Haemophilus influenzae*, a common cause of meningitis in children, are identified by their capsular types. A vaccine, called Hib vaccine, is available for capsular type b.

Capsules can be detected using a *negative stain*, where the bacterial cell and background become stained but the capsule remains unstained. Alternatively, a special capsular staining procedure may be used. Antigen–antibody tests (described in Chapter 9) may be used to identify specific strains of bacteria possessing unique capsular molecules called antigens.

Encapsulated bacteria usually produce colonies on nutrient agar that are smooth (S), mucoid, and glistening, whereas nonencapsulated bacteria tend to grow as dry, rough (R) colonies. Bacterial capsules and slime layers may serve any of several functions. If it is a thin slime layer, the cells may be able to glide, slide, or move on the surface of solid material. Some capsules enable the bacterial species to attach to mucous membranes and tooth surfaces so that they are not flushed away by body secretions. Frequently, encapsulated bacte-

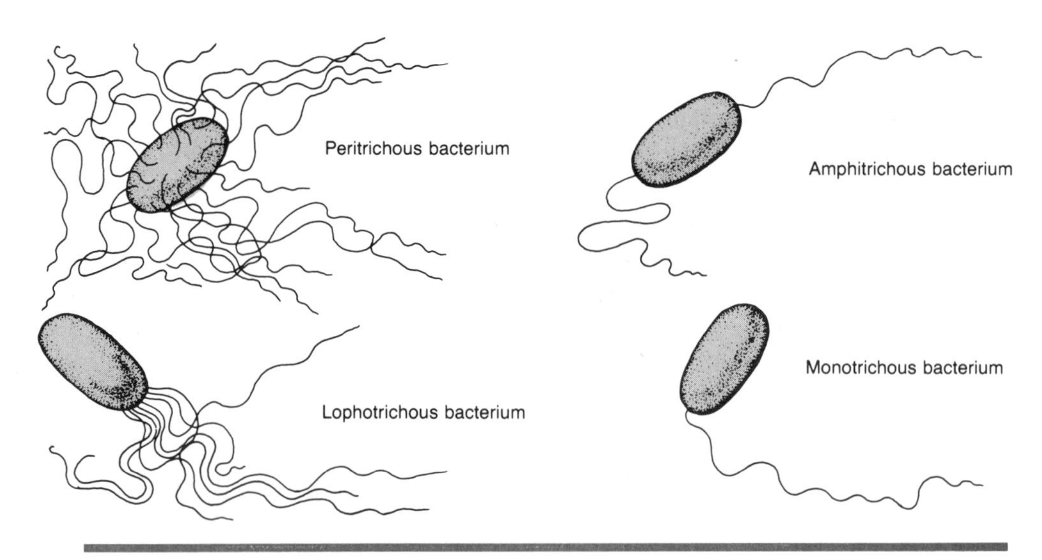

FIGURE 2 – 8
The four basic types of flagella on bacteria: peritrichous flagella all over surface; lophotrichous, a tuft of flagella at one or both ends; amphitrichous, one flagellum at each end; and monotrichous, one flagellum.

ria are not easily ingested by phagocytic white blood cells (phagocytes) and can, therefore, survive longer in the body.

Flagella Many bacteria have flagella, thread-like protein appendages with a whip-like motion that enable bacteria to move. Flagellated bacteria are said to be *motile,* whereas nonflagellated bacteria are usually nonmotile. The number and arrangement of flagella possessed by a certain type of bacterium are characteristics of the species and can therefore be used for classification purposes.

Bacteria possessing flagella over their entire surface (perimeter) are said to be *peritrichous bacteria.* Those with a tuft of flagella at one or both ends are described as being *lophotrichous;* those having one flagellum at each end are *amphitrichous;* and those with a single polar flagellum are *monotrichous bacteria* (Fig. 2 – 8).

Bacterial flagella consist of three, four, or more threads of protein (called *flagellin*) twisted like a rope, unlike eucaryotic flagella and cilia, which have a complex arrangement of microtubules enclosed in a membrane. The flagella of bacteria arise from a basal body in the cell membrane and project outward through the cell wall and capsule, as shown in Figure 2 – 5.

Some *spirochetes* have two flagella-like fibrils called *axial filaments,* one attached to each end of the bacterium. These axial filaments extend toward each other, wrap around the organism between the layers of the cell wall, and overlap in the midsection. As a result, spirochetes can move in a spiral, helical, or inch-worm manner.

Pili or Fimbriae *Fimbriae* (singular, *fimbria*) or *pili* (singular, *pilus*) are hair-like structures most often observed on gram-negative bacteria. They are much thinner than flagella, have a rigid structure, and are not associated with motility. These tiny appendages arise from the cytoplasm and extend through the plasma membrane, cell wall, and capsule. Their functions vary, depending on the bacterial species. They are believed (1) to enable bacteria to attach to other bacteria or to membrane surfaces such as the intestinal lining and red blood cells, (2) to provide a site for the attachment of certain bacterial viruses, and (3) to enable bacteria possessing a *sex pilus* to transfer genetic material to another bacterial cell by a process known as *conjugation* (described in Chapter 4). The pili of *Escherichia coli* can be seen in Figure 2 – 9.

Spores or Endospores A few genera of bacteria (*e.g., Bacillus* and *Clostridium*) are capable of forming endospores (a process called *sporulation*) as a means of survival when their moisture or nutrient supply is low. During sporulation, the genetic material is enclosed in several protein coats that are resistant to heat, drying, and most chemicals. Spores have been shown to survive for many years in soil or dust. When the dried spore lands on a moist,

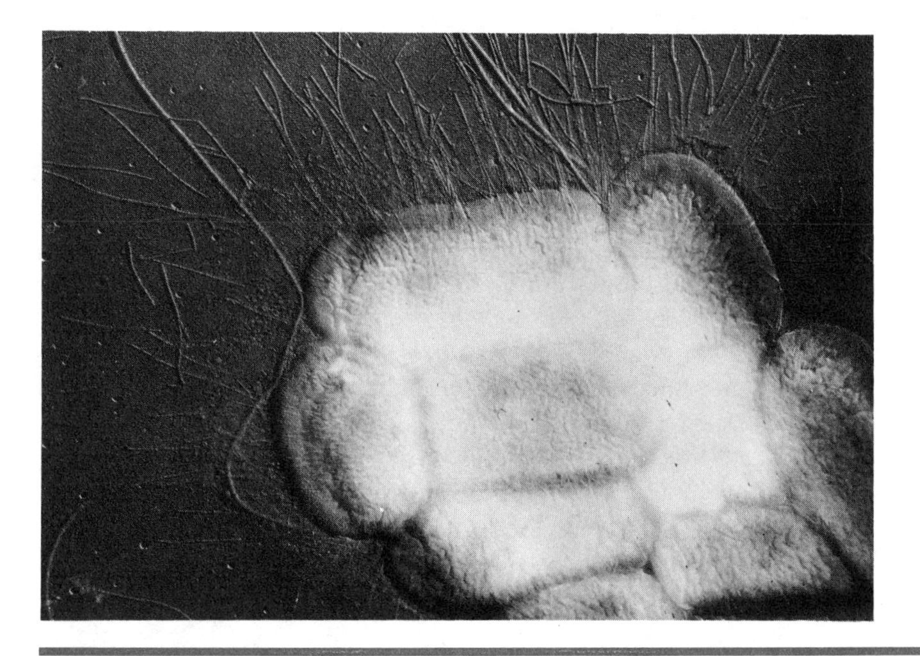

FIGURE 2 – 9
Piliated strain of *Escherichia coli.* Each cell possesses hundreds of pili. Some isolated, broken pili are also seen. A few flagella extend from the cells to the edge of the photograph. (Davis BD, et al.: Microbiology, 4th ed. Philadelphia, Harper & Row, 1987)

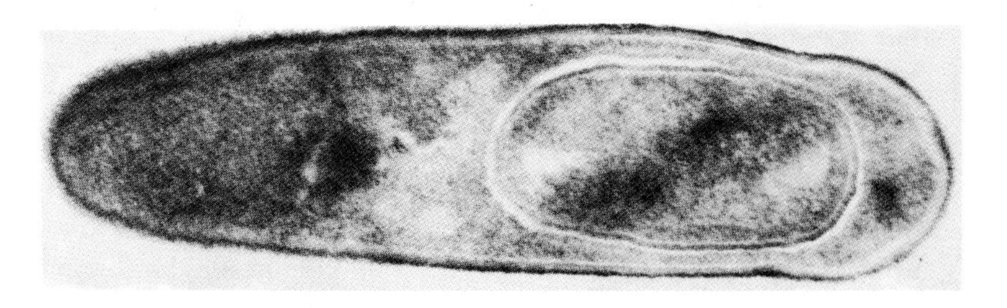

FIGURE 2 – 10
A bacillus with a well-defined endospore. (Lechavalier HA, Pramer D: The Microbes. Philadelphia, JB Lippincott, 1970)

nutrient-rich surface, it germinates, and a new vegetative bacterial cell emerges. Germination of a spore may be compared with germination of a seed. However, spore formation is related to the survival of the bacterial cell, not to reproduction, because usually only one spore is produced in a bacterial cell, and it germinates into only one bacterium (Fig. 2 – 10).

Differences Between Procaryotic and Eucaryotic Cells

Eucaryotic cells are divided into plant and animal types. Animal cells do not have a cell wall, whereas plant cells have a simple cell wall, usually consisting of cellulose. Procaryotic cells have a complex cell wall consisting of proteins, lipids, and polysaccharides. Eucaryotic cells are filled with membrane-bound organelles, such as the endoplasmic reticulum, whereas procaryotic cells have only a few cytoplasmic membranes (mesosomes and photosynthetic membranes), which arise from the plasma membrane. Eucaryotic ribosomes (involved in protein synthesis) are larger and more dense (referred to as 80S ribosomes) than those found in procaryotes (70S ribosomes). The fact that 70S ribosomes are found in the mitochondria and chloroplasts of eucaryotes may indicate that these structures are derived from parasitic procaryotes during their evolutionary development. Other differences are listed in Table 2 – 1.

MICROBIAL CLASSIFICATION

Since Aristotle's time, naturalists have attempted to classify and name plants, animals, and microorganisms in a meaningful way based on their appearance and behavior. Thus, the science of *taxonomy* (biological classification) was devised based on the binomial system developed in the 18th century by the Swedish scientist, Carolus Linnaeus. In the binomial system, each organism is

TABLE 2 – 1
Comparison between Eucaryotic and Procaryotic Cells

	Eucaryotic Cells		
	Plant	Animal	**Procaryotic Cells**
Biological distribution	All plants, fungi, and algae	All animals and protozoa	All bacteria
Nuclear membrane	Present	Present	Absent
Membranous structures other than cell membranes	Generally present	Generally present	Generally absent except mesosomes and photosynthetic membranes
Microtubules and centrioles	Present	Present	Absent
Cytoplasmic ribosomes (density)	80S	80S	70S
Chromosomes	Composed of DNA and proteins	Composed of DNA and proteins	Composed of DNA alone
Flagellla or cilia	When present, have a complex structure similar to centrioles	When present, have a complex structure similar to centrioles	Flagella, when present, have a simple twisted protein structure; no cilia
Cell wall	When present, of simple chemical constitution, usually cellulose	Absent	Of complex chemical constitution, containing peptidoglycan
Active cytoplasmic movements	Present	Present	Absent (not observed)
Photosynthesis (chlorophyll)	Present	Absent	Present in cyanobacteria and some other bacteria

given two names (*e.g., Homo sapiens* for humans). The first name is the *genus* (plural, *genera*), and the second is the *specific epithet.* The first and second names together are referred to as the *species.*

Each organism is categorized into larger groups based on their similarities and differences. According to the currently popular Whittaker classification scheme, all living (and extinct) organisms can be placed in five kingdoms: Animalia for animals, Plantae for plants, Fungi for fungi, Protista for algae and protozoa, and Procaryotae (used in this book) or Monera for cyanobacteria and bacteria. Viruses are not included because they are not truly living cells; they are acellular. Note that four of the five kingdoms consist of eucaryotic organisms. Each kingdom consists of divisions or phyla which, in turn, are divided into classes, orders, families, genera and species. Additional subgroups (such as subspecies) are also used. For an example, see Table 2 – 2. It should be noted

TABLE 2 – 2
Comparison of Human and Bacterial Classification

	Human	Syphilis Pathogen
Kingdom	Animalia	*Procaryotae*
Phylum / Division	Chordata	*Gracilicutes*
Class	Mammalia	*Scotobacteria*
Order	Primate	*Spirochaetales*
Family	Hominidae	*Spirochaetaceae*
Genus } species	*Homo* } species	*Treponema* } species
Specific epithet } species	*sapiens* } species	*pallidum* } species

that not all scientists agree with Whittaker's five kingdom classification system and that other taxonomic classification schemes exist. Protozoa, for example, are sometimes placed into a subkingdom of the Animal Kingdom.

Because written reference is often made to genera and species, biologists throughout the world have adopted a standard method of expressing these names. To express genus, capitalize the first letter of the word and underline or italicize the whole word—for example, *Escherichia.* To express the species, capitalize the first letter of the genus name (the specific epithet is not capitalized) and then underline or italicize the entire species name—for example, *Escherichia coli.* Frequently the genus is designated by a single letter abbreviation; in the example just given, *E. coli* indicates the species. In an essay or article about *Escherichia coli, Escherichia* would be spelled out the first time the organism is mentioned; thereafter, the abbreviated form, *E. coli,* could be used.

BACTERIA

Characteristics

Bacteria were classified into 19 different categories in the single-volume, 8th edition of *Bergey's Manual of Determinative Bacteriology* (1974), which was the standard reference of bacterial classification at the time. In 1984, bacterial classification was restructured because organism identification procedures had become more exact. A more comprehensive edition, renamed *Bergey's Manual of Systematic Bacteriology,* was published in 1984 in four volumes, indicating how the taxonomy of bacteria had expanded. An outline of these volumes can be found in Appendix A. Although about 3,000 different bacteria are described in the 1984 edition, some authorities estimate that these represent only from less than 1% to a few percent of the total number of bacteria in nature. Using computers, microbiologists have established numerical taxonomy systems that not only help to identify bacteria by their characteristics, but also can help establish how closely related these organisms are by comparing the composition of their genetic material and other cell substances.

Many characteristics of bacteria are examined to provide data for identification and classification. These characteristics include the following: (1) morphology, (2) staining, (3) motility, (4) growth, (5) atmospheric requirements, (6) nutritional requirements, (7) biochemical and metabolic activities, (8) pathogenicity, (9) amino acid sequencing of proteins, and (10) genetic composition.

Morphology With the light microscope, the size, shape, and cell arrangement of various bacteria are easily observed. Bacteria vary widely in size, ranging from spheres measuring about 0.2 μm in diameter to spirals 10.0 μm long to even longer filaments. There are three basic shapes (as shown in Figure 2 – 11): (1) spherical or coccoid bacteria—the *cocci* (singular, *coccus*); (2) rod-shaped bacteria—the *bacilli* (singular, *bacillus*); and (3) curved and spiral-shaped bacteria. Cocci are observed in various arrangements (*e.g.*, pairs or *diplococci*, chains or *streptococci*, clusters or *staphylococci*, packets of four or *tetrads*, packets of eight or sarcinae), depending on the particular species and how it divides (Table 2 – 3).

Bacilli (often referred to as rods) may be short or long, thick or thin, pointed or with curved or blunt ends. They may occur singly, in pairs (*diplobacilli*), in chains (*streptobacilli*), in long filaments, or branched. Some rods are very short and, thus, resemble cocci; they are called *coccobacilli. Listeria monocytogenes,* a common cause of neonatal meningitis, is a coccobacillus. Some bacilli stack up next to each other, side by side in a palisade arrangement, which is characteristic of *Corynebacterium diphtheriae* (the cause of diphtheria) and organisms that resemble it in appearance. Vibrios, such as *Vibrio cholerae,* the cause of cholera, are curved (comma-shaped) bacilli.

Spiral-shaped bacteria usually occur singly, but some species may form chains. Different species vary in size, length, rigidity, and the number and amplitude of their coils. Some have rigid cell walls that maintain the shape of a helix or coil. Spirochetes, such as *Treponema pallidum,* the causative agent of syphilis, have a flexible cell wall enabling them to move readily through tissues (Fig. 1 – 10).

Some bacteria may lose their characteristic shape because adverse growth conditions prevent the production of normal cell walls. Such cell-wall–deficient bacteria are called *L-forms.* Some revert to their original shape when placed in favorable growth conditions, whereas others do not. Another group of very small bacteria, the genus *Mycoplasma,* characteristically has no cell wall; thus, microscopically they appear in various shapes (known as *pleomorphism*). Mycoplasmas have no rigid cell shape, reproduce slowly, are relatively fragile, are

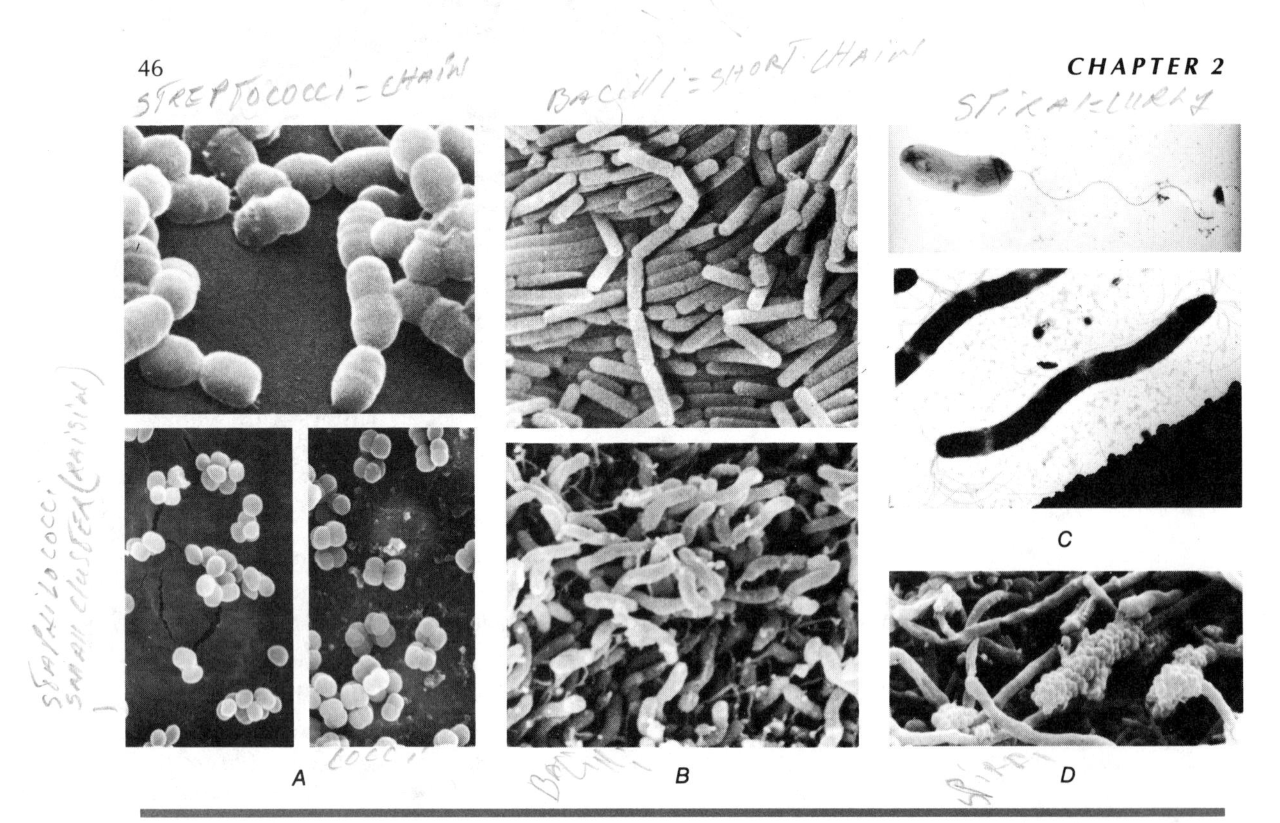

FIGURE 2 – 11
Forms of bacteria. (*A*) Cocci. *Top: Streptococcus mutans,* demonstrating pairs and short chains (original magnification ×9,400). *Bottom left:* Single cells and small clusters of *Staphylococcus epidermidis* (original magnification ×3,000). *Bottom right:* Pairs, tetrads, and regular clusters of *Micrococcus luteus* (original magnification ×3,000). (*B*) Bacilli. *Top:* Single cells and short chains of *Bacillus cereus* (original magnification ×1,700). *Bottom:* Flagellated bacilli (unnamed) associated with periodontitis (original magnification ×3,700). (*C*) *Top:* A cell of *Vibrio cholerae;* note curved cell and single flagellum (original magnification ×8,470). *Bottom:* The spirillum *Aquaspirillum bengal;* note polar tufts of flagella (original magnification ×2,870). (*D*) Variety of organisms (cocci and bacilli) in dental plaque after 3 days without brushing (original magnification ×1,360). (Volk WA, et al.: Essentials of Medical Microbiology, 4th ed. Philadelphia, JB Lippincott, 1991)

very susceptible to changes in osmotic pressure, and are resistant to antibiotics that deter cell wall synthesis.

Staining Various staining methods have been devised to examine bacteria that have been smeared onto a slide and allowed to air dry. Specific stains and techniques are used to observe bacterial morphology (*e.g.,* shape, morphological grouping, type of cell wall, nuclear material, capsules, flagella, endospores, fat globules, and various types of granules).

A simple stain is sufficient to determine bacterial shape and grouping characteristics. For this method, as shown in Figure 2 – 12, a dye (such as methylene

TABLE 2-3
Arrangements of Cocci

Arrangement	Description	Pathogenic Example	Disease
Diplococci	Pairs of cells	*Neisseria gonorrhoeae*	Gonorrhea
Streptococci	Cocci in chains	*Streptococcus pyogenes*	"Strep" throat
Staphylococci	Cocci in clusters	*Staphylococcus aureus*	Boils
Tetrads	Four cocci in a packet	*Micrococcus luteus*	Rarely pathogenic
Octads	Eight cocci in a packet	*Sarcina ventriculi*	Rarely pathogenic

blue) is applied to the air-dried smear, rinsed, dried, and examined using the oil immersion lens of the microscope.

In 1884, Dr. Hans Gram developed a very important staining technique that bears his name—the *Gram stain.* This procedure differentiates between gram-positive and gram-negative bacteria. The color of the bacteria at the end of the Gram-staining procedure (either purple or red) depends on the thickness and

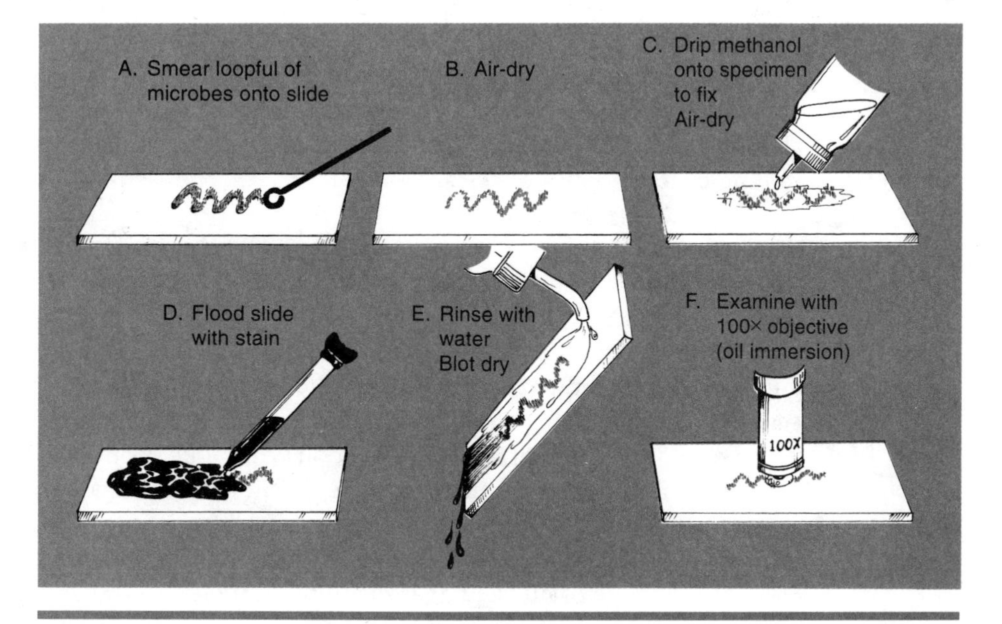

FIGURE 2-12
Simple bacterial staining technique. (*A*) With a flamed loop, smear a loopful of bacteria suspended in broth or water onto a slide. (*B*) Allow slide to air dry. (*C*) Fix the smear with absolute (100%) methanol. (*D*) Flood the slide with the stain. (*E*) Rinse with water. Blot dry. (*F*) Examine the slide with the ×100 microscope objective, using a drop of immersion oil directly on the smear.

chemical composition of the cell wall. This staining process has four steps: (1) flood the fixed and air-dried smear with crystal violet (a purple dye) for 30 seconds; (2) rinse gently with water and cover the smear with Gram's iodine solution; (3) after 30 seconds, wash off the iodine with water and decolorize with ethanol; then (4) counterstain with safranin (a bright red dye) for 1 minute, rinse, dry, and examine under oil. Gram-positive bacteria retain the purple color of the crystal violet. In gram-negative cells, the purple color is removed by the alcohol, and the cells are then stained red by the safranin (see Color Figures 1-9). Some strains of bacteria are neither consistently purple nor red following this procedure; they are referred to as gram-variable bacteria. Examples of such bacteria are members of the genus *Mycobacterium,* such as *M. tuberculosis* and *M. leprae.*

The mycobacteria are more often identified using the *acid-fast stain.* In this method, carbol fuchsin (a bright red dye) is driven into the bacterial cell wall with heat so that the decolorizing agent (a mixture of acid and alcohol) does not remove the red color from the mycobacteria; they are said to be acid-fast. Most other bacteria are decolorized by the acid-alcohol treatment, and are said to be non–acid-fast. The acid-fast stain is especially useful in the "TB lab," where the acid-fast mycobacteria are readily seen as red organisms against a blue or green background in a sputum specimen from a tuberculosis patient (see Color Figures 12, 13).

The procedures for staining bacteria to observe capsules, spores, flagella, fat globules, granules, and nuclear material are described in most microbiology laboratory manuals. Refer to Table 2 – 4 and the Color Figures 1-13 for the staining characteristics of certain pathogens.

Motility The ability of an organism to move by itself is called motility. Bacterial motility is usually associated with the presence of flagella or axial filaments. Most spiral-shaped bacteria and about one half of the bacilli are motile, but cocci are generally nonmotile. Motility can be observed best in a young culture of organisms in a semisolid medium or by the hanging-drop technique. In the latter, motile bacteria suspended in a hanging drop of liquid can be seen darting about in every direction within the drop (Fig. 2 – 13). Some bacteria exhibit gliding motility on secreted slime on solid agar.

Growth The cultural characteristics of colonies of a specific bacterial species vary with the type of nutrient agar medium, nutrients, and dyes present in the medium. The size, color, shape, and consistency of a colony growing on a particular agar medium are characteristic of a given species of bacteria (see Color Figures 14, 15, 16). In a liquid medium, the region in which the organism grows depends on the oxygen needs of that particular species. The rate of growth is also an important characteristic.

COLOR PLATE

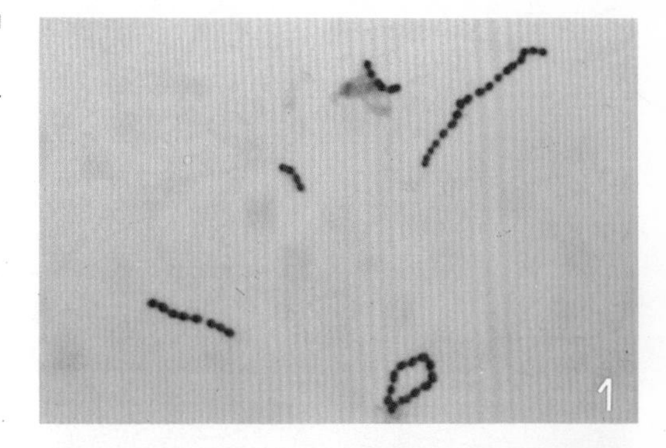

COLOR FIGURE 1
Gram-positive cocci. Streptococci in a Gram-stained smear from a broth culture.

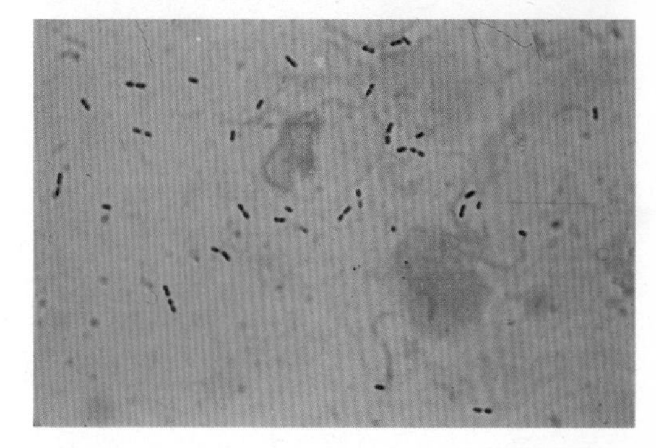

COLOR FIGURE 2
Gram-positive cocci. *Streptococcus pneumoniae* in a Gram-stained smear of a blood culture.

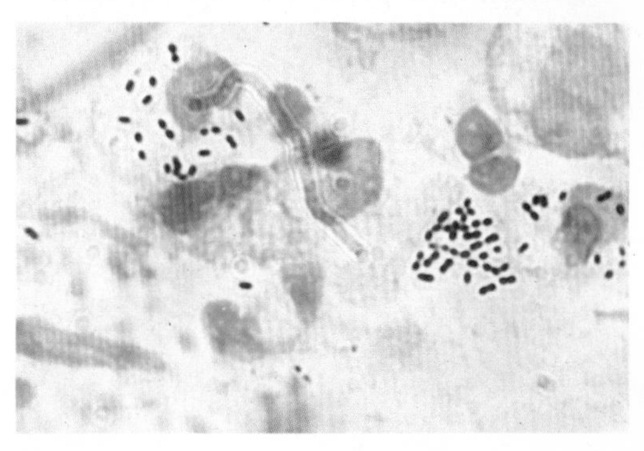

COLOR FIGURE 3
Gram-positive cocci. *Streptococcus pneumoniae* in a Gram-stained smear of a purulent (pus-containing) sputum. Several pink-staining polymorphonuclear leukocytes (PMNs) can be seen.

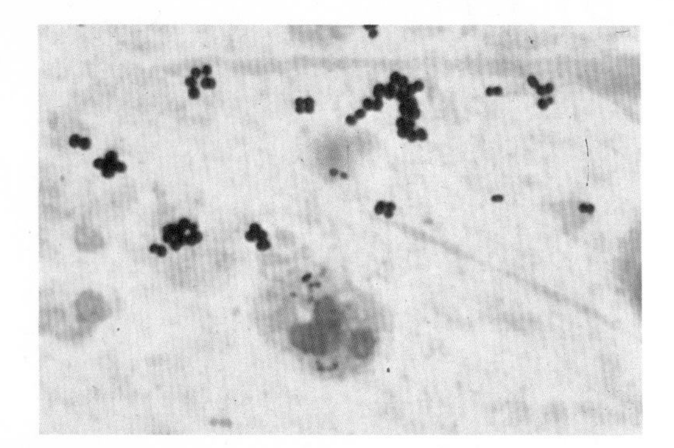

COLOR FIGURE 4
Gram-positive cocci. Staphylococci in a Gram-stained purulent exudate. A pink-staining PMN can be seen.

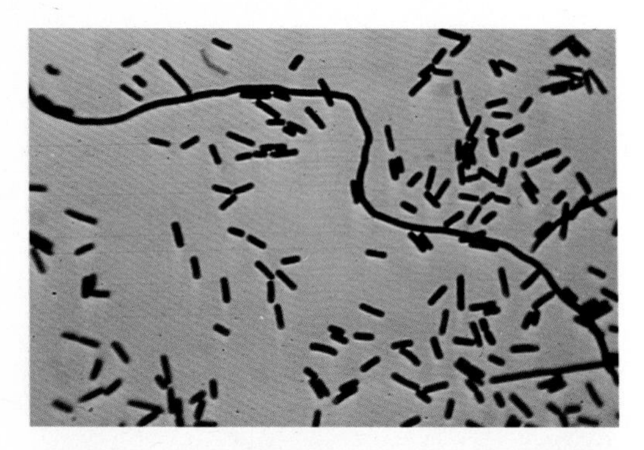

COLOR FIGURE 5
Gram-positive bacilli. *Clostridium perfringens* from a broth culture.

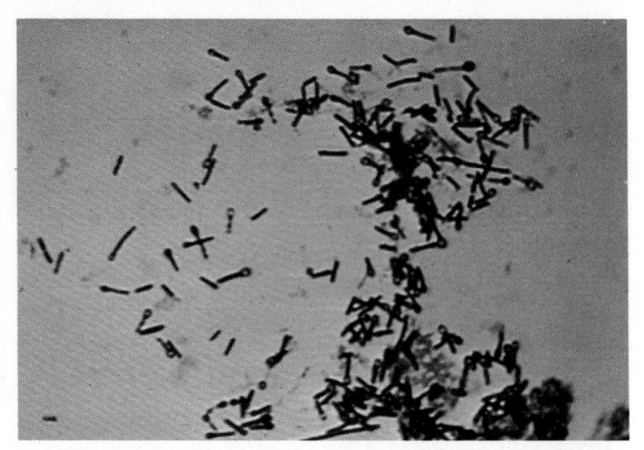

COLOR FIGURE 6
Gram-positive bacilli. *Clostridium tetani* from a broth culture. Terminal spores can be seen on some of the cells.

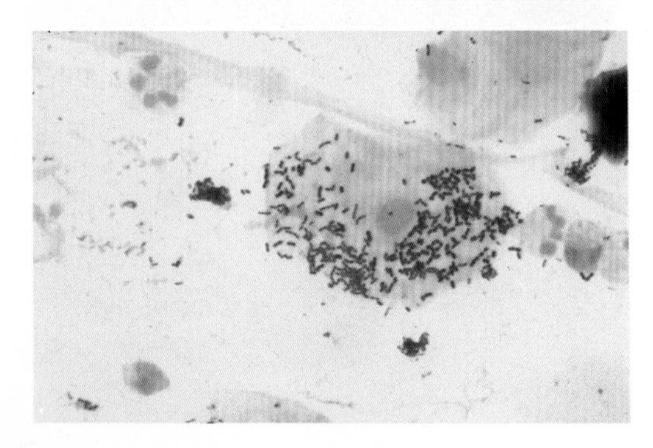

COLOR FIGURE 7
Gram-positive bacteria. Many gram-positive bacteria can be seen on the surface of a pink-stained epithelial cell in this Gram-stained sputum specimen. Several smaller pink-staining PMNs can also be seen.

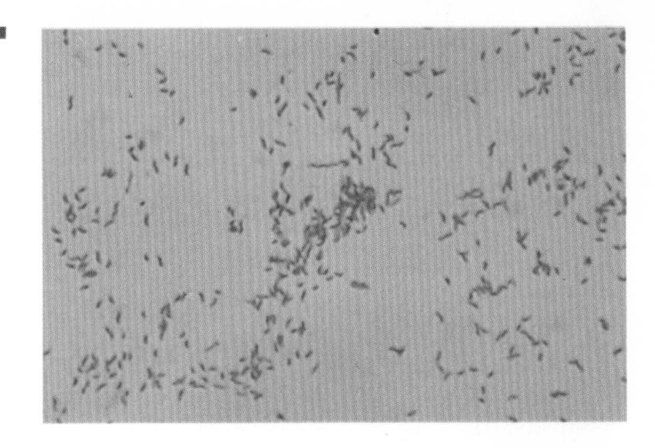

COLOR FIGURE 8
Gram-negative bacilli.

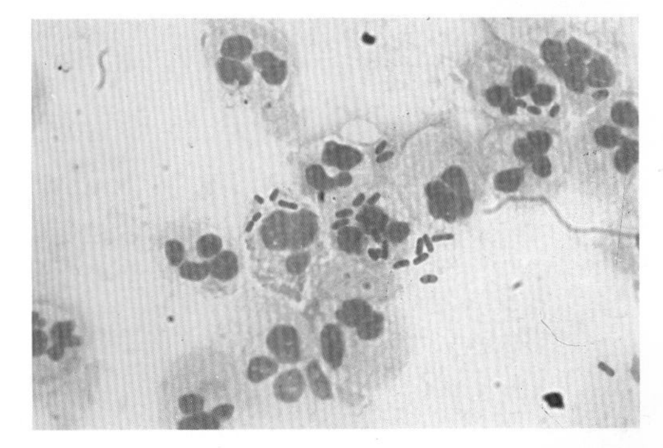

COLOR FIGURE 9
Gram-negative bacilli. Many gram-negative bacilli and many pink-staining PMNs can be seen in this Gram-stained urine sediment from a patient with cystitis.

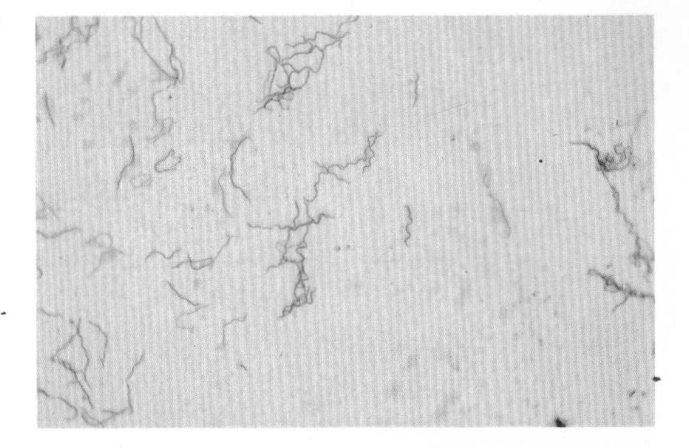

COLOR FIGURE 10
Curved bacilli. *Borrelia burgdorferi,* the etiologic agent of Lyme disease.

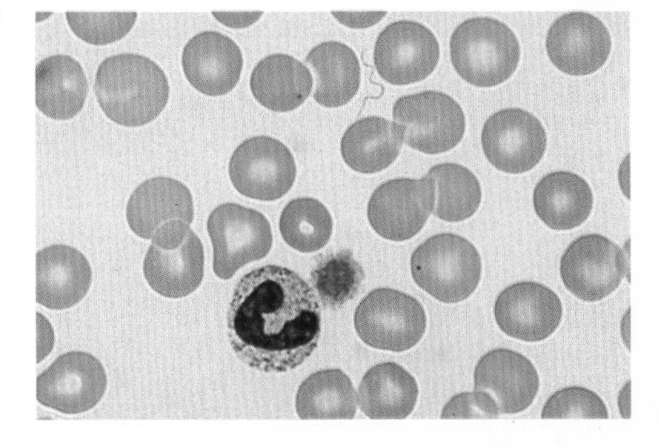

COLOR FIGURE 11
Curved bacilli. Wright's-stained peripheral blood smear from a patient with a *Borrelia* infection. The curved organism can be seen between red blood cells near the top of the photomicrograph. Two white blood cells—a neutrophil and a small lymphocyte—can also be seen.

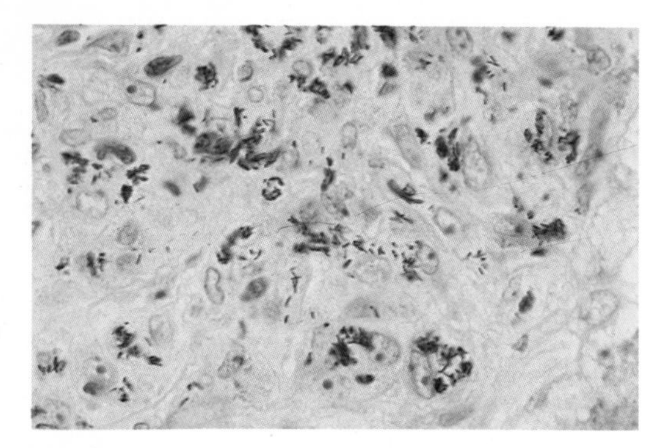

COLOR FIGURE 12
Acid-fast bacteria. Many red acid-fast mycobacteria can be seen in this acid-fast–stained liver biopsy specimen.

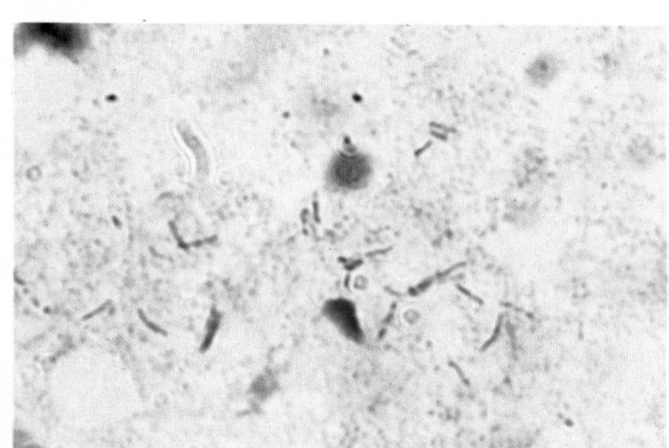

COLOR FIGURE 13
Acid-fast bacteria. Many red acid-fast cells of *Mycobacterium tuberculosis* can be seen in this acid-fast–stained concentrate from a sputum digest.

COLOR FIGURE 14
Beta hemolysis. Colonies of a beta-hemolytic *Streptococcus* species on a blood agar plate.

COLOR FIGURE 15

Colonies on MacConkey agar, a selective and differential medium. It is selective for gram-negative bacteria. Colonies of both lactose-fermenters (pink colonies) and non–lactose-fermenters (clear colonies) can be seen.

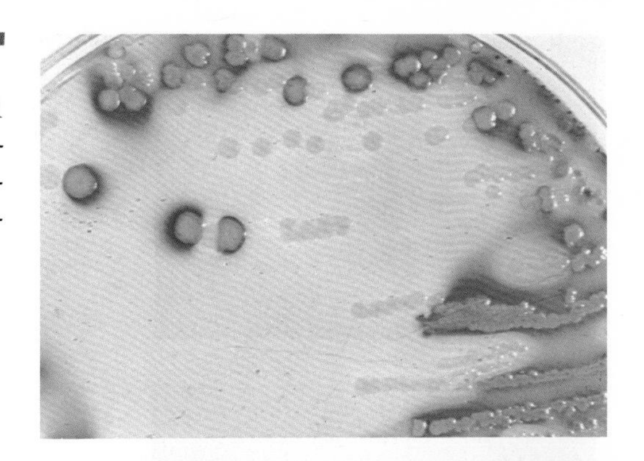

COLOR FIGURE 16

Mannitol salt agar. This agar medium is used to screen for *Staphylococcus aureus.* Any bacteria capable of growing in a 7.5% sodium chloride concentration will grow on this medium, but only *S. aureus* will turn the medium yellow, due to mannitol fermentation.

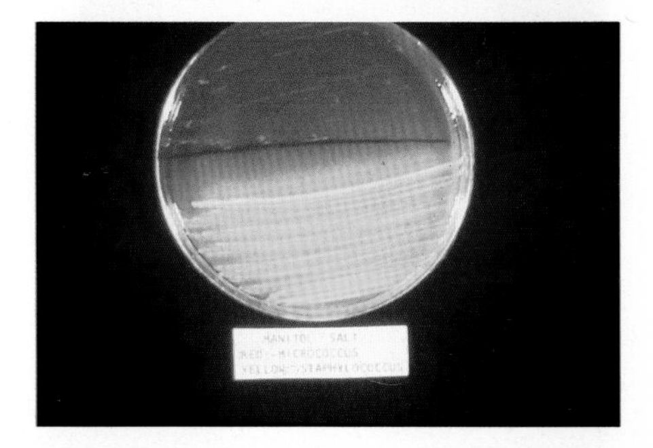

COLOR FIGURE 17

Mini-system for bacterial identification. Shown here is the API 20-E® strip, used to identify members of the family *Enterobacteriaceae.*

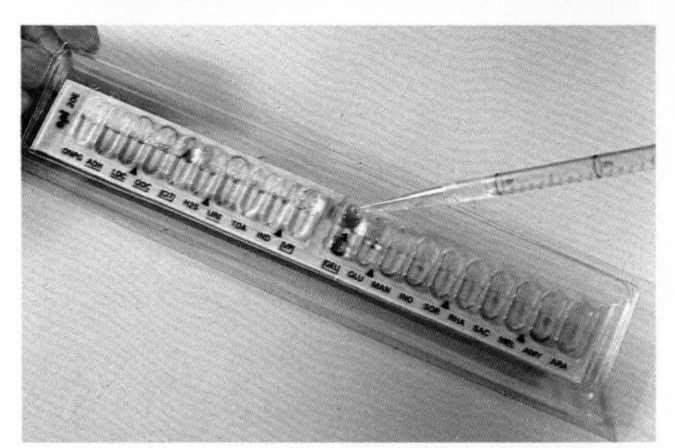

COLOR FIGURE 18

Mini-system for bacterial identification. Shown here is the Enterotube II® system, used to identify members of the family *Enterobacteriaceae.*

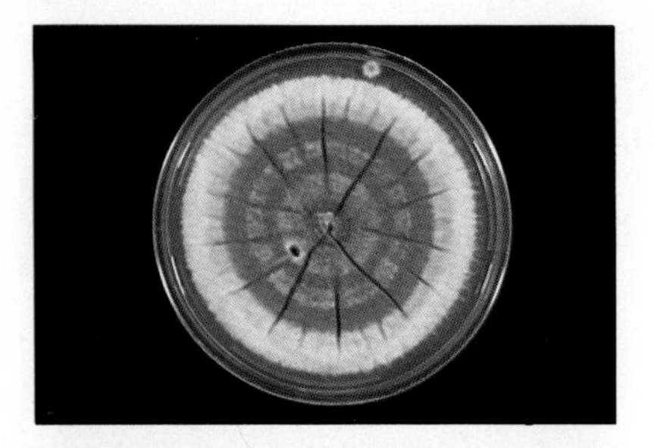

COLOR FIGURE 19

Mold colony. Pictured here is a colony (mycelium) of *Aspergillus fumigatus,* a common cause of pulmonary infections in immunosuppressed patients.

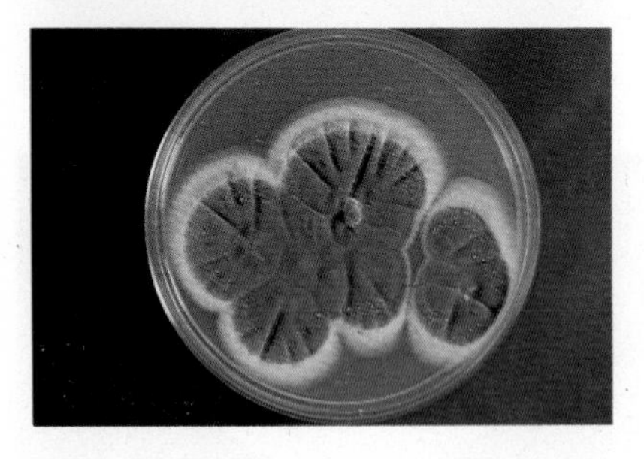

COLOR FIGURE 20

Mold colony. Pictured here is a colony (mycelium) of a *Penicillium* species.

TABLE 2 – 4
Characteristics of Some Important Pathogenic Bacteria

Bacterium	Diseases	Type	Gram-Stain Reaction[a]
Bacillus anthracis	Anthrax	Spore-forming rod	+
Bordetella pertussis	Whooping cough	Rod	−
Brucella abortus and *B. melitensis*	Brucellosis, undulant fever	Rod	−
Chlamydia trachomatis	Lymphogranuloma venereum, trachoma	Coccoid	−
Clostridium botulinum	Botulism (food poisoning)	Spore-forming rod	+
Clostridium perfringens	Gas gangrene, wound infections	Spore-forming rod	+
Clostridium tetani	Tetanus (lockjaw)	Spore-forming rod	+
Corynebacterium diphtheriae	Diphtheria	Rod	+
Escherichia coli	Urinary tract infections	Rod	−
Francisella tularensis	Tularemia	Rod	−
Haemophilus ducreyi	Chancroid	Rod	−
Haemophilus influenzae	Meningitis, pneumonia	Rod	−
Klebsiella pneumoniae	Pneumonia	Rod	−
Mycobacterium leprae	Leprosy	Rod	+/−
Mycobacterium tuberculosis	Tuberculosis	Rod	+/−
Mycoplasma pneumoniae	Atypical pneumonia	Pleomorphic	−
Neisseria gonorrhoeae	Gonorrhea	Diplococcus	−
Neisseria meningitidis	Nasopharyngitis, meningitis	Diplococcus	−
Proteus vulgaris and *P. morgani*	Gastroenteritis, urinary tract infections	Rod	−
Pseudomonas aeruginosa	Respiratory, urogenital, and wound infections	Rod	−
Rickettsia rickettsii	Rocky Mountain spotted fever	Rod	−
Salmonella typhi	Typhoid fever	Rod	−
Salmonella species	Gastroenteritis	Rod	−
Shigella species	Shigellosis (bacillary dysentery)	Rod	−

continued

TABLE 2 – 4
(Continued)

Bacterium	Diseases	Type	Gram-Stain Reaction[a]
Staphylococcus aureus	Boils, carbuncles, pneumonia, septicemia	Cocci in clusters	+
Streptococcus pyogenes	"Strep" throat, scarlet fever, rheumatic fever, septicemia	Cocci in chains	+
Streptococcus pneumoniae	Pneumonia, meningitis	Diplococcus	+
Treponema pallidum	Syphilis	Spirochete	−
Vibrio cholerae	Cholera	Curved rod	−
Yersinia pestis	Plague	Rod	−

[a] + = gram-positive; − = gram-negative; +/− = gram-variable

Atmospheric Requirements In the microbiology laboratory, it is useful to classify bacteria on the basis of their relationship to oxygen (O_2) and carbon dioxide (CO_2). A bacterial isolate can be classified into one of five major groups: *obligate aerobe, microaerophilic aerobe* (microaerophile), *facultative anaerobe, aerotolerant anaerobe, and obligate anaerobe.*

To grow and multiply, obligate aerobes require an atmosphere containing molecular oxygen in concentrations comparable to that found in room air (*i.e.,* 20–21% O_2). Mycobacteria and certain fungi are examples of microorganisms that are obligate aerobes. Microaerophiles (microaerophilic aerobes) also require oxygen for multiplication, but in concentrations lower than that found in

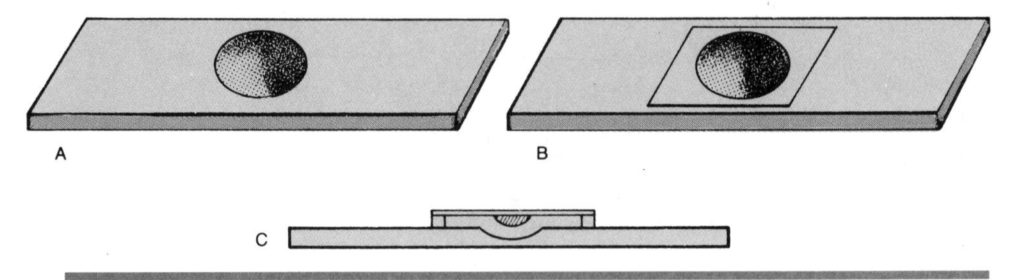

FIGURE 2 – 13
Hanging drop preparation for study of living bacteria. (*A*) Depression slide. (*B*) Depression slide with coverglass over the depression area. (C) Side view of hanging drop preparation, showing the drop of culture hanging from the center of the coverglass above the depression. (Volk WA, Wheeler MF: Basic Microbiology, 5th ed. Philadelphia, JB Lippincott, 1984)

INSIGHT
The Discovery of Anaerobes

Scientists once believed that life in the absence of oxygen was impossible, but we now know differently. We know that there are organisms—obligate anaerobes—that can live in the total absence of oxygen.

Three scientists deserve credit for the discovery—Leeuwenhoek, Spallanzani, and Pasteur. Each made scientific observations that contributed to our knowledge and understanding of anaerobes.

In 1680, Antony van Leeuwenhoek described an experiment he performed using pepper and sealed glass tubes. He wrote that "animalcules developed although the contained air must have been in minimal quantity."

Lazzaro Spallanzani, an Italian scientist, performed similar experiments in the latter half of the 18th century. He drew the air from microbe-containing glass tubes, fully expecting the microbes to die—but some did not. He wrote in a letter to a friend, "The nature of some of these animalcules is astonishing! They are able to exercise in a vacuum the functions they use in free air. . . . How wonderful this is! For we have always believed there is no living being that can live without the advantages air offers it."

It was Louis Pasteur who actually introduced the terms "aerobe" and "anaerobe." In an 1861 paper, he wrote "these infusorial animals are able to live and multiply indefinitely in the complete absence of air or free oxygen. . . . These infusoria can not only live in the absence of air, but air actually kills them. . . . I believe this is . . . the first example of an animal living in the absence of free oxygen." In 1877, Pasteur discovered a pathogenic anaerobe, the bacterium that today is known as *Clostridium septicum.*

room air. *Neisseria gonorrhoeae* (the cause of gonorrhea) and *Campylobacter* species (causes of bacterial diarrhea) are examples of microaerophilic bacteria that prefer an atmosphere containing about 5% oxygen.

Anaerobes can be defined as organisms that do not require oxygen for life and reproduction. However, they vary in their sensitivity to oxygen. The terms obligate anaerobe, aerotolerant anaerobe, and facultative anaerobe are used to describe the organism's relationship with molecular oxygen. An obligate anaerobe is an anaerobe that grows only in an anaerobic environment. It will not grow in a microaerophilic environment (about 5% O_2), in a CO_2 incubator (about 15% O_2), or in air.

An aerotolerant anaerobe does not require oxygen and grows better in the absence of oxygen, but can survive in atmospheres containing molecular oxygen (such as air and a CO_2 incubator). Facultative anaerobes are capable of surviving in either the presence or absence of oxygen; anywhere from 0% O_2 to 20–21% O_2. Many of the bacteria routinely isolated from clinical specimens are facultative anaerobes (*e.g.,* members of the family *Enterobacteriaceae,* streptococci, staphylococci).

Some bacteria grow better in the presence of increased concentrations of CO_2. Such organisms are referred to as *capnophiles* (capnophilic organisms). Some anaerobes (*e.g., Bacteroides* and *Fusobacterium* species) are capnophiles, as are some aerobes (*e.g.*, certain *Neisseria* and *Campylobacter* species). Capnophilic aerobes will grow in a candle extinction jar, but not in room air. A candle extinction jar (or candle jar, as it is sometimes called) generates a final atmosphere containing 12–17% O_2 and 3–5% CO_2.

Nutritional Requirements All bacteria need some form of the elements carbon, hydrogen, oxygen, sulfur, phosphorus, and nitrogen for growth. Special elements such as potassium, calcium, iron, manganese, magnesium, cobalt, copper, zinc, and uranium are needed by certain bacteria. Some have specific vitamin requirements; others need organic substances secreted by other living microorganisms during their growth. The nutritional needs are characteristic of the species of bacteria. These nutritional requirements are discussed in Chapter 4.

Biochemical and Metabolic Activities As bacteria grow, they produce many waste products and secretions, some of which are enzymes that enable them to invade their host and cause disease. The pathogenic strains of many bacteria, such as staphylococci and streptococci, can be tentatively identified by the enzymes they secrete. Also, in particular environments, certain bacteria are characterized by the production of carbon dioxide, hydrogen sulfide, oxygen, or methane.

Pathogenicity The disease-producing abilities of pathogens are important to understand. Many pathogens are able to cause disease because they have capsules or endotoxins (a part of the cell wall of gram-negative bacteria) or because they secrete exotoxins and enzymes that damage cells and tissues (described in Chapter 7). Frequently, pathogenicity (the ability to cause disease) is tested by injecting the organism into mice. Some common pathogenic bacteria are listed in Table 2 – 4.

Amino Acid Sequencing of Proteins Some proteins found within a bacterium are specific for that species. Thus, by comparing the amino acid sequence of certain bacterial proteins, the species and how closely related it is to other bacteria can be determined.

Genetic Composition The composition of the genetic material (DNA) is unique to each species. Thus, by determining the base composition and by comparing the cytosine / guanine ratio with the total amount of bases, a numerical ratio can be calculated (see Chapter 3). Also, by identifying or hybridizing a sequence of bases in portions of DNA or RNA, the researcher can determine the degree of relationship between two different bacteria and, perhaps, the particular species or strain of bacteria.

RUDIMENTARY FORMS OF BACTERIA

The rickettsias, chlamydias, and mycoplasmas are bacteria, but they do not possess all of the attributes of typical bacterial cells. Because they are so small and difficult to isolate, they were formerly thought to be viruses.

Rickettsias and Chlamydias

Rickettsias and chlamydias are coccoid, rod-shaped, or pleomorphic (irregular) gram-negative bacteria with a bacterial-type cell wall; unlike viruses, they contain both DNA and RNA. Most known forms are obligate intracellular parasites that are pathogenic to humans and other animals. As the name implies, an obligate intracellular parasite is a parasite that must live within a host cell.

Because they appear to have leaky cell membranes, most rickettsias must live inside another cell to retain all the necessary cellular substances. They are usually transmitted by arthropod *vectors* (carriers). An exception is *Coxiella burnetii,* the cause of Q fever, which may also be airborne or foodborne.

Arthropods like lice, fleas, and ticks are frequently vectors of rickettsial diseases; they transmit pathogens from one host to another by their bites or waste products. Diseases caused by arthropod-borne rickettsias include Rocky Mountain spotted fever and typhus fevers. Closely related organisms called *Bartonella* species are associated with cat scratch disease, bacteremia, and endocarditis.

Chlamydias are probably the most primitive of all bacteria because they lack the enzymes required to perform many essential metabolic activities, particularly production of ATP. They are obligate intracellular parasites that are transferred by direct contact between hosts, not by arthropods.

Chlamydias have two forms in their life cycle (Fig. 2 – 14). The infectious form, called the elementary body, attaches to the host cell. After it is engulfed by the host cell, the elementary body reorganizes into the larger, less-infectious form, called the reticulate body. This form finally divides to produce many small infectious elementary bodies that are released to spread and infect surrounding host cells or other individuals. Chlamydias are easily transmitted during sexual contact and cause infections of the urethra (urethritis), bladder, (cystitis), fallopian tubes (salpingitis), prostate (prostatitis), or other complications. They cause many diseases in vertebrate animals (cats, dogs, sheep, cattle, birds, humans), including trachoma (an eye disease), lymphogranuloma venereum (LGV; a sexually transmitted disease), and psittacosis (a respiratory infection in humans and birds, often called "parrot fever").

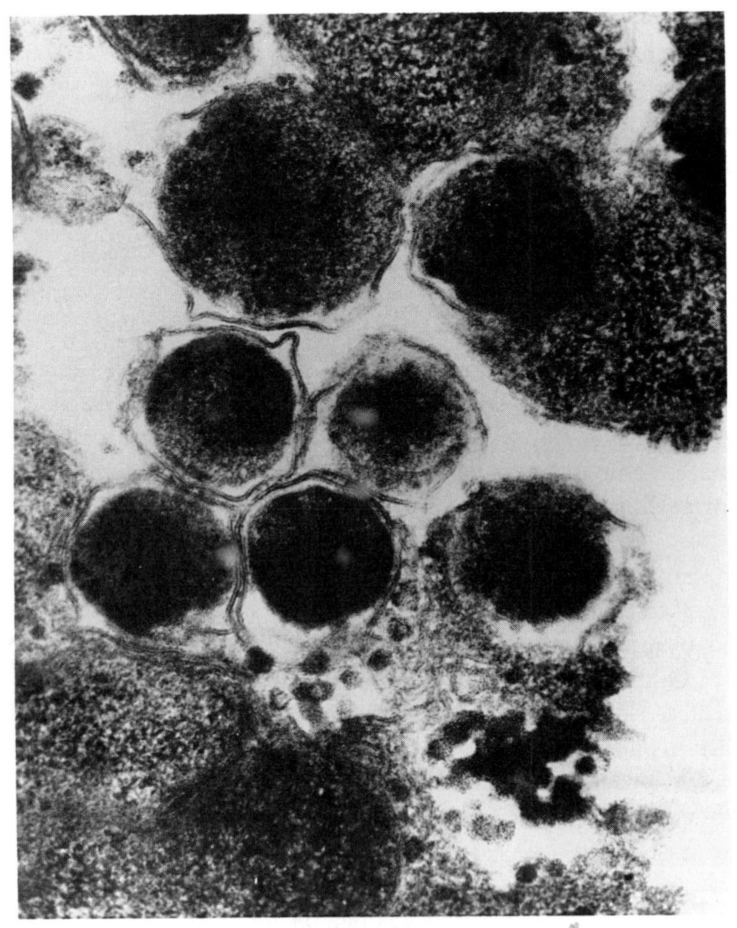

FIGURE 2 – 14
Elementary bodies of chlamydia. (Courtesy of S. Koester)

Mycoplasmas

Mycoplasmas are the smallest of the cellular microbes (Fig. 2 – 15). Because they lack cell walls, they assume many shapes, from coccoid to filamentous. Sometimes they are confused with the L-forms of bacteria that were described earlier; however, even in the most favorable growth media, mycoplasmas are not able to produce cell walls, which is not true for L-forms. Mycoplasmas were formerly called pleuropneumonia-like organisms (PPLO) because they were first isolated from cattle with lung infections. These organisms may be free-living or parasitic and are pathogenic to many animals and some plants. In humans, pathogenic mycoplasmas cause primary atypical pneumonia and many secondary infections. Because they have no cell wall, they are resistant to treatment with penicillin and other antibiotics that work by inhibiting cell wall synthesis.

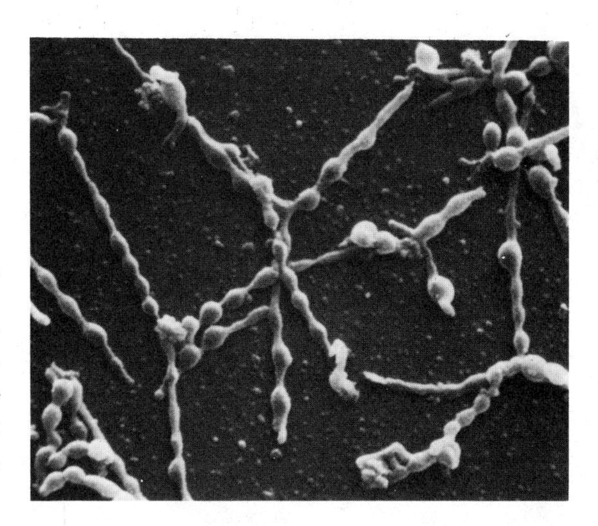

FIGURE 2 – 15
Scanning electron micrograph of *Mycoplasma pneumoniae;* note the coccoid structures within the filaments (original magnification ×10,000). (Volk WA, et al.: Essentials of Medical Microbiology, 4th ed. Philadelphia, JB Lippincott, 1991)

PROTOZOA

Protozoa are eucaryotic, usually single-celled, animal-like microorganisms, ranging in length from 3 to 2000 μm. Most of them are free-living organisms found in soil and water. They have no chlorophyll; therefore, they cannot make their own food by photosynthesis. Some ingest whole algae, yeasts, bacteria, and other smaller protozoans as their source of nutrients (Fig. 2 – 16); others live on dead, decaying organic matter. Parasitic protozoa break down and absorb nutrients from the body of the host in which they live. Many parasitic protozoa are pathogens, such as those that cause malaria, giardiasis, and amebic dysentery (see Chapter 10). Other parasitic protozoa co-exist with the host animal in a type of symbiotic (living together) relationship, wherein both organisms benefit. A typical example of symbiosis is the termite and its intestinal protozoa, which can digest the wood eaten by the termite; thus, both organisms absorb the nutrients necessary for life. Without the parasitic protozoa, the termite cannot digest wood, its main source of food, and thereby starves to death.

Classification

Protozoa are divided into groups (phyla or classes) according to their method of locomotion or the presence or absence of cilia (short, hair-like structures) and flagella (Table 2 – 5). The *amebas* or amebae (subphylum *Sarcodina* in the phylum *Sarcomastigophora*), lack cilia and flagella. They move by extending a bit of cytoplasm (a *pseudopodium,* or false foot) and slowly flowing into it; this process is called ameboid movement. An ameba engulfs food by surrounding the food

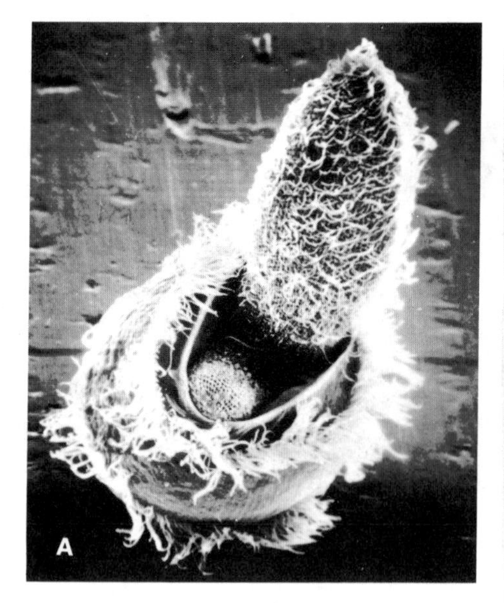

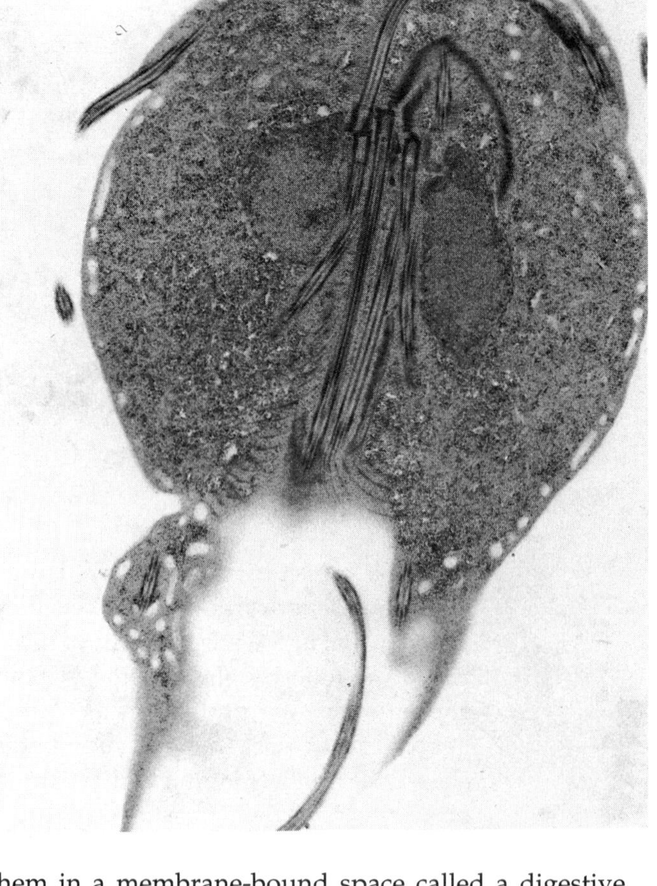

FIGURE 2 – 16
Protozoa. (*A*) *Didinium nasutum* with partially ingested prey. (Lechavalier HA, Pramer D: The Microbe. Philadelphia, JB Lippincott, 1970). (*B*) TEM longitudinal section of *Giardia lamblia.* (From S. Koester and P. Engelkirk)

particles and enclosing them in a membrane-bound space called a digestive vacuole, where the particles are digested and used as a source of nutrients. Phagocytosis is the engulfment of large particles, whereas *pinocytosis* is the intake of fluids that may contain dissolved nutrients. Some of our white blood cells move through tissues and ingest materials in the same manner as amebae. Phagocytosis is further discussed in Chapter 9. One important pathogenic ameba is *Entamoeba histolytica,* which causes amebic dysentery and additional extraintestinal (outside the intestine) abscesses.

The *flagellates* (subphylum Mastigophora in the phylum *Sarcomastigophora*) move by means of one or more flagella. Some species are pathogenic. For example, *Trypanosoma brucei gambiense* is transmitted by the tsetse fly and causes African sleeping sickness in humans; *Trichomonas vaginalis* causes persistent infections (trichomoniasis) of the male and female genital tracts; and *Giardia lamblia* causes a persistent intestinal infection (giardiasis).

Ciliates (phylum *Ciliophora*) move about as a result of large numbers of cilia on their surfaces. They are the most complex of all the protozoa. A pathogenic ciliate, *Balantidium coli,* causes a severe intestinal infection in pigs and humans.

TABLE 2–5
Characteristics of Major Protozoa

Phylum	Means of Movement	Selected Differentiating Properties (Method of Reproduction)		Representatives
		Asexual	Sexual	
Ciliophora	Cilia	Transverse fission	Conjugation	*Balantidium coli, Paramecium, Stentor, Tetrahymena, Vorticella*
Sarcodina	Pseudopodia (false feet)	Binary fission	When present, involves flagellated sex cells	*Amoeba, Naegleria, Entamoeba histolytica*
Mastigophora	Flagella	Binary fission	None	*Chlamydomonas, Giardia lamblia, Trichomonas, Trypanosoma*
Sporozoea	Generally nonmotile except for certain sex cells	Multiple fission	Involves flagellated sex cells	*Plasmodium, Toxoplasma gondii, Cryptosporidium, Pneumocystis carinii*

In underdeveloped countries, it is usually transmitted to humans via swine feces. *B. coli* is the only ciliated protozoan that causes disease in humans.

The nonmotile protozoa are classified as sporozoa (phylum *Sporozoea*). The most important pathogens are those of the *Plasmodium* species that cause malaria in many areas throughout the world. One of these, *Plasmodium vivax,* causes a relatively mild form of malaria in some parts of the United States. These pathogens are carried and transmitted by the female *Anopheles* mosquito. Two sporozoans, not previously recognized as serious pathogens, *Pneumocystis carinii* and *Cryptosporidium,* cause severe secondary infections in immunosuppressed patients, especially those with acquired immunodeficiency syndrome (AIDS) (see Chapter 10). *Pneumocystis* causes a type of pneumonia (called PCP), and *Cryptosporidium* causes a diarrheal disease (cryptosporidiosis). A 1993 epidemic in Milwaukee, Wisconsin, caused by *Cryptosporidium* oocysts in drinking water resulted in more than 400,000 cases of cryptosporidiosis.

FUNGI

Fungi are found almost everywhere on earth—living on organic matter in water and soil and living on animals and plants. They may be harmful or beneficial. Fungi also live on many unlikely materials, causing deterioration of leather and plastics and spoilage of jams, pickles, and many other foods. Beneficial fungi are important in the production of cheese, yogurt, beer, and wine and other foods and beverages as well as certain drugs and antibiotics.

Characteristics

Fungi are eucaryotic organisms that include mushrooms, molds, and yeasts. As saprophytes, their main source of food is dead and decaying organic matter. Fungi are the "garbage disposers" of nature, the "vultures" of the microbial world. By secreting digestive enzymes into dead plant and animal matter, they decompose this material into absorbable nutrients for themselves and other living organisms; thus they are the original "recyclers." Imagine living in a world without saprophytes, stumbling through endless piles of decaying debris! Some fungi also live as parasites, on living animals and plants.

Fungi are often incorrectly referred to as plants. They differ from plants and algae in that they are not photosynthetic—they have no chlorophyll or other photosynthetic pigments. Whereas the cell walls of plant cells contain cellulose, fungal cell walls contain a substance called chitin. Although many fungi are unicellular during some phase of their life cycle (*e.g.*, yeasts), others grow as filaments called *hyphae* (singular, *hypha*), which intertwine to form a mass called a *mycelium* (plural, *mycelia*); thus, they are different from saprophytic bacteria. Remember that bacteria are procaryotic, whereas fungi are eucaryotic.

Reproduction Depending on the particular species, fungal cells can reproduce asexually by budding, by hyphal extension, or by the formation of spores (Fig. 2 – 17).

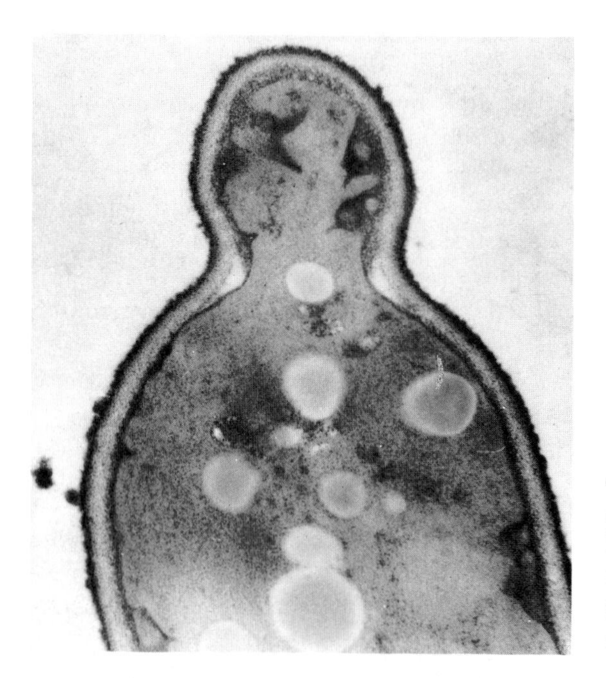

FIGURE 2 – 17
Longitudinal section of a budding yeast cell (original magnification ×15,500). (Lechavalier HA, Pramer D: The Microbes. Philadelphia, JB Lippincott, 1970)

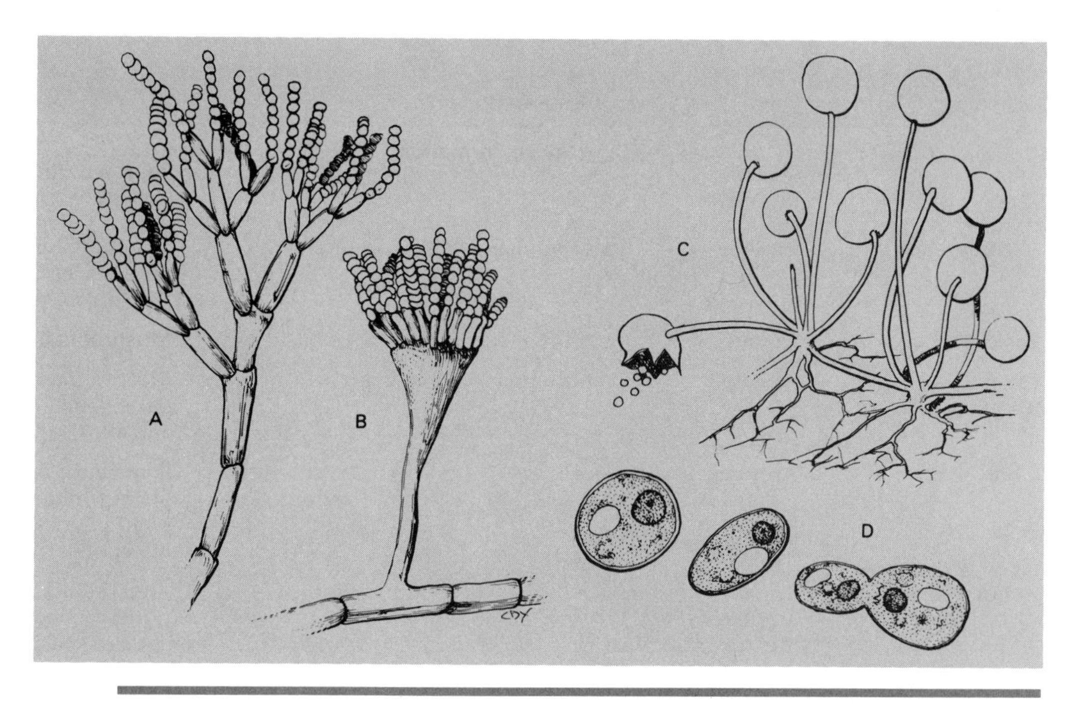

FIGURE 2 – 18
Typical fungi. (*A*) *Penicillium* mold, blue-green spores arranged like a brush. (*B*) *Aspergillus*, blue-green with yellow areas. (*C*) *Rhizopus*, white to dark gray with root-like rhizoids. (*D*) *Saccharomyces cerevisiae*, a yeast.

Spores Sexual reproduction in many fungi involves the nuclear fusion of two gametes and their subsequent division into many sexual spores. Some species of fungi produce both asexual and sexual spores or more than one type of asexual spore. Asexual spores are also called *conidia* (singular, *conidium*). Fungi are classified in accordance with the type of sexual spore that they produce or the type of structure upon which spores are produced. Figure 2 – 18 illustrates some typical fungi. Fungal spores are very resistant structures that are carried great distances by wind. They are resistant to heat, cold, acids, bases, and other chemicals.

Classification of True Fungi

Mycologists, scientists who study fungi, have separated the so-called "true fungi" into five classes: Basidiomycetes, Oomycetes, Zygomycetes, Ascomycetes, and Deuteromycetes. These classes are based on mode of reproduction and types of mycelia, spores, and gametes that are produced. The characteristics of each of these classes are shown in Table 2 – 6.

TABLE 2–6
Selected Characteristics of the Major Classes of Fungi

Class	Type of Mycelium	Site of Spore Formation		Representative Groups
		Asexual	Sexual	
Ascomycetes	Septate	Tips of hyphae	Within sacs	Common antibiotic-producing fungi, *Penicillium,* yeasts
Basidiomycetes	Septate	Tips of hyphae	Surface of basidium	Mushrooms, rusts
Deuteromycetes (fungi imperfecti)	Septate	Tips of hyphae	None present	Most human pathogenic molds and yeasts
Oomycetes	Aseptate	In sacs	Within a unicellular female sex organ (oogonium)	Some aquatic forms, mildew, plant blights, fish infections
Zygomycetes	Usually aseptate	In sacs	In mycelium	Bread mold (*Rhizopus nigricans*), aquatic species

Mushrooms Mushrooms (Basidiomycetes) are a class of true fungi that consist of a network of filaments or strands (mycelium) that grow in the soil or in a rotting log, and the fruiting body (mushroom) that forms and releases spores. A spore, much like the seed of a plant, germinates into a new organism. Many mushrooms are delicious to eat, but many that resemble edible fungi are extremely toxic and may cause permanent brain damage or death.

Molds Molds are the fungi often seen in water and soil and on food. They grow in the form of filaments or hyphae that make up the mycelium of the mold (see Color Figures 19, 20). Reproduction is by spore formation, either sexually or asexually, on the reproductive hyphae. Various species of molds are found in each of the classes of fungi except Basidiomycetes (mushrooms). An interesting mold in class Oomycetes is *Phytophtera,* the potato blight mold that caused a famine in Ireland in the mid-19th century. The black bread mold, *Rhizopus,* is a zygomycete. Both of these genera are primitive molds with aseptate hyphae; this means the hyphae are not divided into individual cells. Although there may be many nuclei in aseptate hyphae, they are not separated by cell walls (septa). Among the Ascomycetes and Deuteromycetes classes are found many antibiotic-producing molds, such as *Penicillium* and *Cephalosporium.*

Molds have great commercial importance. They are the main source of antibiotics. Many new antibiotics are developed by growing soil cultures and isolating any molds that inhibit the growth of bacteria. The antibiotic penicillin was accidentally discovered when a *Penicillium notatum* mold contaminated an agar plate containing a *Staphylococcus* and inhibited growth of the bacteria. Today, antibiotics can be synthesized and chemically altered in the biochemistry laboratory, as has been done with the various penicillins. Some molds are also used to produce large quantities of enzymes (such as amylase, which converts starch to glucose) and of citric acid and other organic acids that are used commercially. The flavor of cheeses such as bleu, Roquefort, Camembert, and Limburger are the result of molds that grow on them.

Molds also can be harmful. The rusts and smuts of crop plants, grains, corn, and potatoes not only destroy crops, but some are also toxic. The aflatoxin from an *Aspergillus* mold on peanuts and cottonseed and the ergot (a type of smut) on rye and wheat are extremely toxic to humans and farm animals. Aflatoxins have been shown to be carcinogenic (cancer causing) as well.

Yeasts Yeasts are microscopic, eucaryotic, single-celled Ascomycetes or Deuteromycetes that lack mycelia. They usually reproduce by budding (Fig. 2 – 17), but occasionally do so by spore formation. Yeasts are found in soil and water and on the skins of many fruits and vegetables. People produced wine, beer, and alcoholic beverages for centuries before Pasteur discovered that naturally occurring yeasts on the skin of grapes and other fruits and grains were responsible for these fermentation processes. The common yeast *Saccharomyces cerevisiae* ferments sugar to alcohol under anaerobic conditions (without oxygen). In aerobic conditions (with oxygen), this yeast breaks down simple sugars to carbon dioxide and water; for this reason, it has long been used to leaven light bread. Yeasts are also a good source of nutrients for humans because they produce many vitamins and proteins. Some yeasts are human pathogens. *Candida albicans* is the yeast most frequently isolated from human clinical specimens.

Dimorphism

A few fungi, usually pathogens, can live either as molds or as yeasts depending on growth conditions. When they are isolated from living tissues at body temperature (37° C), they appear in a unicellular parasitic form as yeasts. If they are grown at room temperature (25° C) or isolated from soil or dust, they grow in the saprophytic form as a mold with hyphae and spores. *Histoplasma,* which causes histoplasmosis; *Sporothrix,* which causes sporotrichosis; *Coccidioides,* which causes coccidioidomycosis; and *Blastomyces,* which causes blastomycosis, are examples of dimorphic fungi (Fig. 2 – 19).

Fungal Diseases

Considering the large number of fungal species, very few are pathogenic for humans, and most of those are found in the class Deuteromycetes (see Table 2 –

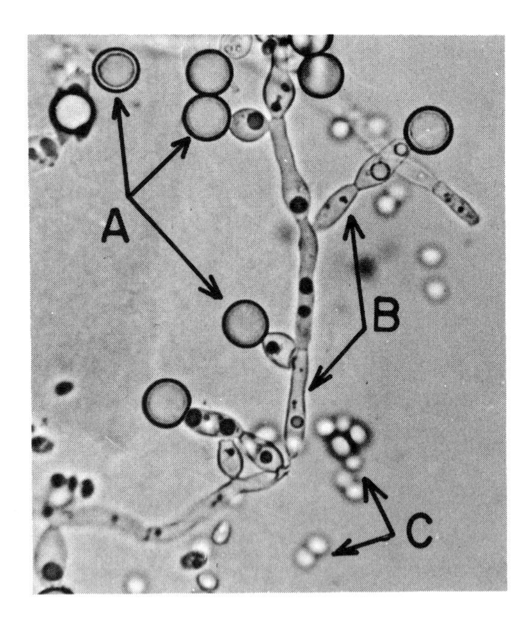

FIGURE 2 – 19
A culture of *C. albicans* showing (*A*) chlamydospores, (*B*) pseudohyphae (elongated yeast cells, linked end to end), and (*C*) budding yeast cells (blastospores) (original magnification ×450). (Davis BD, et al.: Microbiology, 4th ed. Philadelphia, Harper & Row, 1987)

6). Diseases caused by fungi are called *mycoses* (singular, *mycosis*) and are categorized as superficial, cutaneous, subcutaneous, or systemic mycoses. In some cases the infection may progress through all of these stages.

Superficial and Cutaneous Mycoses Superficial and cutaneous fungal infections are caused by dermatophytes or other fungi that live on skin, hair, fingernails, toenails, or mucous membranes at body openings. Dermatophytes (fungi living on or within the skin) cause forms of "ringworm" or tinea infections like athlete's foot and lesions of the nails, scalp, and hair follicles.

Candida albicans is an opportunistic yeast that lives harmlessly on the skin and mucous membranes of the mouth, intestine, and reproductive tract. However, when the chemical balance (homeostasis) is upset and the number of indigenous bacteria is reduced, this yeast flourishes to cause infections of the mouth (oral thrush), skin, and vagina (yeast vaginitis). This type of local infection may become a focal site from which the organisms invade the bloodstream to become a generalized or systemic infection in many internal areas.

Subcutaneous and Systemic Mycoses Spores of some pathogenic fungi may be inhaled with dust from contaminated soil and dried bird and bat feces, or they may enter through wounds of the hands and feet. If the spores are inhaled into the lungs, they may germinate there to cause a respiratory infection similar to tuberculosis. This type of deep-seated pulmonary infection is caused by species of *Coccidioides* (coccidioidomycosis), *Histoplasma* (histoplasmosis), *Blastomyces* (blastomycosis), and *Cryptococcus* (cryptococcosis). In each case, the pathogens may invade further to cause systemic infections, especially in immunosuppressed individuals.

Skin tests for most of these infections are available, but the diseases are difficult to cure. Mycotic diseases are most effectively treated with antifungal agents like nystatin, amphotericin B, or 5-fluorocytosine. Because these chemotherapeutic agents may be toxic to humans, they are prescribed with due consideration and caution.

ALGAE

Algae are photosynthetic organisms that may be multicellular or unicellular. They are arranged in colonies or strands and are found in water and soil and on trees, plants, and rocks. Algae produce their energy by photosynthesis, using energy from the sun, carbon dioxide, water, and inorganic nutrients from the soil to build cellular material. However, a few species use organic nutrients, and others survive with very little sunlight. The organisms previously known as blue-green algae are simple procaryotic cells, now classified in the kingdom Procaryotae as cyanobacteria. The green, brown, and red algae are eucaryotic organisms in the kingdom Protista along with protozoa. Only one genus of algae (*Prototheca*) is a true pathogen, although a few other genera secrete substances that are toxic to humans, fish, and other animals (Table 2 – 7).

It is easy to find algae. They include the large seaweed and kelp found along ocean shores, the green scum floating on ponds, and the slippery material on wet rocks. There are also many microscopic forms in pond water that differ from the colorless, motile protozoa in that they are photosynthetic, colored, and may move very slowly. Some algae (*e.g., Euglena* and *Volvox*) have characteristics that cause them to be classified as protozoa by some taxonomists.

Algae are an important source of food, iodine and other minerals, fertilizers, emulsifiers for pudding, and stabilizers for ice cream and salad dressings; they are also used as a gelling agent for jams and nutrient media for bacterial growth. The agar that is used to produce solid laboratory culture media is derived from a red marine alga. Damage to water systems is frequently caused by algae clogging filters and pipes where many nutrients are present. Some typical algae are shown in Figure 2 – 20.

ACELLULAR INFECTIOUS AGENTS

Viruses

Mature virus particles, called *virions,* are so small and simple in structure that they do not fit the living cell classification. Most range in size from 10 to 300 nm

TABLE 2 – 7
Characteristics of Algae (Major Groups)

Division	Common Name	General Structural Arrangement	Stored Materials	Motility	Method of Reproduction	Cell Wall Composition	Habitat
Chlorophyta	Green algae	Unicellular to multicellular	Starch, oils	Mostly nonmotile	Asexual by multiple fission; spores sexual	Cellulose and pectin	Fresh water, salt water, soil, lichens
Chrysophyta	Golden algae (includes diatoms)	Mainly unicellular	Oils	Unique movement in diatoms; others have flagella	Asexual and sexual	Pectin, some with silica or calcium	Fresh water, salt water, soil
Euglenophyta	*Euglena*	Unicellular	Fats	Motile by means of flagella	Asexual only by binary fission	None	Fresh water
Phaeophyta	Brown algae	Multicellular	Fats	Motile	Asexual by motile zoospores; sexual by motile gametes	Cellulose, pectin, algin	Salt water (cool environment)
Pyrrophyta	Dinoflagellates	Unicellular	Starch, oils	Motile	Asexual; sexual rare	Cellulose and pectin	Fresh water, salt water
Rhodophyta	Red algae	Multicellular	Starch, oils	Nonmotile	Asexual by spores; sexual by gametes	Cellulose, pectin, agar, carrageenan	Salt water (warm environment)

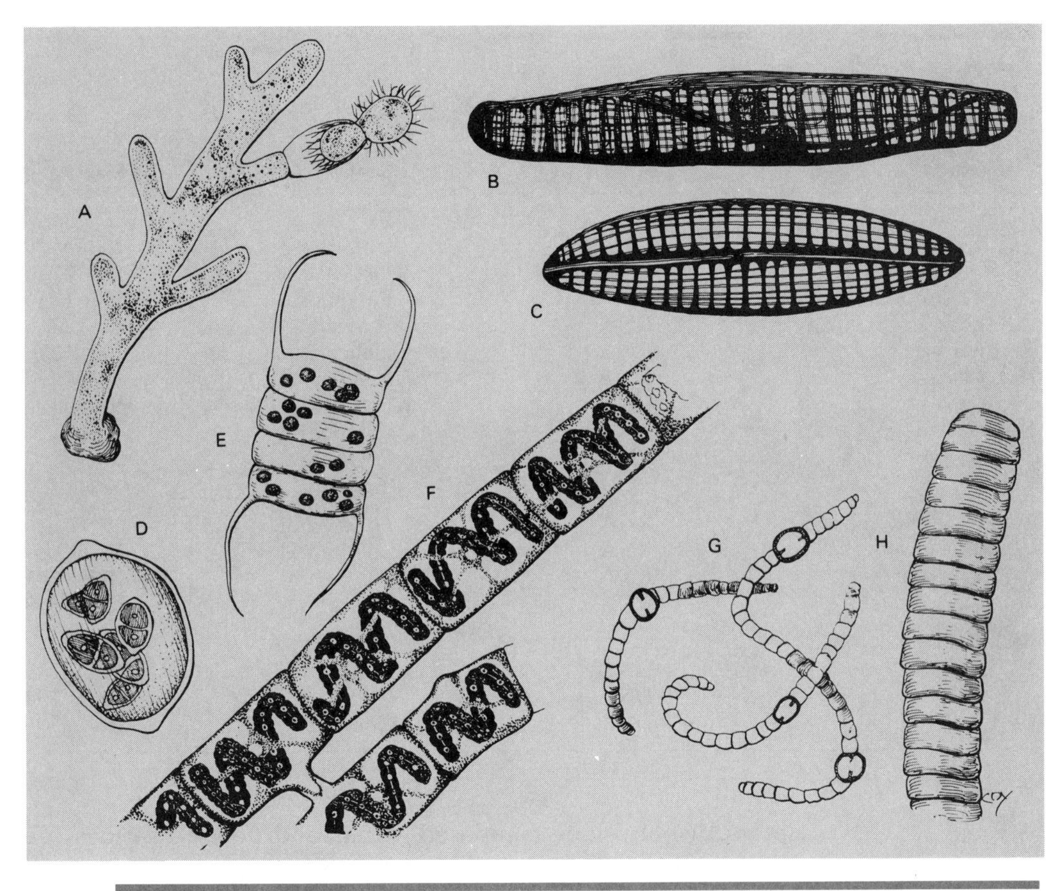

FIGURE 2 – 20
Typical algae. (*A*) *Vaucheria.* (*B*) Diatom. (*C*) *Navicula.* (*D*) *Oocystis.* (*E*) *Scenedesmus.* (*F*) *Spirogyra.* (*G*) *Nostoc.* (*H*) *Oscillatoria.*

in diameter. The smallest is about the size of the large hemoglobin molecule of a red blood cell. Some viruses produce diseases or genetic changes in animals, plants, algae, fungi, protozoa, and bacterial cells (Table 2 – 8).

Recently, virions have been described as having five specific properties that distinguish them from living cells: (1) they possess *either* DNA or RNA, never both; (2) their replication (duplication) is directed by the viral nucleic acid within a host cell; (3) they do not divide by binary fission or mitosis; (4) they lack the genes and enzymes necessary for energy production; and (5) they depend on the ribosomes, enzymes, and nutrients of the infected cells for protein production.

A typical virus particle consists of a genome (nuclear material) of DNA or RNA surrounded by a *capsid* (protein coat), which is composed of many small protein units called *capsomeres* (Fig. 2 – 21). Some virions (enveloped viruses)

TABLE 2-8
Relative Sizes and Shapes of Some Viruses

Viruses	Nucleic Acid Type	Shape	Size Range (nm)
Animal Viruses			
Vaccinia	DNA	Complex	200 by 300
Mumps	RNA	Helical	150–250
Herpes simplex	DNA	Polyhedral	100–150
Influenza	RNA	Helical	80–120
Retroviruses	RNA	Helical	100–120
Adenoviruses	DNA	Polyhedral	60–90
Reoviruses	RNA	Polyhedral	60–80
Papovaviruses	DNA	Polyhedral	40–60
Polioviruses	RNA	Polyhedral	28
Plant Viruses			
Turnip yellow mosaic	RNA	Polyhedral	28
Wound tumor	RNA	Polyhedral	55–60
Alfalfa mosaic	RNA	Polyhedral	18 by 36–40
Tobacco mosaic	RNA	Helical	18 by 300
Bacteriophages			
T2	DNA	Complex	65 by 210
λ	DNA	Complex	54 by 194
$\phi\chi$-174	DNA	Complex	25

have a protective envelope composed of fats and polysaccharides. Bacterial viruses may also have a sheath and tail fibers. There are no ribosomes for protein synthesis or sites of energy production; hence, the virus must take over a functioning cell to produce new virus particles.

Viruses are classified by the following characteristics: (1) type of genetic material (either DNA or RNA); (2) shape of the capsid; (3) number of capsomeres; (4) size of capsid; (5) presence or absence of an envelope; (6) host that it infects (plant, animal, or microorganism); (7) type of disease produced; (8) target cell; and (9) immunological properties.

The genetic material of most viruses is either double-stranded DNA or single-stranded RNA, but a few have single-stranded DNA or double-stranded RNA. Viral genomes are usually circular molecules, but some are linear (with two ends). Capsids of viruses have various shapes and symmetry. They may be polyhedral (many sided), helical (coiled tubes), bullet shaped, spherical, or a complex combination of these shapes. Polyhedral capsids have 20 sides or facets, geometrically referred to as icosahedrons. Each facet consists of several capsomeres; thus, the size of the virus is determined by the size of each facet and the number of capsomeres in each. Frequently, the envelope around the capsid makes the virus appear spherical or irregular in shape in electron micro-

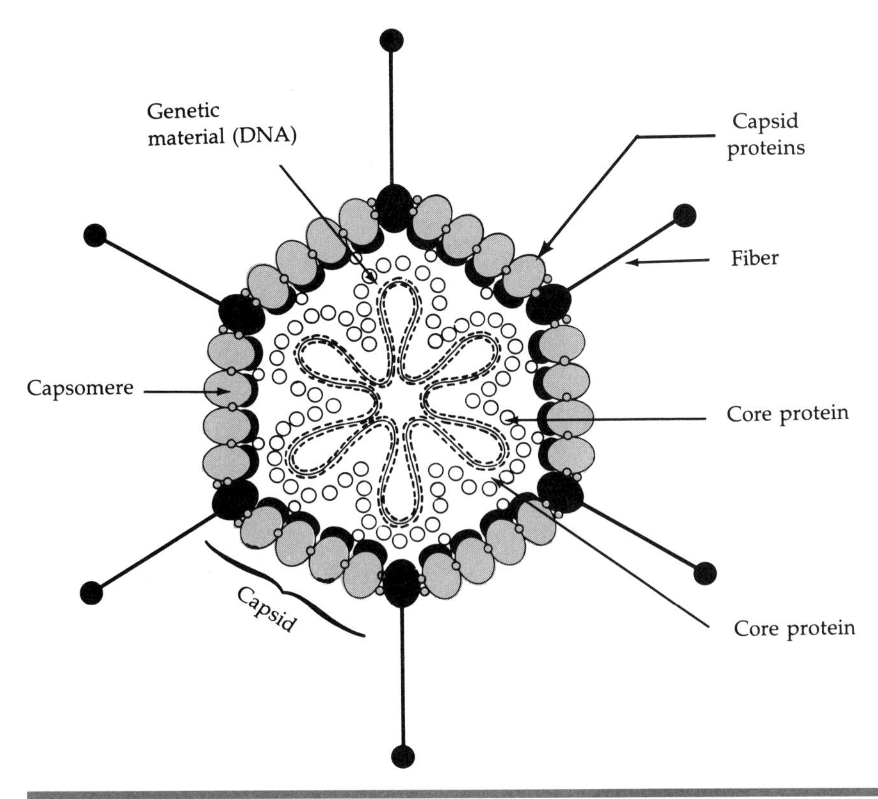

FIGURE 2 – 21
Model of an icosahedral virus: adenovirus. (Volk WA, et al.: Essentials of Medical Microbiology, 4th ed. Philadelphia, JB Lippincott, 1991)

graphs. The envelope is usually acquired by certain animal viruses from the host nuclear or cellular membrane as the new virus particle leaves the cell. Apparently, viruses are able to alter these membranes and add protein fibers, spikes, and knobs that enable the virus to recognize the next host cell to be invaded. A list of some viruses, their characteristics, and diseases they cause is presented in Table 2 – 9. Sizes of viruses are depicted in Figure 2 – 22.

The virion usually infects a cell by injecting its DNA or RNA into the cell (Fig. 2 – 23) or by cellular phagocytosis (Fig. 2 – 24). The viral genetic material may remain latent or inactive in the cell and be transferred to each daughter cell when it divides. This is the *lysogenic cycle* of viral infection. However, the viral genetic material may be induced by heat (fever), ultraviolet light, or certain chemical agents to take over the cell and make viruses.

When the genetic material from the virus takes over the metabolic "machinery" of the cell to produce viruses, it has entered the *lytic cycle* of virus infection. It breaks up the cellular DNA and produces viral DNA or RNA. It then

TABLE 2 – 9
Selected Important Groups of Viruses and Viral Diseases

Virus Type	Viral Characteristics	Virus	Disease
Poxviruses	Large, brick shape with envelope, d.s. DNA	Variola Vaccinia	Smallpox Cowpox
Polyoma-Papilloma	d.s. DNA, polyhedral	Papillomavirus Polyomavirus	Warts Some tumors, some cancer
Herpesviruses	Polyhedral with envelope, d.s. DNA	Herpes simplex I Herpes simplex II Herpes zoster Varicella	Cold sores or fever blisters Genital herpes Shingles Chickenpox
Adenoviruses	d.s. DNA, icosahedral, no envelope		Respiratory infections, pneumonia, conjunctivitis, some tumors
Picornaviruses (the name means small RNA viruses)	s.s RNA, tiny icosahedral, with envelope	Rhinovirus Poliovirus Hepatitis types A and B Coxsackievirus	Colds Poliomyelitis Hepatitis Respiratory infections, meningitis
Reoviruses	d.s. RNA, icosahedral with envelope	Enterovirus	Intestinal infections
Myxoviruses	RNA, helical with envelope	Orthomyxoviruses types A & B Myxovirus parotidis Paramyxovirus Rhabdovirus	Influenza Mumps Measles (rubeola) Rabies
Arbovirus	Arthropod-borne RNA, cubic	Mosquito-borne type B Mosquito-borne types A and B Tick-borne, corona-virus	Yellow fever Encephalitis (many types) Colorado tick fever
Retrovirus	d.s. RNA, helical with envelope	RNA tumor virus HTLV virus HIV (Human immunodeficiency virus)	Tumors Leukemia AIDS (Acquired immunodeficiency syndrome)

d.s. = double-stranded
s.s. = single-stranded

synthesizes viral protein capsids and assembles many new virus particles until the cell bursts (lyses) and releases new viruses into the area to infect neighboring cells.

A good example of the lysogenic (latent) and lytic cycles is the ordinary cold sore, which is caused by herpes simplex virus. Persons who develop cold sores had the latent or temperate virus genetic material in a nearby nerve ganglion in the lysogenic cycle. When a fever develops or the individual is exposed to

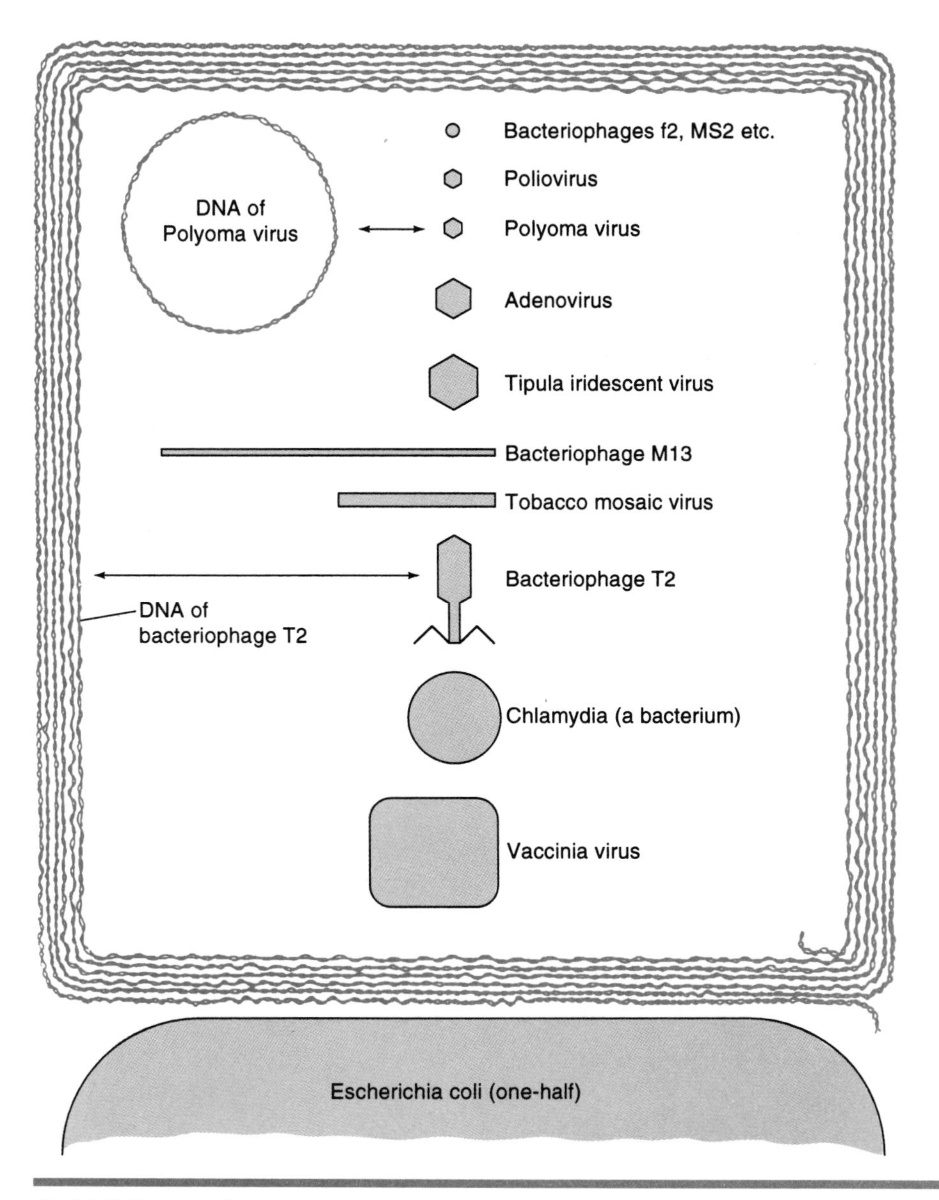

FIGURE 2 – 22
Comparative sizes of virions, their nucleic acids, and bacteria. (adapted from Davis BD, et al.: Microbiology, 4th ed. Philadelphia, JB Lippincott, 1990)

excessive ultraviolet light of the sun, the viral genetic material may be stimulated to take over the cells to produce more viruses (lytic cycle) and, in the process, a cold sore develops. Such viral infections are usually limited by the body's defenses—the phagocytes and antiviral proteins called interferons that are produced by virus-infected cells.

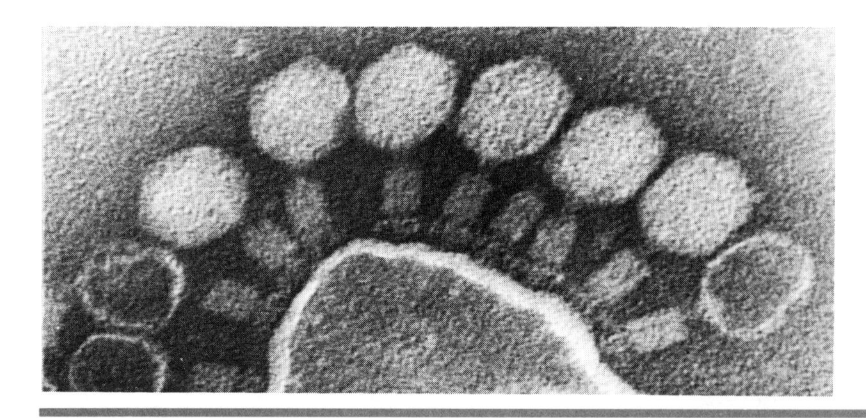

FIGURE 2 – 23
A partially lysed cell of *Vibrio cholerae* with attached virions of phage CP-T1. Note the empty capsids, full capsids, contracted tail sheaths, base plates, and spikes (original magnification ×257,000). (Courtesy of R. W. Taylor and J. E. Ogg, Colorado State University, Fort Collins, Colorado)

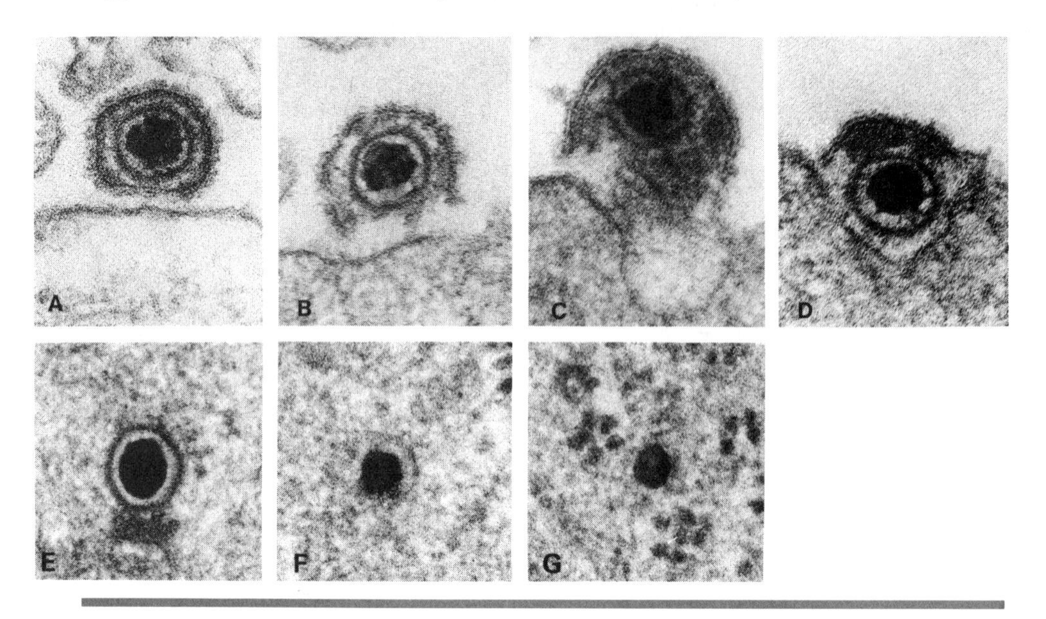

FIGURE 2 – 24
Adsorption (*A*), penetration (*B–D*), and digestion of the capsid (*E–G*) of herpes simplex on HeLa cells, as deduced from electron micrographs of infected cell sections. Penetration involves local digestion of the viral and cellular membranes (*B,C*), resulting in fusion of the two membranes and release of the nucleocapsid into the cytoplasmic matrix (*D*). The naked nucleocapsid is intact in (*E*), is partially digested in (*F*), and has disappeared in (*G*), leaving a core containing DNA and protein. (Morgan C, et al.: J Virol 2:507, 1968)

Antibiotics are not effective against viral infections because these drugs work by inhibiting certain metabolic reactions within living pathogens. Because viruses are not independent metabolizing cells, agents that affect the synthesis of new virus particles may also cause damage to host cells. For colds and influenza, antibiotics should be given only to prevent secondary bacterial infections that might occur after the virus infection. However, several new chemicals (antiviral agents) have been developed that interfere with virus-specific enzymes and virus production by either disrupting critical phases in viral cycles or inhibiting the synthesis of viral DNA or RNA. Refer to Chapter 6 for a discussion of chemotherapy.

Many tumors and cancers are apparently caused by viruses that change the genetic composition of cells and cause uncontrolled growth of abnormal cells under certain environmental conditions. Some chemotherapeutic drugs used to treat cancer are chemicals that interfere with DNA and RNA synthesis in rapidly dividing cells, thus inhibiting or destroying tumor cells and also some human cells (hair, blood, and sperm).

Remnants or collections of viruses are often seen in infected cells and are used as a diagnostic tool to identify certain diseases. These are called *inclusion bodies* and may be found in the cytoplasm (referred to as cytoplasmic) or within the nucleus (intranuclear), depending on the disease. In rabies, the cytoplasmic inclusion bodies are called Negri bodies. The inclusion bodies of AIDS and the Guarnieri bodies of smallpox are also cytoplasmic. Herpes and poliomyelitis viruses cause intranuclear inclusion bodies. In each case, the inclusion bodies may be merely aggregates or collections of viruses. Some important human viral diseases include the common cold, influenza, mumps, measles, chickenpox, smallpox, rabies, cold sores, genital herpes, warts, poliomyelitis, encephalitis, and AIDS.

Bacteriophages Viruses that infect bacteria are called *bacteriophages* or simply phages. Most bacteriophages are species- and strain-specific. Those that infect the coliform (intestinal) bacteria, such as *Escherichia coli*, are called coliphages. Refer to Figures 2 – 23 and 2 – 25 to observe the complexity of the phages. The end plate, tail fibers, and spikes are used to attach to bacteria, often at the site of a pilus; the phage DNA is then injected into the bacteria much like water from a syringe (Fig. 2 – 24).

The phages that *lyse* host bacterial cells during the production of new bacteriophages are referred to as *virulent bacteriophages.* However, many DNA phages are *temperate bacteriophages;* their viral genome does not take over the host cell. The temperate phage DNA is injected into the bacterium but causes no damage to the cell. Then, each time the host cell divides, the viral genome also replicates and is passed to each daughter cell. Thus, each daughter cell is

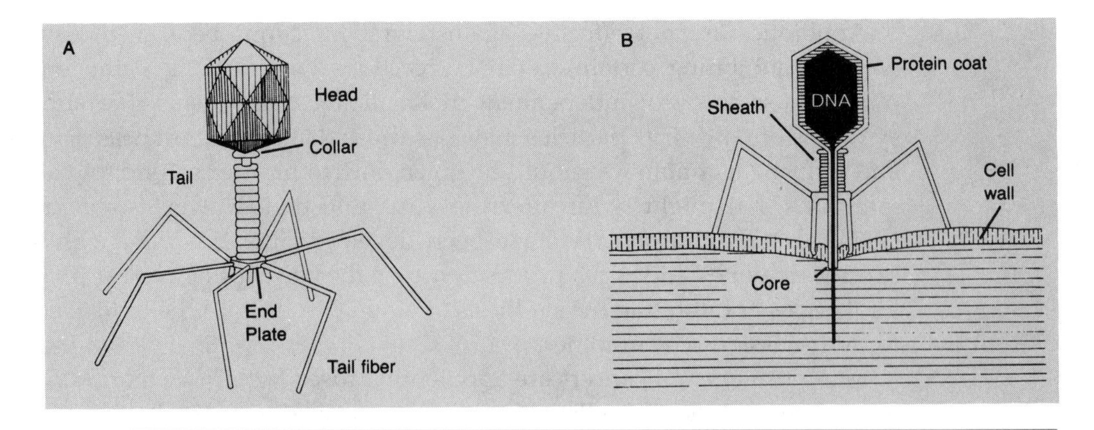

FIGURE 2 – 25
(*A*) The bacterial virus T4 is an assembly of protein components. The head is a protein membrane with 20 facets, filled with DNA. It is attached to a tail consisting of a hollow core surrounded by a sheath and based on a spiked end-plate to which six fibers are attached. (*B*) The sheath contracts, driving the core through the cell wall, and viral DNA enters the cell. (Volk WA, Wheeler MF: Basic Microbiology, 5th ed. Philadelphia, JB Lippincott, 1984)

infected with the phage DNA. This relationship between the phage and its host cell is called *lysogeny.* The infected bacteria are said to be *lysogenic bacteria,* and the phages that infect them are temperate phages. The latent viral genome is called a *prophage.*

In some cases, under appropriate environmental conditions (heat, ultraviolet light, certain chemicals), the prophage may be stimulated to produce new complete phages and lyse the bacterial cell. This involves a process called induction of the lytic cycle.

VIRUSES AND GENETIC CHANGES

Much research in genetics has been accomplished using bacteria and bacteriophages. Some pathogens are identified by the specific bacteriophages that will infect them. When bacteria are injected with the genetic material from a bacteriophage, the genetic constitution of those bacteria is changed by the addition of the phage DNA. This process is called *lysogenic conversion.* An example of such a conversion is the pathogenic diphtheria bacterium, which is pathogenic only when a certain phage gene is present. Evidently, the gene (called the tox gene) that enables diphtheria bacteria (*Corynebacterium diphtheriae*) to produce a lethal toxin (diphtheritoxin) is injected with the phage DNA. Other genetic changes produced in bacteria by bacteriophages are discussed in Chapter 4.

Viroids and Prions

Although viruses are very small, nonliving infectious agents, viroids and prions are even smaller and less-complex infectious agents. Viroids consist of circular, single-stranded RNA that can infect the nucleus of a plant cell. They are transmitted between plants by mechanical means (birds, insects) or via pollen. It is believed that they cause more than 10 plant diseases, such as potato spindle-tuber disease, chrysanthemum stunt disease, and exocortosis of citrus trees.

Prions are small infectious proteins that apparently cause nervous system diseases in livestock and humans, such as scrapie in sheep and goats, spongiform encephalopathy in cattle, and kuru and Creutzfeldt-Jakob disease in humans.

SUMMARY

Differences that exist among the various types of microorganisms determine their classification into different kingdoms, phyla, orders, classes, families, genera, and species. When a variety of characteristics are used to categorize each of these organisms, some appear to be more primitive than others. A study of the various types of cells, especially the unicellular microorganisms, helps us understand why many scientists believe that microbes were probably the first kinds of cells on earth. As we learn more about the differences between procaryotic and eucaryotic cells, we start to understand why some chemotherapeutic drugs, including antibiotics, destroy or inhibit certain pathogens but do not kill the patient.

This survey of the types of microorganisms emphasizes the characteristics that enable them to carry on beneficial ecological functions. Without these various types of microbes, the earth would not be habitable.

PROBLEMS AND QUESTIONS

1. Which microbes are eucaryotes?
2. Which microorganisms are procaryotes?
3. What function do mitochondria and mesosomes have in common?
4. Where does protein synthesis occur in eucaryotic cells? in procaryotic cells?
5. Which microorganisms do not have a nuclear membrane?
6. List five differences between eucaryotic animal cells and procaryotic bacterial cells.
7. What is the difference between bacterial and protozoal flagella?
8. Describe peritrichous, amphitrichous, lophotrichous, and monotrichous flagellation in bacteria.
9. Describe a symbiotic relationship of a parasitic protozoan.
10. What is the role of fungi in recycling nutrients?
11. What are the differences between protozoa and green algae?
12. List five properties of viruses that distinguish them from living cells.
13. Describe the lytic and lysogenic cycles of viruses.

Self Test

After you have read Chapter 2, examined the objectives, reviewed the chapter outline, studied the new terms, and answered the problems and questions above, complete the following self test.

Matching Exercises

Complete each statement from the list of words provided with each section.

DESCRIPTIVE TERMS

amphitrichous	spirochete	monotrichous
diplococci	bacilli	streptobacilli
procaryotes	lophotrichous	diplobacilli
axial filaments	staphylococci	peritrichous
eucaryotes	cocci	streptococci

1. Cyanobacteria and bacteria are ________________.
2. Bacteria with one flagellum on each end are said to be ________________.
3. Rod-shaped bacteria are called ________________.
4. Some spirochetes have two flagella-like fibrils attached at each end, called ________________.

5. Fungi, protozoa, green algae, plants, and animal cells are ________.
6. Bacteria that have a tuft of flagella at one or both ends are said to be ________.
7. *Treponema pallidum* is an example of a ________.
8. Cocci that are found in pairs are called ________.
9. Cells that have no true nuclear membrane are procaryotes.
10. Plant cells have cellulose-containing cell walls, and animal cells have no cell wall, but both are ________.
11. The spherical or round bacteria are called cocci.
12. *Salmonella* organisms are motile because they have flagella covering their entire surfaces; they are said to be ________.
13. Cells that have invaginations of the cell membrane, called mesosomes, for cellular respiration are ________.
14. Cocci that appear as grape-like clusters on a stained smear are staphylococci
15. Cells that have one "naked" chromosome composed of DNA are ________.
16. Bacilli that appear in pairs are called ________.
17. Some bacteria that live in the intestine have only one flagellum; they are said to be ________.
18. Bacilli that form long chains are ________.
19. Bacteria are classified as ________.
20. Cells that have membrane-bound organelles, such as mitochondria, endoplasmic reticulum, and Golgi bodies are ________.
21. Cocci that form chains are called ________.

CHARACTERISTICS OF MICROORGANISMS

algae	protozoa	fungi
viruses	rickettsias	mycoplasmas
chlamydias	*Escherichia coli*	

1. Most of the saprophytic organisms, which live on decaying organic materials, are ________.
2. All the organisms in this group are photosynthetic; they are ________.
3. Single-celled microbes that are classified by their means of locomotion are protozoa.
4. Bacteria that must obtain their energy (ATP) from their host cells are chlamydias.
5. Bacteria that lack a cell wall are ________.
6. Bacteria that are usually found as a part of the indigenous microflora of the human intestine are ________.
7. Bacteria that must be transmitted from person to person by arthropod vectors such as fleas, lice, and ticks are ________.
8. The smallest microorganisms, consisting mainly of the genetic material and a protein coat, are called ________.

DISEASES AND MICROORGANISMS

Match the following diseases with the type of microorganisms that cause them, using the following words:

virus	protozoa	algae
mycobacteria	staphylococci	mycoplasmas
streptococci	rickettsias	spirochete
chlamydias	fungi	curved rods

1. "Strep" sore throat ________________
2. Typhus ________________
3. Cold sores ________________
4. Boils, wound infections ________________
5. Atypical pneumonia, sinusitis ________________
6. Cholera ________________
7. Rabies ________________
8. Syphilis ________________
9. Smallpox ________________
10. Tuberculosis ________________
11. Poliomyelitis ________________
12. Lymphogranuloma venereum ________________
13. Influenza ________________

True or False (T or F)

___ 1. Microorganisms, with the exception of viruses, exhibit the properties found in all living systems.

___ 2. Mitochondria are the powerhouses that generate energy for eucaryotic cells.

___ 3. The chromosomes of procaryotic cells consist of histone proteins and DNA.

___ 4. Lysosomes contain lysozyme and other enzymes to aid in the digestion of ingested materials.

___ 5. Nuclei of cyanobacteria have a well-defined nuclear membrane.

___ 6. Bacterial flagella are structurally like protozoan flagella and cilia.

___ 7. Viruses contain both DNA and RNA.

___ 8. An icosahedron is a 20-sided figure.

___ 9. Bacteriophages are viruses that infect bacteria.

___ 10. Fungi may reproduce sexually or asexually.

___ 11. Pathogenic rickettsias differ from viruses in that they require living cells for growth and reproduction.

___ 12. Molds differ from bacteria in that they are multicellular and reproduce by spores.

___ 13. During the cyst stage, protozoa resemble bacterial spores because they have thick walls, are dormant, and resist drying.

___ 14. Bacteria reproduce by budding, and yeasts reproduce by transverse fission.

___ 15. Vibrio organisms (like *Vibrio cholerae*) are curved, rod-shaped bacteria.

___ 16. PPLO, like the L-forms, have a rigid cell wall.

___ 17. Rickettsial diseases are usually transmitted by an arthropod vector.

Multiple Choice

1. The semipermeable structure controlling the entry and exit of materials between the cell and its environment is the
- **a.** cell wall
- **b.** protoplast
- **c.** cytoplasm
- **d.** plasma membrane

2. Procaryotic cells reproduce by
- **a.** gamete production
- **b.** budding
- **c.** mitosis
- **d.** binary fission

3. Centrioles are
- **a.** cylindrical organelles
- **b.** involved in cell division
- **c.** found in eucaryotes
- **d.** structurally similar to cilia and flagella
- **e.** all of the above

4. The group of bacteria that lack rigid cell walls and take on irregular shapes is
- **a.** rickettsias
- **b.** chlamydias
- **c.** *Clostridium*
- **d.** mycoplasmas

5. Sporulation in bacteria is
- **a.** a means of reproduction
- **b.** degeneration of organelles
- **c.** a means of survival
- **d.** development of a cell wall

6. Lysogenic bacteria may be induced to enter a lytic cycle by exposure to
- **a.** ultraviolet light
- **b.** sunlight
- **c.** heat
- **d.** certain chemicals
- **e.** all of the above

7. Gram-positive bacteria stain
- **a.** purple
- **b.** red
- **c.** yellow
- **d.** green

8. Rickettsias are the causative agent of
- **a.** vaginitis
- **b.** Rocky Mountain spotted fever
- **c.** trichomoniasis
- **d.** typhoid fever

9. Infections of the male and female genital tracts may be caused by *Trichomonas vaginatis,* which is a
- **a.** protozoan
- **b.** virus
- **c.** yeast
- **d.** fungus

10. During the lysogenic cycle, the virus
- **a.** lyses the host cell
- **b.** is latent in the cell
- **c.** has not yet infected the cell
- **d.** induces the production of more viruses

CHAPTER 3

Introduction to the Chemistry of Life

OBJECTIVES

After studying this chapter, you should be able to

- Differentiate among elements, atoms, molecules, and compounds
- Describe an acid-base reaction
- Discuss the importance of water in biochemical reactions
- List the characteristics of monosaccharides, disaccharides, and polysaccharides
- Describe four main types of biochemical molecules
- Distinguish between organic and inorganic compounds
- Discuss the structure of carbohydrates, fats, proteins, and nucleic acids and their breakdown products
- Describe the role of enzymes in metabolism
- Discuss how DNA directs cellular activities

CHAPTER OUTLINE

Basic Chemistry
Atoms, Molecules, Elements, Compounds
Chemical Bonding

Importance of Water in Living Cells and Systems
Solutions
Acids
Bases
Salts
pH

Organic Chemistry
Carbon Bonds
Cyclic Compounds

Biochemistry
Carbohydrates
Monosaccharides
Disaccharides
Polysaccharides
Lipids
Waxes
Compound Lipids
Derived Lipids
Proteins
Amino Acid Structure
Protein Structure
Enzymes
Nucleic Acids
Function
Structure
DNA Structure
Replication
Protein Synthesis

NEW TERMS

Acid
Adenosine triphosphate
Amino acids
Anion
Anticodon
Atom
Atomic number
Atomic weight
Base
Biochemistry
Carbohydrates
Catalyst
Cation
Cistron
Codon
Coenzyme
Cofactor
Covalent bond
Compound
Dehydration synthesis
Dehydrolysis
Dipeptide
Disaccharide
DNA polymerase
DNA replication
Electrolyte
Element
Enzyme
Fatty acid
Genetic code
Glucose
Glycogen
Hydrocarbon
Hydrogen bonding
Hydrolysis
Inorganic chemistry
Ion
Ionic bond
Isotopes
Lipids
Messenger RNA (mRNA)
Metabolite
Molecule
Monosaccharides
Mutation
Nucleic acids
Nucleotides
Organic compounds
Organic chemistry
Polar bonding
Polarity
Polymer
Polypeptide
Polysaccharide
Proteins
Purine
Pyrimidine
Ribosomal RNA (rRNA)
RNA polymerase
Salt
Solute
Solution
Solvent
Starch
Substrate
Transcription
Transfer RNA (tRNA)
Translation
Tripeptide

The various ways microorganisms function and survive in their environment depend on their chemical make-up. A microbe can be thought of as a "bag" of chemicals that interact with each other in a variety of ways; even the "bag" itself is composed of chemicals. To understand microbial cells and how they work, one must have a basic knowledge of the chemistry of atoms, molecules, and macromolecules (large and complex molecules). As explained in Chapter 2, even procaryotic cells consist of very large macromolecules of DNA, RNA, *proteins, lipids,* and *polysaccharides* as well as many combinations of these macromolecules that combine to make up the capsule, cell wall, cell membrane, mesosomes, cytoplasm, and flagella. These macromolecules can be broken down into smaller units, or molecules, of *nucleotides, amino acids,* glycerol, fatty acids, and *monosaccharides* (or simple sugars). Each of these molecules, in turn, may be broken down into inorganic molecules of water, carbon dioxide, ammonia, sulfides, and phosphates, and finally into the atoms of carbon, hydrogen, oxygen, nitrogen, sulfur, and phosphorus. Basic *inorganic chemistry* is introduced in this chapter, followed by a discussion of the *organic chemistry* and *biochemistry* of the major macromolecules found in living cells. Organic chemistry is the study of compounds that contain carbon and hydrogen; biochemistry is the chemistry of living cells; inorganic chemistry includes all other chemical reactions.

Only when all of these units are in place and working together properly can the cell function like a well-managed industrial plant. As in industry, the cell must have the appropriate structures and parts, the regulatory molecules (*enzymes*) to control its activities, the fuel (nutrients or light) to provide energy, and raw materials (nutrients) for manufacturing end products.

BASIC CHEMISTRY

Atoms, Molecules, Elements, Compounds

All substances, whether gas, liquid, or solid, have certain fundamental characteristics in common. If one could break down any substance into its smallest elemental units, it would be composed of *atoms.* Although it is difficult to observe atomic structure, atoms are known to consist of a mass of positively charged protons, noncharged (or neutral) neutrons, tiny negatively charged electrons, and other even smaller subparticles. An atom has equal numbers of electrons and protons; thus it has a net charge of zero (neither positive nor negative). When an atom loses or gains electrons it becomes a positively or negatively charged *ion* (*e.g.,* Na^+, Cl^-). The protons and neutrons are found in a central nucleus, and the electrons travel around the nucleus, like negatively charged satellites attracted to a positively charged planet. Figure 3 – 1 illustrates an oxygen atom. When a multitude of atoms with identical numbers of

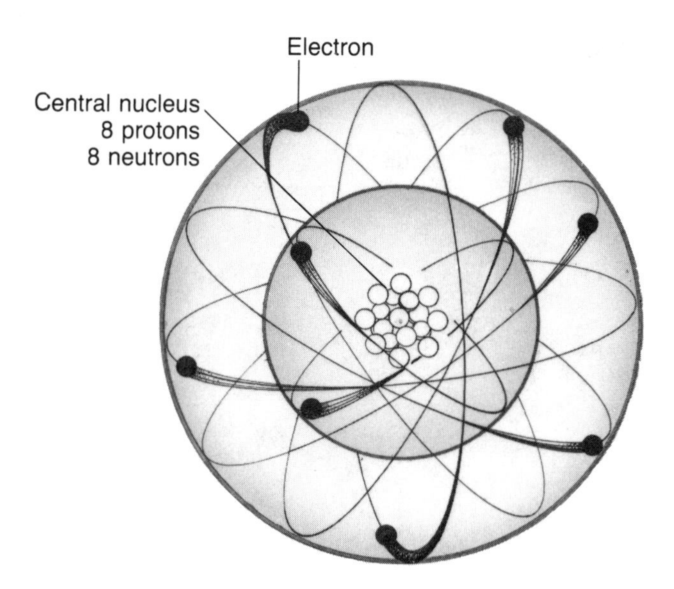

FIGURE 3 – 1
Representation of an oxygen atom. Eight protons and eight neutrons are tightly bound in the central nucleus, around which the eight electrons revolve.

electrons and protons and the same chemical properties exist together, the substance is known as an *element,* for example, carbon (C), hydrogen (H), oxygen (O), nitrogen (N), sulfur (S).

The atoms of each element have their own special weight (or mass) called their *atomic weight,* which is the total mass of the number of protons and neutrons, each of which is assigned a value of 1. Electrons are negligible in weight. The atomic weight of a specific element is an average of the weights of all *isotopes* of that element as they occur in nature. Isotopes are atoms of the same element, with identical chemical characteristics but with different atomic weights. They differ in weight because they have various numbers of neutrons within the atoms. For example, carbon-12 has six protons and six neutrons for an atomic weight of 12, whereas carbon-14 has six protons and eight neutrons for an atomic weight of 14; both are isotopes of the element carbon. Elements are also identified by a number that indicates the number of positive charges, or protons, in the nucleus. This is called the *atomic number.* Thus, the atomic number of carbon is 6. Atomic weights and atomic numbers can be found in a periodic table of the elements in any chemistry book; some atomic weights and numbers are shown in Table 3 – 1.

Two or more atoms of the same or different elements that combine into a single stable unit form a *molecule;* examples include carbon dioxide (CO_2), water (H_2O), or sulfur dioxide (SO_2). When many molecules exist together in a substance that can be seen or weighed, the substance is called a *compound, e.g.,* carbon dioxide gas (CO_2). Although there are only 106 known elements, the number of known compounds is already in the millions, and new ones are being created and discovered every year. The component parts of living cells and systems are composed of many macromolecules that interact to produce

TABLE 3 – 1
The First 20 Elements

Element	Symbol	Atomic Number	Atomic Weight
Hydrogen	H	1	1
Helium	He	2	4
Lithium	Li	3	6.9
Beryllium	Be	4	9
Boron	B	5	10.8
Carbon	C	6	12
Nitrogen	N	7	14
Oxygen	O	8	16
Fluorine	F	9	19
Neon	Ne	10	20.1
Sodium	Na	11	23
Magnesium	Mg	12	24.3
Aluminum	Al	13	27
Silicon	Si	14	28.1
Phosphorus	P	15	31
Sulfur	S	16	32.1
Chlorine	Cl	17	35.5
Argon	Ar	18	40
Potassium	K	19	39.1
Calcium	Ca	20	40.1

and maintain life. You will learn about many of these macromolecules, such as polysaccharides, structural proteins, regulatory enzymes, and DNA, although many of the most complex are still not fully identified.

Chemical Bonding

In general, atoms form molecules by gaining or losing electrons (forming *ionic bonds*) or by sharing electrons (forming *covalent bonds*). Ionic bonds hold molecules together in minerals such as sodium chloride (NaCl, common table salt) and many others that ionize in water. When sodium chloride dissolves in water, the sodium atom (Na) loses a negative electron and assumes a positive charge, thus becoming a positive ion (Na^+). Likewise, the chlorine atom (Cl) accepts one electron from the donor sodium atom to become a chloride ion (Cl^-). The ionic bonds between sodium and chlorine atoms are easily broken; however, they re-form in water because the unlike charges are attracted to each

Types of Chemical Bonds
Ionic Bonds
Covalent Bonds
Peptide Bonds
Glycosidic Bonds
Hydrogen Bonds

other. They may temporarily form NaCl molecules and then dissociate into ions again (Fig. 3 – 2). Positively charged ions are called *cations,* whereas negatively charged ions are called *anions.*

Other molecules are held together by covalent bonds. A covalent bond is formed by atoms sharing one or more pairs of electrons. In Figure 3 – 3, the methane (CH_4) molecule has four hydrogen atoms that each share two electrons with the carbon atom. This is a much stronger bond than an ionic bond; thus, the molecule is much more stable. Carbon compounds are especially good examples of covalent bonding. All of organic chemistry is based on stable covalently bonded carbon molecules and how they react with other molecules. Biochemistry, the chemistry of living cells and systems, is a study of large macromolecules composed of covalently bonded carbon, hydrogen, oxygen, nitrogen, phosphorus, and sulfur atoms. Loosely bound electrons on the surface of macromolecules enable them to react and bind with other molecules in living organisms. Peptide bonds, which hold amino acids together in a protein, and glycosidic bonds, which bind the sugar groups of polysaccharides together, are also covalent bonds.

Hydrogen bonding occurs when a hydrogen atom that is covalently bonded to an oxygen or nitrogen atom is attracted to another oxygen or nitrogen atom on a different molecule or group, as seen in the α-helix of DNA and protein structure.

IMPORTANCE OF WATER IN LIVING CELLS AND SYSTEMS

Water is the most abundant molecule in living cells and is essential for the functioning of living cells. A water molecule, consisting of two hydrogen atoms on one side of an oxygen atom, has *polarity;* that is, a water molecule has positive and negative areas (Fig. 3 – 4). This characteristic makes water an ideal solvent, or suspending medium, for other ionic or charged molecules. The attraction between water molecules is due to *polar bonding.* This polarity of the

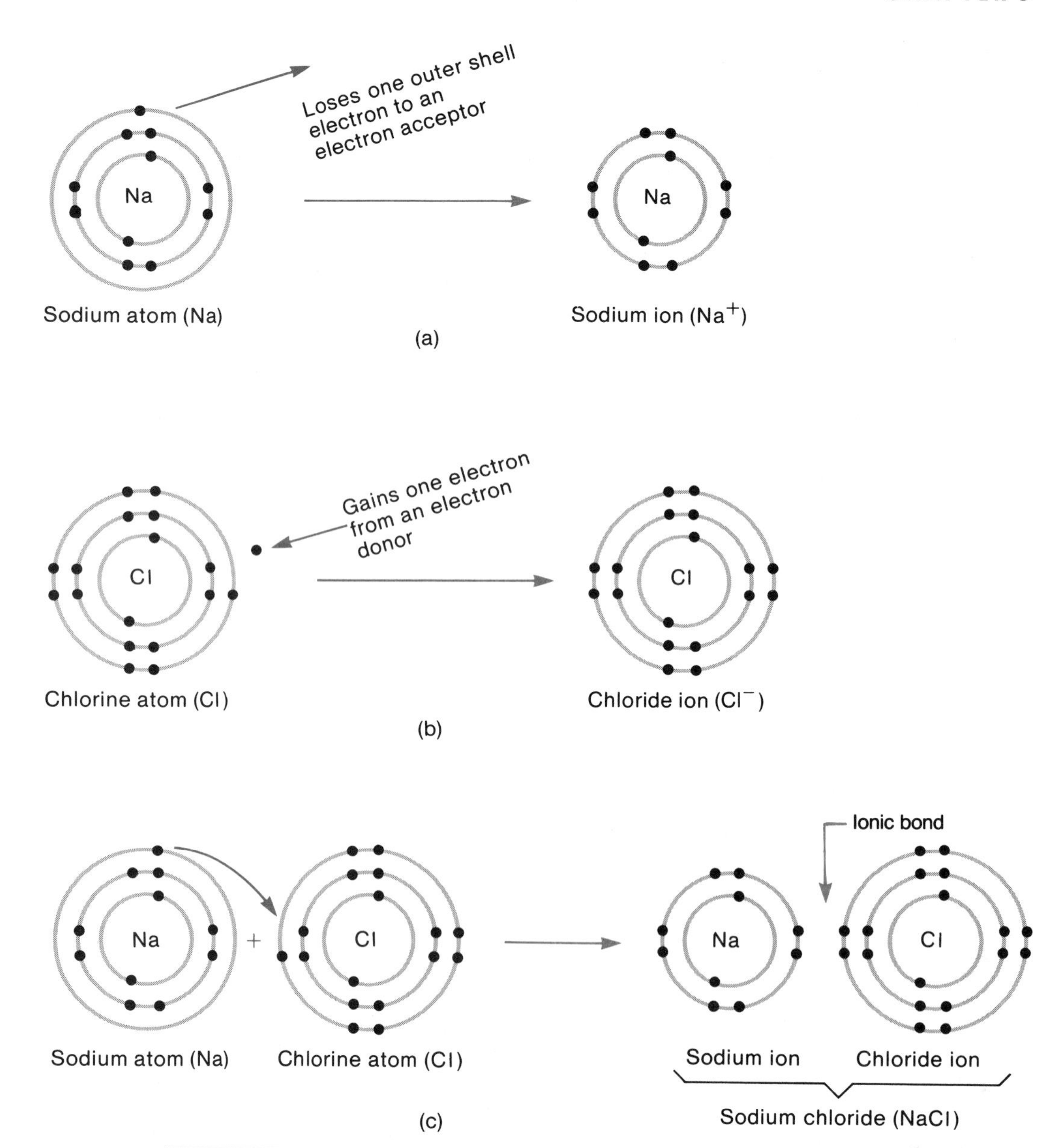

FIGURE 3 – 2

Ionic bond formation. (*A*) A sodium atom loses an electron to become a positively charged sodium ion (Na^+). (*B*) A chlorine atom gains an electron to become a negatively charged chloride ion (Cl^-). (*C*) The positive sodium ion is attracted to the negative chloride ion to form a sodium chloride molecule, which is held together by a weak ionic bond.

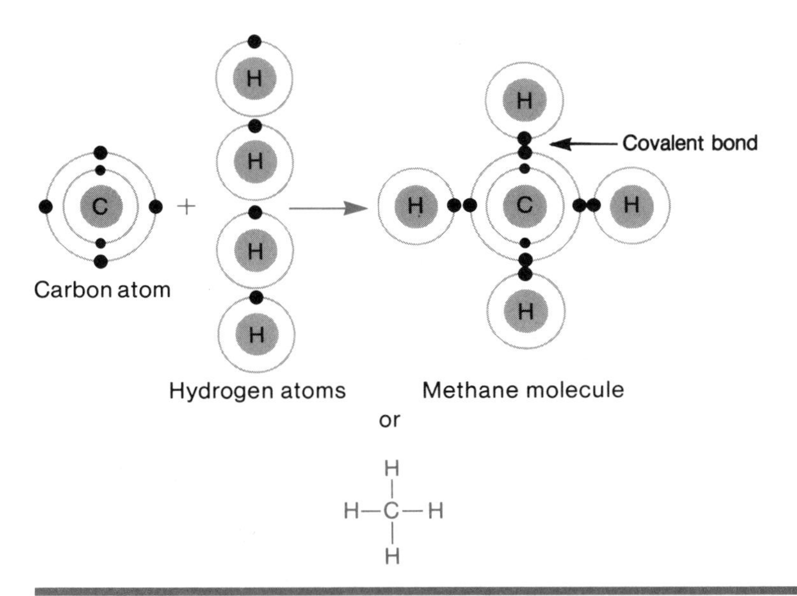

FIGURE 3 – 3
Covalent bond formation. A methane molecule with four hydrogen atoms sharing electrons with a carbon atom, forming four covalent bonds.

water molecule accounts for the following four major characteristics that make it an essential part of living cells.

1. Water is an excellent *solvent* (a liquid in which molecules of one or more other chemicals, called *solutes,* are dissolved, thus forming a *solution*). Nutrients and waste materials may move in and out of a cell, crossing the cell membrane, because they are dissolved in water.
2. Its polarity makes the molecules adhere to each other, producing surface tension and capillary action; hence, water can move within and among cells and tissues.
3. It also serves as an excellent buffer against heat changes because it can absorb and hold large amounts of heat; thus, the cell is protected from temperature fluctuations.

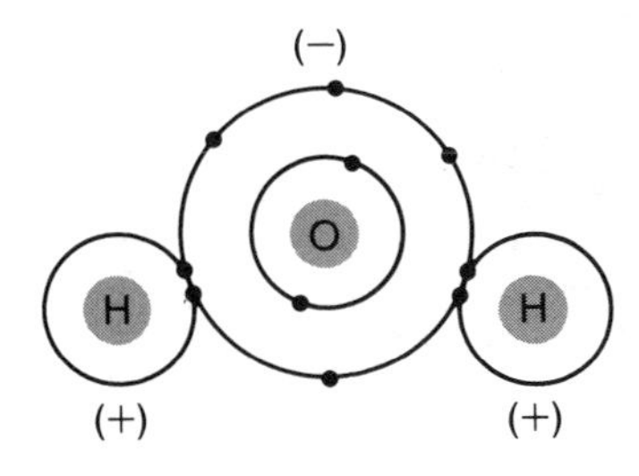

FIGURE 3 – 4
A water molecule showing two hydrogen atoms covalently bonded to an oxygen atom. The arrangement gives the molecule a more positive charge near the hydrogen protons and a negative charge near the oxygen electrons, thus providing an attraction between water molecules (polar bonding).

4. Water enters into *hydrolysis* reactions (breakdown or digestion of large macromolecules like cellulose by breaking apart the molecular units) and *dehydrolysis* reactions (in which macromolecules are synthesized by removal of water molecules).

Solutions

Solutions of various types result from the mixing of elements with elements, elements with compounds, or compounds with compounds. It is these homogeneous (evenly distributed throughout) mixtures that form almost all existing substances. The most common solutions are those in which a compound is dissolved in the solvent water. Other solvents, such as alcohol, are more commonly used with *organic compounds.* There are two types of compounds—*acids* and *bases*—that have characteristics that set them apart from all others. One interesting feature of these two groups is that any acid will react with any base, almost without exception. This certain reactivity is not true for other classes of compounds. Also, it is relatively easy to measure the acidity or alkalinity of solutions of acids and bases.

INSIGHT
The Oxygen Holocaust

In the beginning, all the world was anaerobic—there was no oxygen. Scientists tell us that the first organisms were anaerobic microorganisms that evolved some 3 to 4 billion years ago. Life on earth remained anaerobic for hundreds of millions of years.

Then, about 2 billion years ago, the first-worldwide pollution crisis occurred. "The oxygen holocaust" (as described by Margulis and Sagan) came about as the result of the evolution of the purple and green photosynthetic microbes. These organisms were able to make use of the hydrogen in water, by photosynthesis, leaving a waste product called oxygen. Yes, the oxygen that humans consider so precious was originally a gaseous poison dumped into the atmosphere.

As stated by Lovelock, "the first appearance of oxygen in the air heralded an almost fatal catastrophe for early life." Many anaerobic microbes were immediately destroyed. The microbes able to survive were those that responded to the crisis by developing ways to detoxify and eventually exploit the dangerous pollutant. These were the organisms that developed the ability to produce enzymes—like catalase, peroxidase, and superoxide dismutase—that break down and neutralize the various toxic reduction products of oxygen.

Those organisms lacking such enzymes either died or were forced to retreat to ecological niches devoid of oxygen such as soil and mud and deep within the bodies of animals. Those anaerobes that constitute part of our own indigenous microflora, for example, lead a rather pampered existence. We provide them with warmth and nutrients and a safe haven from their worst enemy—oxygen.

(To learn more about this subject, refer to *Microcosmos* by Lynn Margulis and Dorion Sagan [1986] and *Gaia—A New Look at Life on Earth* by J. E. Lovelock [1979]).

TABLE 3 – 2
A Few Common Acids and Their Formulas

Acid	Formula	Location or Use
Hydrochloric acid	HCl	Acid in stomach
Sulfuric acid	H_2SO_4	Industrial mineral acid and battery acid
Boric acid	H_3BO_3	Eye wash
Nitric acid	HNO_3	Industrial oxidizing acid
Acetic acid	$HC_2H_3O_2$	Acid in vinegar
Formic acid	$HCHO_2$	Acid in some insect venom
Carbonic acid	H_2CO_3	Acid in carbonated drinks

Acids

The sour tastes of lemons, grapefruit, and vinegar are familiar sensations. Acids can be recognized by this sour taste. However, this is *not* a safe method of identification because some acids are poisonous or very destructive to human tissues. Acids are identified by a more reliable, safer technique using chemical indicators or dyes that exhibit a distinct color in an acid solution. Some common acids are shown in Table 3 – 2.

All acids behave in a similar way because they all share a common feature: the ability to produce hydrogen ions when in solution. An acid molecule has at least one hydrogen atom. As an illustration, consider hydrogen chloride or hydrochloric acid, which has the formula HCl. Hydrochloric acid in pure form is a gas, but when bubbled into water, it releases a hydrogen ion (H^+), by donating an electron to form the chloride ion (Cl^-). The result is a cation (the positively charged hydrogen ion) and an anion (the negatively charged chloride ion), as illustrated in the reaction equation shown in Fig. 3 – 5.

Note that HCl is an *electrolyte,* that is, it is a substance with charged particles that can conduct electricity. All acids produce hydrogen ions, which are the particles that contribute to the sour taste of acids and also affect the color of

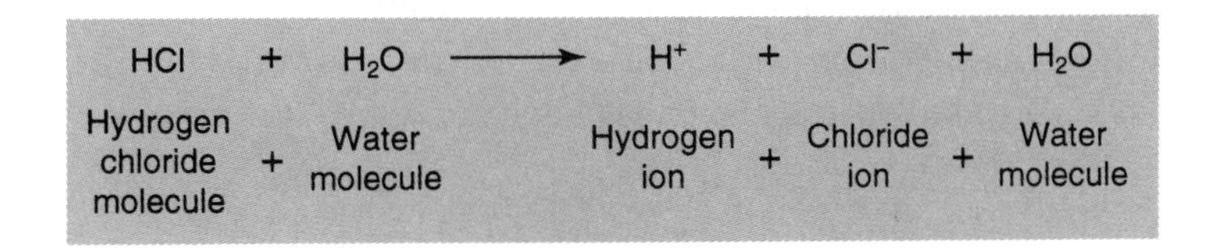

FIGURE 3 – 5
The ionization of hydrochloric acid in water.

TABLE 3 – 3
Some Common Bases

Base	Formula	Location or Use
Sodium hydroxide	$NaOH$	Caustic soda
Potassium hydroxide	KOH	Caustic potash
Ammonium hydroxide	NH_4OH	Household cleaners
Calcium hydroxide	$Ca(OH)_2$	Limewater
Magnesium hydroxide	$Mg(OH)_2$	Antacid drugs

indicators. Indicator dyes are acidic or basic compounds that have distinctive colors when in acidic or basic solutions.

Bases

If you have had the misfortune to taste soap, you know the bitter taste of bases. However, due to their potentially toxic nature, this is not a wise method to use for identification of a base. These substances have a hydroxide ion (OH^-) as their common feature. When dissolved in water, they release the hydroxide ion and a cation and become electrolytes. Table 3 – 3 lists some common bases. Notice the hydroxide ion in their formulas.

As mentioned before, all acids will react with all bases. When they react, they form one common product—water (H_2O). To illustrate, three reactions between different acids and bases are shown in Fig. 3 – 6.

Acid–base reactions can be summarized by a general statement: *An acid plus a base produces a salt and water.*

Salts

Salts are another general class of electrolytic compounds. Chemical formulas are often written in a manner to assist in determining whether a substance is an acid, base, or salt. For example, acids are written with the "H" at the left, as in HCl and H_2SO_4 (sulfuric acid). Bases have an "OH" at the right, such as $NaOH$

FIGURE 3 – 6
General equation: acid + base → salt + water

$$HCl + NaOH \longrightarrow NaCl + H_2O$$
$$H_2SO_4 + Mg(OH)_2 \longrightarrow MgSO_4 + 2\ H_2O$$
$$H_2CO_3 + NaOH \longrightarrow NaHCO_3 + H_2O$$

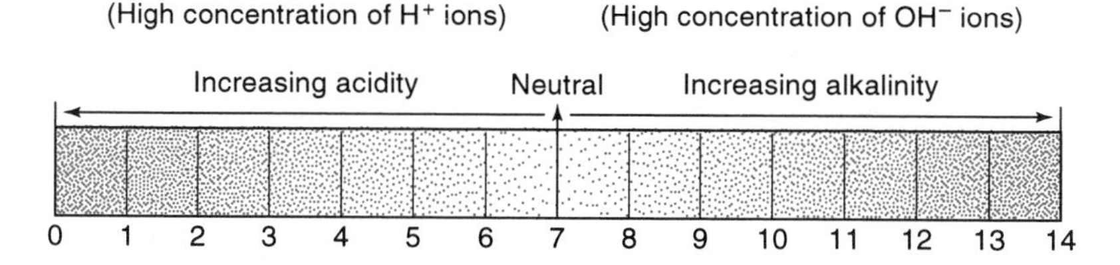

FIGURE 3 – 7
The pH scale is derived by using a complex mathematical formula to determine the free hydrogen ion (H^+) concentration.

(sodium hydroxide) and $Mg(OH)_2$ (magnesium hydroxide). The formulas for salts have neither an H at the left nor an OH at the right. Sodium chloride (NaCl) and sodium bicarbonate ($NaHCO_3$) are examples.

It is important for all living creatures, including microorganisms, to have and maintain the appropriate acid, base, and salt balance to metabolize and function properly. When a nutrient medium is prepared for growing microbial cells in the laboratory, the correct acid–base–salt balance is as important as the availability of the necessary nutrients.

pH

pH is a measure of acidity. Naturally, acids are "acidic," and bases are "basic" or "alkaline" in nature. The pH scale used to indicate acidity or alkalinity ranges from 0 to 14, with 7 as the neutral point (Fig. 3 – 7). Pure water has a pH of 7. If an acid is added to water, the pH *decreases* to between 0 and 7. Adding a base to water *increases* the pH to between 7 and 14. Table 3 – 4 lists some common substances and the pH for each.

TABLE 3 – 4
The pH of Some Common Substances

Substance	**pH**
Household ammonia cleaner	11.9
Blood (human)	7.35–7.45
Water	7
Milk	6.9
Black coffee	5
Orange juice	2.9
Gastric juice (stomach acid)	1.5

The pH of a substance can be measured with a pH meter, with indicator solutions, or with chemically treated papers that turn a definite color within certain pH ranges. These papers are called indicator papers. In the human body the pH of various fluids (blood, lymph) must remain within very narrow ranges to keep the delicately balanced metabolic processes working properly; as with microbial cells, if the pH varies too far from optimum, metabolism is disrupted and death may result.

ORGANIC CHEMISTRY

Organic chemistry is the branch of chemistry that deals primarily with the element carbon and its covalent bonds. This definition makes organic chemistry a broad and important branch of the science of chemistry.

Carbon Bonds

In our current understanding of life, carbon is the primary requisite for all living systems. The element carbon exists in three forms: diamond, graphite, and carbon or carbon black. These three forms have dramatically different physical properties, and it is difficult to believe that they are truly the same element. In addition to these unique physical differences, carbon has a valence of 4, which means that its atoms bond to four other atoms. For convenience, the carbon atom is illustrated in this text with the symbol C and four bonds:

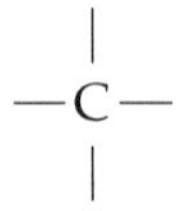

```
  |
—C—
  |
```

The uniqueness of carbon lies in the ability of its atoms to bond to each other to form a multitude of compounds. The variety of carbon compounds increases still more when atoms of other elements also attach in different ways to the carbon atom.

When two carbon atoms bond together there are three types of resulting covalent bonds: single bonds, double bonds, and triple bonds. Each line represents a bond between the carbon atoms, which is formed from a pair of shared electrons.

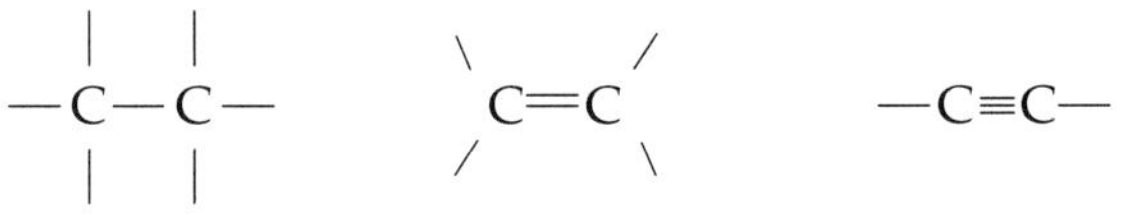

```
  |  |          \     /
—C—C—           C=C          —C≡C—
  |  |          /     \

Single         Double        Triple
bond           bond          bond
```

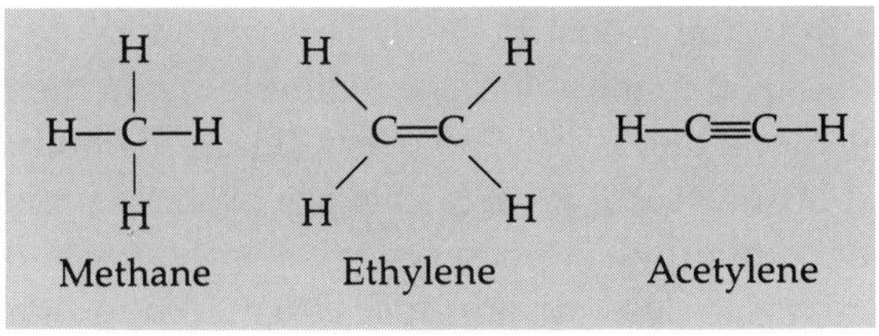
Methane
Ethylene
Acetylene

FIGURE 3 – 8
Simple hydrocarbons.

When atoms of other elements attach to additional available bonds of the carbon atoms, stable compounds are formed. For example, if hydrogens are bonded to the available bonds, compounds called *hydrocarbons* are formed. Just a few of the many hydrocarbon compounds are shown in Figure 3 – 8.

When more than two carbons are linked together, longer molecules are formed. A series of many carbon atoms bonded together is logically called a chain. The long-chain hydrocarbons are usually liquid or solid; the short-chain hydrocarbons, such as the ones in Figure 3 – 8, are gases.

Cyclic Compounds

Carbon atoms may link to carbon atoms to close the chain, forming ring or cyclic compounds. An example is benzene, which has six carbons and six hydrogens, as shown in Figure 3 – 9.

BIOCHEMISTRY

Biochemistry is the study of living matter and its biochemical changes or metabolism. Large biochemical molecules, called macromolecules, are present in every cell in animals, plants, and microorganisms. These macromolecules are of many different types and are classified as *carbohydrates,* lipids, proteins, and *nucleic acids.* Vitamins, enzymes, hormones, and high-energy molecules, such as *adenosine triphosphate* (ATP), are also included among these biochemicals.

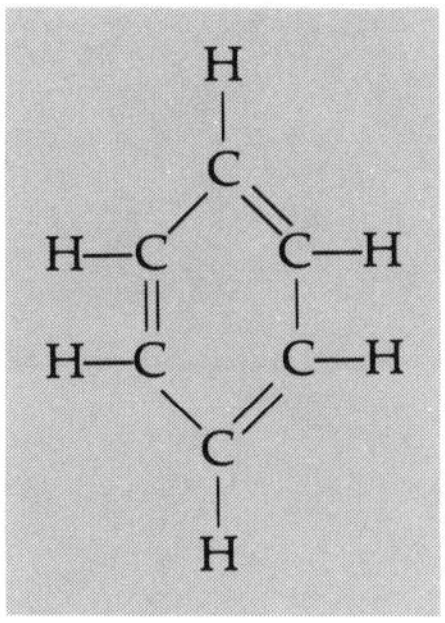

FIGURE 3 – 9
The benzene ring.

Humans obtain their nutrients from the foods they eat. The carbohydrates, fats, nucleic acids, and proteins contained in these foods are digested, and their components are absorbed into the blood and carried to every cell in the body. In the body cells, components from the food biochemicals are absorbed and rearranged. In this way, the proper compounds necessary for cell structure and function are synthesized. Microorganisms also absorb their essential nutrients into the cell by various means to be described in Chapter 4. These nutrients are then used in metabolic reactions as sources of energy and as building blocks for enzymes, structural macromolecules, and genetic materials.

Carbohydrates

Carbohydrates are composed of carbon (C), hydrogen (H), and oxygen (O) in the ratio of 1:2:1, or simply CH_2O.

Monosaccharides The simplest carbohydrates are sugars, and the smallest of these sugars are called monosaccharides (Gr. "mono" = "one"; "sakcharon" = "sugar"). The most important monosaccharide in nature is *glucose* ($C_6H_{12}O_6$), which may occur as a chain or in alpha or beta ring configurations, as shown in Figure 3 – 10. Monosaccharides may contain from three to seven carbon atoms. A three-carbon monosaccharide is called a triose; one containing four carbons is called a tetrose; five, a pentose; six, a hexose; and seven, a heptose. Glucose is a hexose.

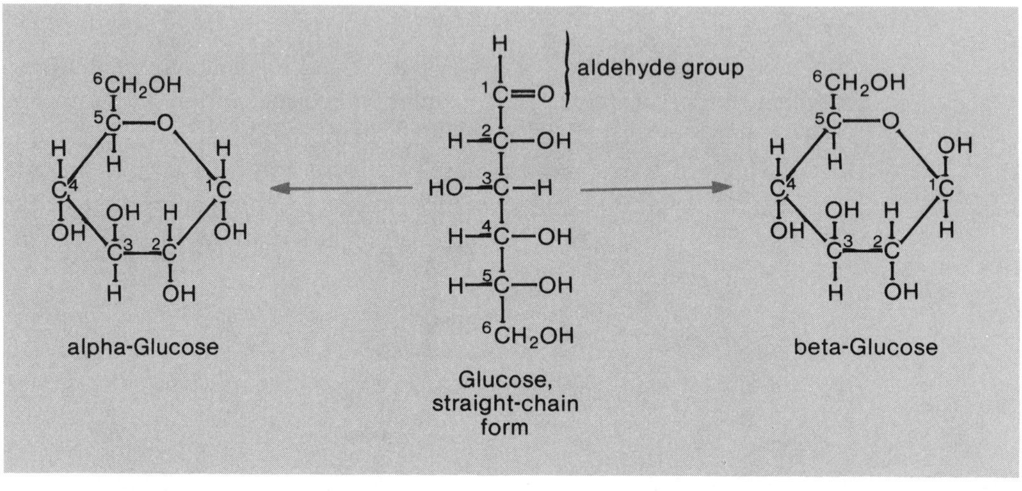

alpha-Glucose

aldehyde group

Glucose, straight-chain form

beta-Glucose

FIGURE 3 – 10
Glucose. The straight-chain and α and β forms may all exist in equilibrium in solution.

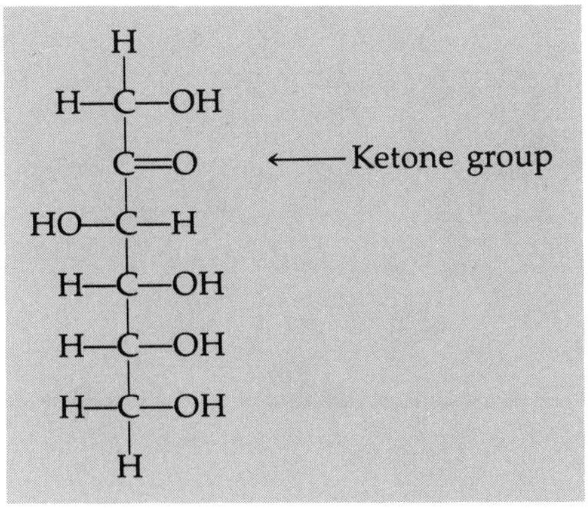

H
H—C—OH
C=O ←— Ketone group
HO—C—H
H—C—OH
H—C—OH
H—C—OH
H

FIGURE 3 – 11
Fructose in straight-chain form. (Fructose may also exist in the ring form as shown in Figure 3 – 12.)

The main source of energy for body cells, glucose (also called dextrose), is found in most sweet fruits and in blood. The glucose carried in the blood to the cells is oxidized to produce the high-energy molecule ATP with its high-energy phosphate bonds. This ATP molecule is the main source of energy used to drive most metabolic reactions. Other monosaccharides are galactose, ribose (a pentose), and fructose (Fig. 3 – 11). Fructose, the sweetest of the monosaccharides, is found in fruits and honey.

Disaccharides Disaccharides are sugars that result from the combination of two monosaccharides. The synthesis of disaccharides from two monosaccharides with removal of a water molecule is called *dehydration synthesis* (Fig. 3 – 12). Sucrose (table sugar) is a sweet *disaccharide* made up of a glucose molecule and a fructose molecule. Sucrose comes from sugar cane, sugar beets, and maple sugar. Lactose and maltose are also disaccharides. Lactose is the sugar in milk, and maltose is the sugar in malt.

Disaccharides react with water in a process called hydrolysis, which causes them to break down into two monosaccharides:

$$\text{disaccharide} + H_2O \rightleftharpoons 2 \text{ monosaccharides}$$
$$\text{maltose} + H_2O \rightleftharpoons \text{glucose} + \text{glucose}$$
$$\text{lactose} + H_2O \rightleftharpoons \text{glucose} + \text{galactose}$$
$$\text{sucrose} + H_2O \rightleftharpoons \text{glucose} + \text{fructose}$$

Polysaccharides Polysaccharides, such as starch, cellulose, and glycogen, are essentially made up of hundreds of repetitive glucose units, that is, they consist of long chains of glucose molecules, which are held together by glycosidic bonds. They are *polymers,* meaning that they consist of many similar subunits. Some of these molecules are so large that they are insoluble in water. In the presence of the proper enzymes or acids, polysaccharides may be hydrolyzed or broken down into disaccharides and then finally into monosaccharides (Fig. 3 – 13).

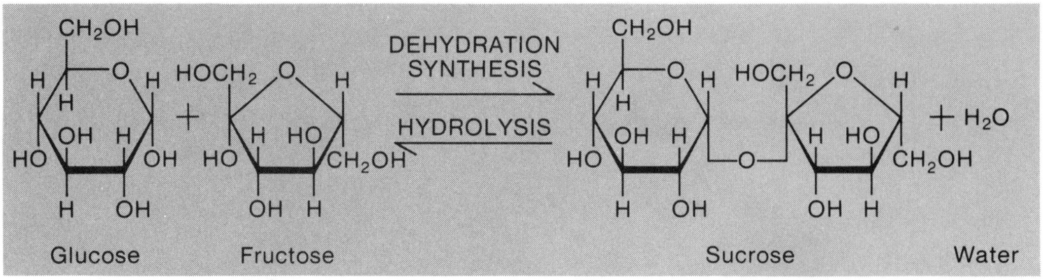

FIGURE 3 – 12
The dehydration synthesis and hydrolysis of sucrose.

Polysaccharides serve two main functions. One is to store energy that can be used when the external food supply is low. The common storage molecule in animals is *glycogen,* which is found in the liver and in muscles. In plants, glucose is stored as *starch* and is found in potatoes and other vegetables and seeds. Some algae store starch, whereas bacteria contain glycogen granules as a reserve nutrient supply. The other function of polysaccharides is to provide a "tough" molecule for structural support and protection. Many bacteria secrete polysaccharide capsules for protection against drying and phagocytosis. Plant and algal cells have cellulose cell walls to provide support and shape as well as protection against the environment.

Cellulose is insoluble in water and indigestible for humans and most animals. Some protozoa, fungi, and bacteria have enzymes that will break the β-glycosidic bonds linking the glucose units in cellulose. These microorganisms

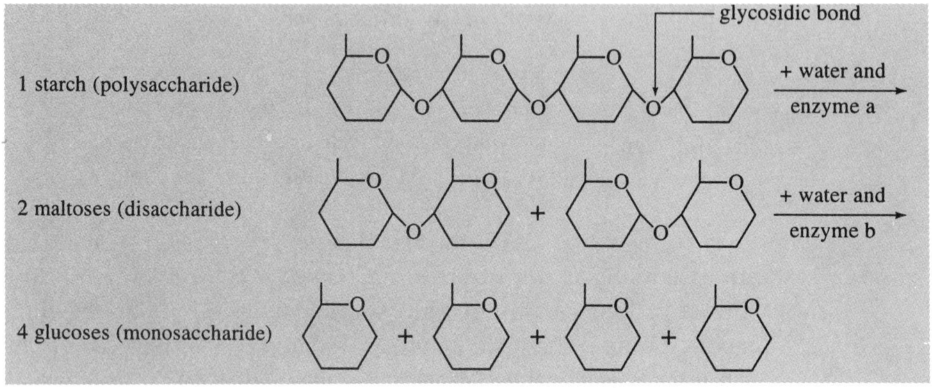

FIGURE 3 – 13
The hydrolysis of starch.

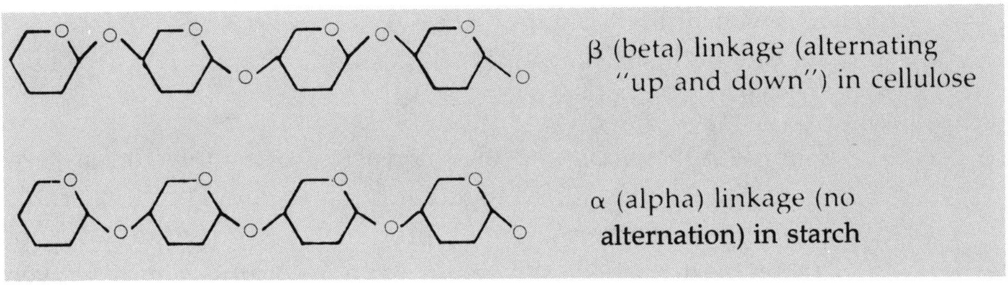

β (beta) linkage (alternating "up and down") in cellulose

α (alpha) linkage (no alternation) in starch

FIGURE 3 – 14
The difference between cellulose and starch.

(saprophytes) are able to disintegrate dead plants in the soil and live in the digestive organs of herbivores (plant eaters). Protozoa in the gut of termites digest the cellulose in the wood that the termites eat. Starch and glycogen are easily digested by animals because they have the digestive enzyme that hydrolyzes the α-glycosidic bonds that link the glucose units into long, helical, or branched polymers (Fig. 3 – 14). Fibers of cellulose extracted from certain plants are used to make paper, cotton, linen, and rope. These fibers are relatively rigid, strong, and insoluble because they consist of 100 to 200 parallel strands of cellulose.

When polysaccharides combine with other chemical groups (amines, lipids, and amino acids), extremely complex macromolecules are formed that serve specific purposes. Glucosamine and galactosamine (amine derivatives of glucose and galactose, respectively) are important constituents of the supporting polysaccharides in connective tissue fibers, cartilage, and chitin. Chitin is the main component of the hard outer covering of insects, spiders, crabs, and fungi. The main portion of the rigid cell wall of bacteria consists of amino sugars and short polypeptide chains that combine to form the peptidoglycan layer.

Lipids

Lipids are a class of biochemical compounds consisting mainly of fats and oils. Most of them are insoluble in water but soluble in fat solvents such as ether, chloroform, and benzene. Lipids are essential constituents of almost all living cells. They may be classified into the following categories:

1. Simple lipids (contain C, H, O)
   - **a.** Fats and oils (butter, vegetable oils)
   - **b.** Waxes (beeswax, lanolin)
2. Compound lipids (contain C, H, O, N, P)
   - **a.** Phospholipids (in cell membranes)
   - **b.** Glycolipids (in nerve cells)

3. Derived lipids (contain C, H, O)
   a. Steroids (sex hormones, cholesterol, vitamin D)
   b. Fat-soluble vitamins A, E, K

Simple lipids (fats and oils) consist of one molecule of glycerol and three *fatty acid* molecules joined together by the removal of three molecules of water (dehydration synthesis), which produces a triglyceride fat or oil (Fig. 3 – 15).

If the hydrocarbon side chains have no double bonds between carbons, the lipid is called a saturated fatty acid because every carbon atom is saturated with hydrogen. When there are double bonds between the carbons (C=C), it is an unsaturated fatty acid. Sources of unsaturated fats are peanut, olive, corn, soybean, and cottonseed oils. Fats are liquids or low-melting point solids at room temperature, depending on the relative composition of the fatty acids. Unsaturated fatty acids with short carbon chains are usually liquids. Many people are on low-saturated fat diets to help prevent lipid cholesterol deposits in their hearts and arteries.

A nutritious human diet should include at least the two essential fatty acids that the body cannot synthesize. These are linoleic acid ($C_{17}H_{31}COOH$) from fish liver oils and vegetable oils and arachidonic acid ($C_{19}H_{31}COOH$) from egg yolk, liver, kidney tissues, and fish liver oils. However, different fatty acids are produced by and are necessary for the growth of bacteria and other microorganisms, depending on the species of the organism.

Saponification is the hydrolysis of a fat into glycerol (glycerin) and its three fatty acids. When NaOH or KOH (strong alkalis) are used to hydrolyze fats, the results are sodium or potassium salts of the fatty acids, which are soaps. Detergents, bile salts, and fat-digestive enzymes also saponify fats.

Waxes Waxes are chemically different from fats because they are long-chain or complex alcohols, other than glycerol, with fatty acids attached. Waxes are

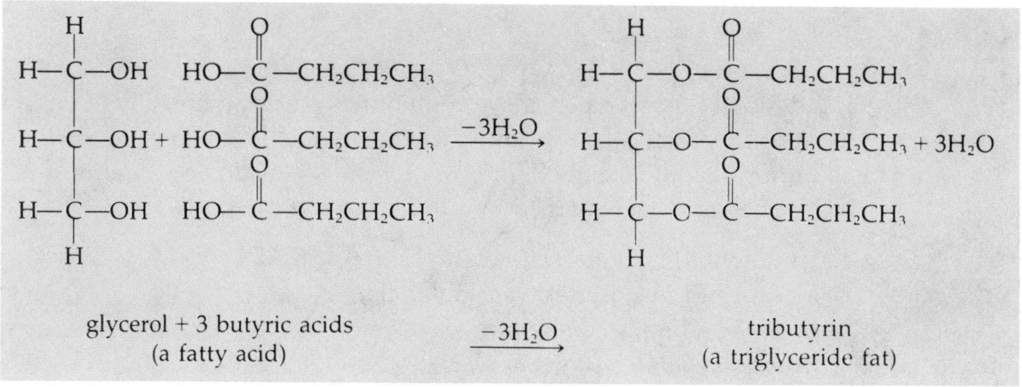

$$\begin{array}{l} \text{H—C(H)—OH} \quad \text{HO—C(=O)—CH}_2\text{CH}_2\text{CH}_3 \qquad \text{H—C(H)—O—C(=O)—CH}_2\text{CH}_2\text{CH}_3 \\ \text{H—C—OH} + \text{HO—C(=O)—CH}_2\text{CH}_2\text{CH}_3 \xrightarrow{-3H_2O} \text{H—C—O—C(=O)—CH}_2\text{CH}_2\text{CH}_3 + 3H_2O \\ \text{H—C(H)—OH} \quad \text{HO—C(=O)—CH}_2\text{CH}_2\text{CH}_3 \qquad \text{H—C(H)—O—C(=O)—CH}_2\text{CH}_2\text{CH}_3 \end{array}$$

glycerol + 3 butyric acids (a fatty acid) $\xrightarrow{-3H_2O}$ tributyrin (a triglyceride fat)

FIGURE 3 – 15
The synthesis of a fat.

semi-solid substances that serve as protective coatings on the surfaces of leaves, stems, fruits, insects, and some bacteria (*e.g.*, mycobacteria).

Compound Lipids Compound lipids include the phospholipids of cell membranes and the glycolipids of nerve and brain cells. Even when an organism is starving, the amount of these nonfat lipids in the cells does not vary.

Derived Lipids Derived lipids are classified as lipids because they are also soluble in fat solvents. These complex molecules of four interlocking carbon rings are called steroids. Many steroid anti-inflammatory medications are produced by fungi and bacteria and are used in the treatment of arthritis and cancer. Other examples of steroids include cholesterol, male and female hormones, adrenal cortex hormones, and vitamin D. Vitamins A, E, and K are not steroids but are fat-soluble vitamins.

Cholesterol is a necessary body *metabolite.* It is found in most eucaryotic membranes and in nervous tissues and aids in fatty acid adsorption. It is also essential for the production of sex hormones, vitamin D, and adrenal cortex hormones. However, an excess of cholesterol may be deposited in either the arteries, causing heart disease, or in the gallbladder, becoming gallstones.

Vitamin D helps maintain the calcium and phosphorus balance in the body. Good sources of vitamin D are fish oils and fortified foods. Vitamin A can be obtained from liver, eggs, butter, cheese, carrots, and green leafy vegetables; a lack of this vitamin can lead to night blindness or hardening of the mucous membranes.

Vitamin E (from plant oils, leafy vegetables, and eggs) reduces sterility in animals and the wasting effects of aging. Vitamin K, which is synthesized in the body, is necessary for normal blood clotting. Some colon bacteria (specifically *Escherichia coli*) also secrete vitamin K, which is absorbed into the blood.

Proteins

Proteins are among the most essential chemicals in all living cells. Some are the structural components of membranes, cells, and tissues, whereas others are enzymes and hormones that chemically control the metabolic balance within both the cell and the entire organism. All proteins are polymers of amino acids; however, they vary widely in the number of amino acids present and in the sequence of amino acids, as well as in their size, configuration, and functions.

Proteins are synthesized by plants, algae, and some bacteria from carbon dioxide, water, nitrates, and sulfates through the process called photosynthesis. In this process, the chemicals combine with the aid of the sun's energy. Humans can synthesize proteins, but they must ingest eight essential amino acids to synthesize the other amino acids necessary in the building of proteins. Nonphotosynthetic microorganisms also must absorb their essential amino

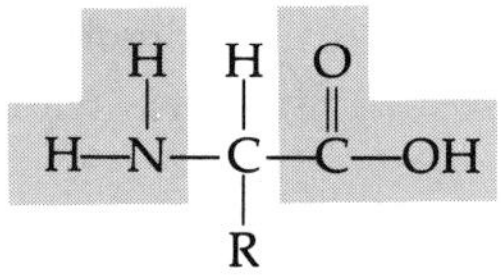

Basic amine group H—N—C—C—OH Acid carboxyl group

FIGURE 3 – 16
The basic structure of amino acids.

acids that are not synthesized or stored in living cells. All of the amino acids (ingested and synthesized) must be present during the protein synthesis process in cells. Any excess can be used to produce energy or eliminated with the soluble wastes.

Amino Acid Structure There are 20 different amino acids. Each is composed of carbon, hydrogen, oxygen, and nitrogen; three of the amino acids also have sulfur atoms in the molecule. The general formula for amino acids is shown in Figure 3 – 16. In this figure, the "R" group represents any of the 20 groups that may be substituted into that position to build the various amino acids. For instance, "H" in place of the "R" represents glycine, and "CH_3" in that position results in a structural formula representing alanine.

The thousands of different proteins in the body are composed of a great variety of amino acids in various arrangements and amounts. The number of proteins that can be synthesized is virtually unlimited. Proteins are not limited by the number of different amino acids, just as the number of words in a written language is not limited by the number of letters in the alphabet. The actual number of proteins produced by an organism and the amino acid sequence of those proteins are determined by the particular genes present on the organism's chromosome(s).

Protein Structure When water is removed, by dehydration synthesis, amino acids become linked together by a peptide bond as shown in Figure 3 – 17. A

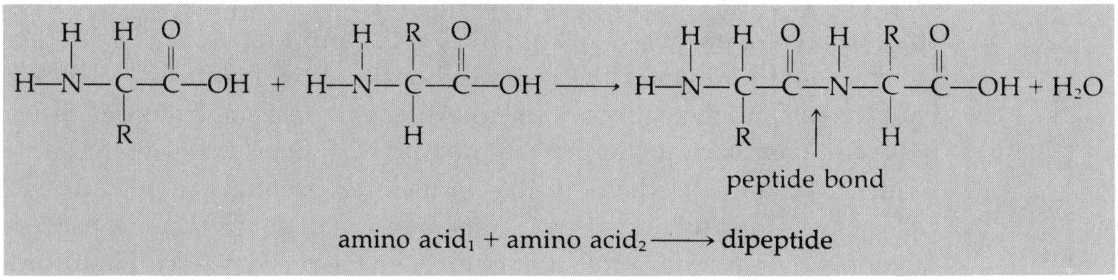

H—N—C—C—OH + H—N—C—C—OH ⟶ H—N—C—C—N—C—C—OH + H_2O

peptide bond

amino $acid_1$ + amino $acid_2$ ⟶ **dipeptide**

FIGURE 3 – 17
The formation of a dipeptide (R = any amino acid side-chain group).

dipeptide is formed by bonding two amino acids; three amino acids form a *tripeptide.* A long chain or polymer of amino acids is referred to as a *polypeptide.* Polypeptides are said to have primary protein structure, a sequence of amino acids in a chain (Fig. 3 – 18). Some fibrous proteins are made of strands of polypeptides bonded to each other by hydrogen bonds and wound together into a helix. Collagen, found in connective tissue, scars, and tendons is another fibrous protein, consisting of three polypeptides in a triple helix, forming a very strong structure.

Most polypeptide chains naturally twist into spirals or helices as a result of the charged side chains protruding from the carbon–nitrogen backbone of the molecule. This helical configuration is referred to as secondary protein structure and is found in fibrous proteins such as the keratin fibers of hair, nails, and skin and the microtubules and microfilaments of cells.

Because a long coil can become entwined by folding back on itself, a poly-

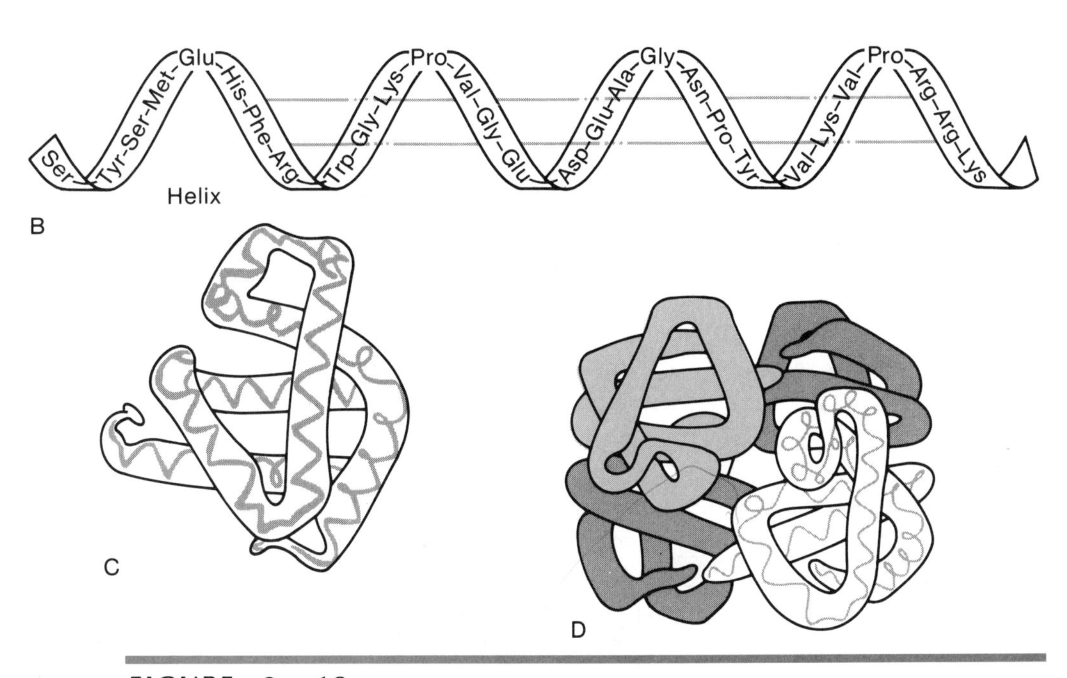

FIGURE 3 – 18
Basic protein structure. (*A*) Primary sequence of amino acids. (*B*) Secondary helix. (*C*) Tertiary globular structure. (*D*) Quarternary structure with four polypeptide chains.

peptide helix may become globular (Fig. 3 – 18). In some areas the helix is retained, but other areas curve randomly. This globular, tertiary protein structure is stabilized, not only by hydrogen bonding, but also by disulfide bond cross-links between two sulfur groups (S–S). This three-dimensional configuration is characteristic of enzymes, which work by fitting on and into specific molecules (see the next section).

When two or more polypeptide chains are bonded together by hydrogen and disulfide bonds, the resulting structure is referred to as quaternary protein structure (Fig. 3 – 18). For instance, hemoglobin consists of four globular myoglobins. The size, shape, and configuration of a protein is specific for the function it must perform. If the amino acid sequence and, thus, the configuration of hemoglobin in red blood cells is not perfect, the red blood cells may become distorted and assume a sickle shape (as in sickle cell anemia). In this state they are unable to carry oxygen, which is necessary for cellular metabolism. Myoglobin, the oxygen-binding protein found in skeletal muscles, was the first protein to have its primary, secondary, and tertiary structure defined.

Enzymes Enzymes are protein molecules produced by living cells as "instructed" by the DNA or genes on the chromosomes. Enzymes are essential as *catalysts* in the biochemical reactions of metabolism; in other words, they speed up these reactions. Almost every reaction in the cell requires the presence of a specific enzyme. All enzymes are proteins; many consist of protein only, whereas others require a nonprotein *cofactor,* such as metal ions (Ca^{2+}, Fe^{2+}, Mg^{2+}, Cu^{2+}) or organic compounds (*coenzymes*) such as vitamin C. Although enzymes influence the direction of the reaction and increase its rate of reaction, they do not provide the energy needed to activate the reaction.

Enzymes are usually named by adding the ending "-ase" to the word indicating the compound or types of compounds on which an enzyme exerts its effect. For example, proteases, carbohydrases, and lipases are enzymes specific for proteins, carbohydrates, and lipids, respectively. The specific molecule upon which an enzyme acts is referred to as that enzyme's *substrate.* Each enzyme has a particular substrate upon which it exerts its effect; thus, enzymes are said to be very specific.

Poisons and toxins usually cause damage to the body by interfering with the action of certain necessary enzymes. For example, cyanide poison binds to the iron and copper ions in the cytochrome systems of the mitochondria of eucaryotic cells. As a result, the cells cannot use oxygen to synthesize ATP, which is essential for energy production, and they soon die.

Proteins, including enzymes, may be denatured (changed) by heat or certain chemicals. In a denatured protein, the bonds that hold the molecule in a tertiary structure are broken. With the bonds broken, the protein is no longer functional. Enzymes are discussed further in Chapter 4.

Nucleic Acids

The fourth major group of macromolecules in living cells are the nucleic acids: deoxyribonucleic acid (DNA) and ribonucleic acid (RNA). DNA is the macromolecule that makes up the major portion of the chromosomes. Its important task is to carry the genetic information for each cell of an organism. The information from the DNA must be carried to the rest of the cell for the cell to function properly; this is accomplished by RNA. Thus, RNA is found not only in the nucleus, but in the cytoplasm as well.

Function Nucleic acids have two very important functions. First, they determine precisely all the proteins that are synthesized, many of which control the metabolism of the organism. DNA's second role is to act as the genetic "molecular blueprint of life." The genetic information must be passed from one generation to the next, from each parent cell to each daughter cell via the DNA.

Structure In addition to carbon, hydrogen, oxygen, and nitrogen, DNA and RNA contain the element phosphorus, whereas proteins (described previously) may contain sulfur. The building blocks of these nucleic acid polymers are called nucleotides. These are more complex monomers (single molecular units that can be repeated to form a polymer) than amino acids, which are the building blocks of proteins. Nucleotides consist of three subunits: a nitrogen-containing base, a five-carbon sugar, and a phosphate group, joined together, as shown in Figure 3 – 19.

As previously stated, there are two kinds of nucleic acids in cells: DNA and RNA. DNA contains deoxyribose as its five-carbon sugar (a pentose), whereas RNA contains the sugar ribose. There are three types of RNA, which are named for the function they serve: *messenger RNA* (*mRNA*), *ribosomal RNA* (*rRNA*), and *transfer RNA* (*tRNA*). The five most important nitrogen-containing bases are adenine (A), guanine (G), thymine (T), cytosine (C), and uracil (U). Thymine occurs in DNA, and uracil occurs in RNA. The other three bases (A, G, C) occur in both DNA and RNA. Adenine and guanine are molecules called *purines*, whereas the other three nitrogen bases are *pyrimidines*.

The nucleotides join together between their sugar and phosphate groups to form very long polymers, 100,000 or more monomers long, as shown in Figure 3 – 20.

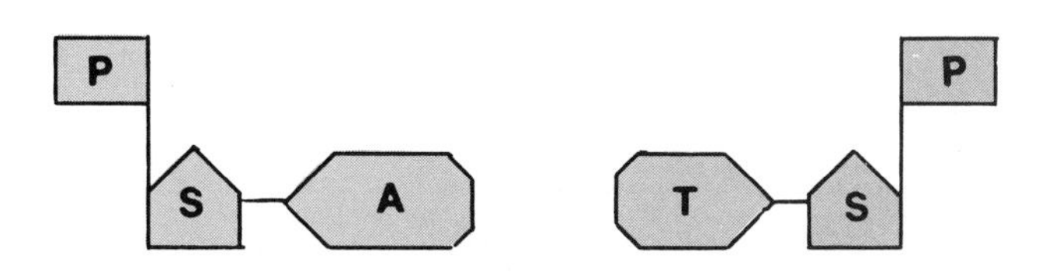

FIGURE 3 – 19
Two nucleotides, each consisting of a nitrogen base (A or T), a five-carbon sugar (S), and a phosphate group (P).

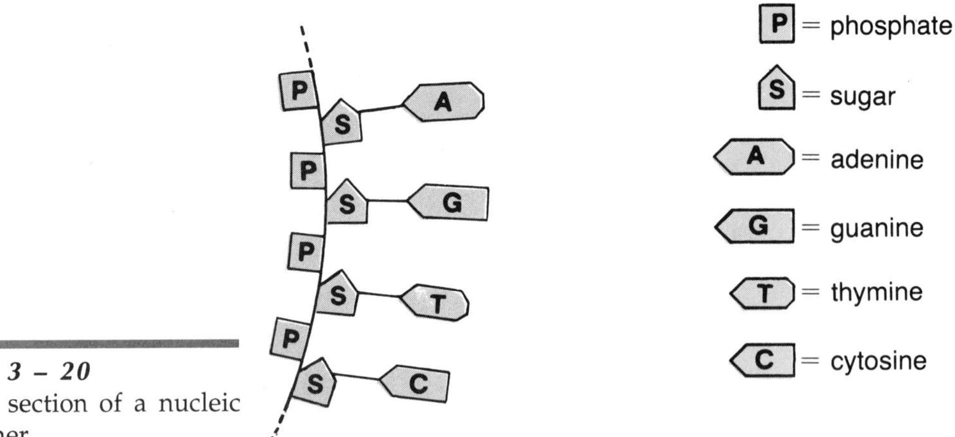

FIGURE 3 – 20
One small section of a nucleic acid polymer.

DNA Structure In 1953, James Watson and Francis Crick proposed a double-stranded helical structure for DNA to indicate how it could copy (replicate) itself exactly to pass the genetic information to each daughter cell. For a double-stranded DNA to form, the bases on the two separate strands must bond together. It was found that adenine always bonds with thymine (with two hydrogen bonds) and guanine with cytosine (with three hydrogen bonds) because of the size and bonding attraction between the molecules. The bonding forces of the double-stranded polymer cause it to assume the shape of a double α-helix, which is similar to a right-handed spiral staircase (Fig. 3 – 21).

Replication When a cell is preparing to divide, all of the DNA molecules in the chromosomes of that cell must duplicate, thereby ensuring that the same genetic information is passed to both daughter cells. This process is called *DNA replication.* It occurs by separation of the DNA strands and the building of complementary strands by the addition of the correct nucleotides, as indicated

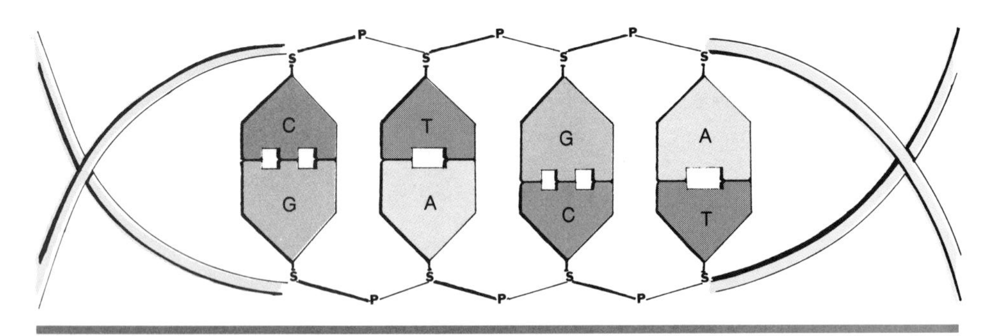

FIGURE 3 – 21
The DNA double helix.

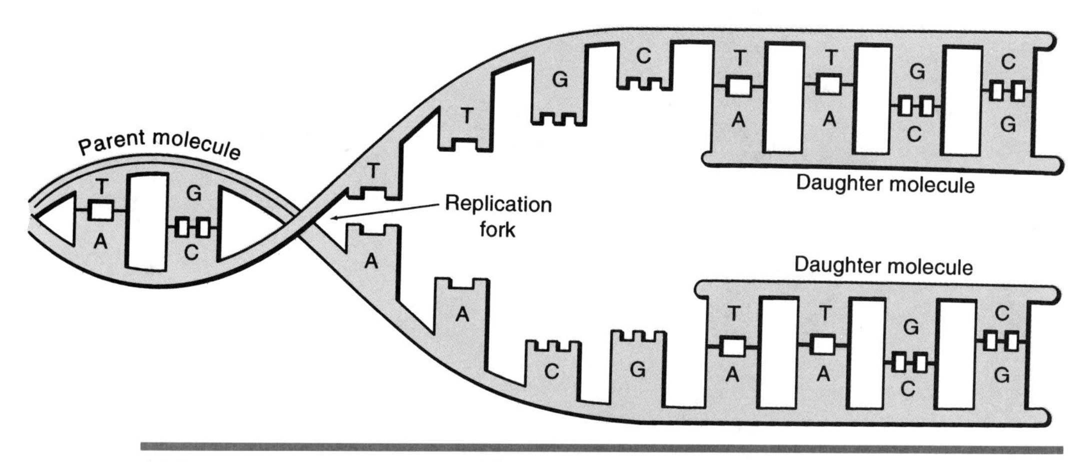

FIGURE 3 – 22
Replication of DNA before cell division.

in Figure 3 – 22. The duplicated DNA of the chromosomes can then be separated during ordinary cell division so that the same number of chromosomes, the same genes, and the same amount of DNA as in the parent cell are found in each daughter cell (except during meiosis, the reduction division to produce egg and sperm cells). The enzyme required for DNA replication to occur is called *DNA polymerase* (or DNA-dependent DNA polymerase).

Protein Synthesis In a normally functioning living cell, the DNA of the chromosomes controls all the metabolic activities. It accomplishes this by directing the synthesis of the protein enzymes that regulate the chemical reactions occurring in metabolism.

The translation and interpretation of the information carried by the DNA molecule to produce the proper proteins is a complex procedure. It is the sequence of the four nitrogen-containing bases (A, T, C, and G) that carries the code for the amino acid sequences of the protein to be synthesized in the cytoplasm of the cell. Protein synthesis is shown in Figure 3 – 23.

You should remember that the chromosomes, which are located in the nuclear region of the cell, consist of many genes that carry the genetic information for the inherited traits of the organism. Each gene consists of one or more *cistrons,* which are the portions of the chromosomal DNA that code for *one* particular protein or enzyme. The terms cistron and gene are often used synonymously.

When a cell is stimulated (by need) to produce a particular protein, such as insulin, the DNA of the appropriate cistron is activated to unwind temporarily from its helical configuration. This unwinding exposes the bases, which then attract the bases of free nucleotides, and a messenger RNA (mRNA) begins to build on one strand, the activated strand, of the opened DNA. Thus, the DNA

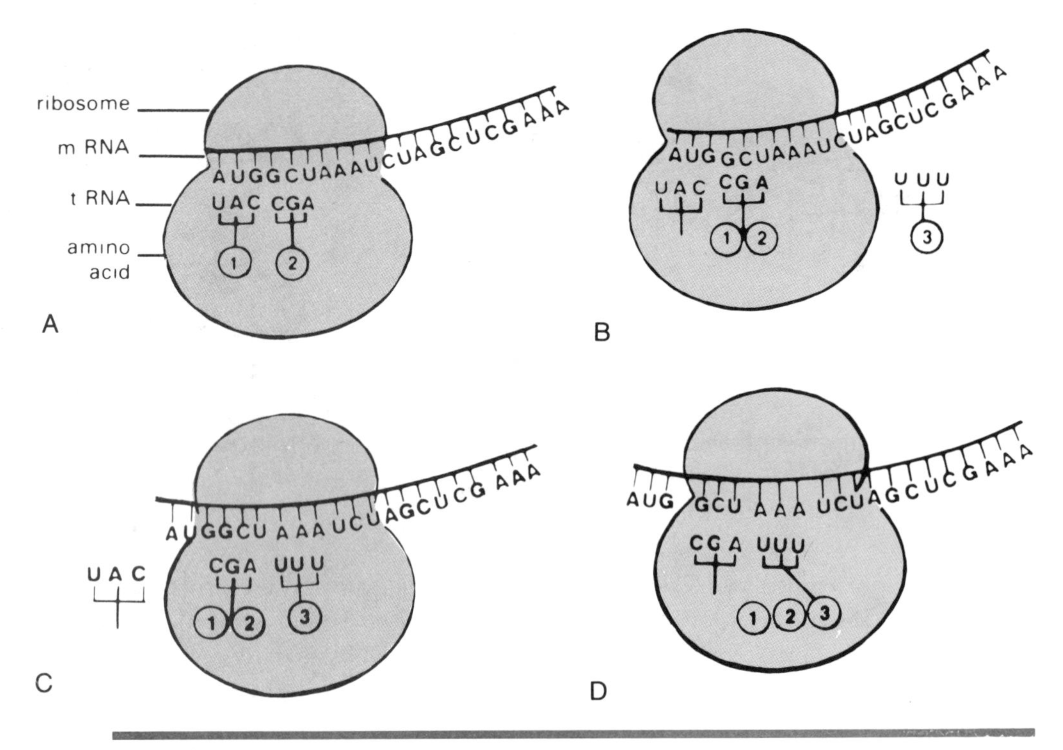

FIGURE 3 – 23
Outline of protein synthesis. (*A*) Messenger RNA bound to ribosome. Amino acids 1 and 2 are bonded to their transfer RNAs. The t-RNA-amino acid 1 is bound to the peptidyl binding site. The t-RNA-amino acid 2 is bound to the amino acid binding site. (*B*) Amino acid 1 is released from its t-RNA and is joined to amino acid 2 through a peptide bond. (*C*) The ribosome moves along the m-RNA, and t-RNA 1 leaves the peptidyl binding site. The t-RNA 2 with the attached dipeptide moves to the peptidyl site and t-RNA amino acid 3 binds to the amino acid binding site. (*D*) Amino acid 2 is bonded to amino acid 3 as in step *B*. The process repeats until the protein is completed.

has served as a template, or pattern, and has coded for a complementary mirror image of its structure in the RNA. On the growing mRNA molecule, an A will be introduced opposite a T on the DNA molecule, a G opposite a C, a C opposite a G, and a U opposite an A. This procedure is called *transcription* because the message from the DNA is transcribed onto the mRNA. After the mRNA has been synthesized over the length of the cistron, it is released from the active DNA strand to carry the message to the cytoplasm and direct the synthesis of that particular protein, which is insulin in this case. The enzyme that is required for transcription to occur is called *RNA polymerase* (or DNA-dependent RNA polymerase).

In eucaryotes, the released mRNA passes through the pores of the nuclear membrane into the cytoplasm and takes its position on the protein "assembly line," that is, the ribosomes containing ribosomal RNA (rRNA), which attract the mRNA. Ribosomes usually reside on the endoplasmic reticulum membranes in eucaryotic cells.

In procaryotes, the ribosomes may attach to the mRNA as it is being transcribed at the DNA because no nuclear membrane is present; thus, transcription and translation may occur simultaneously.

The base sequence of the mRNA is read or interpreted in groups of three bases, called *codons.* The sequence of a codon's three bases is the code that determines which amino acid is inserted in that position in the protein being synthesized. Amino acids must first be activated by attaching to an appropriate transfer RNA (tRNA) molecule that then carries each amino acid from the cytoplasmic matrix to the site of protein assembly. The three-base sequence of the codon determines which tRNA brings its specific amino acid to the ribosome because the tRNA has an *anticodon,* three-base sequence, which is complementary to, or attracted to, the codon of the mRNA. For example, the tRNA with the anticodon base sequence UUU carries the amino acid lysine to the mRNA codon AAA. Similarly, the mRNA codon CCG codes for the tRNA anticodon GGC, which carries the amino acid proline at the opposite end of the tRNA molecule. This information system is called the *genetic code.* A chart can be prepared to show the sequence of three bases in the DNA that codes for a particular codon in mRNA, which, in turn, attracts a particular anticodon on the tRNA carrying a specific amino acid:

DNA	mRNA	tRNA	Amino Acid
G	C	G	
G	C	G	Proline
C	G	C	

The process of translating the message carried by the mRNA, whereby particular tRNAs bring amino acids to be bound together in the proper sequence to make a specific protein (*e.g.,* insulin), is called *translation.* This process is summarized in Figure 3 – 23. It should be noted that a eucaryotic cell is constantly producing mRNAs in its nucleus that direct the synthesis of all the proteins, including metabolic enzymes necessary for the normal functions of that specific type of cell. Also, mRNA and tRNA are short-lived nucleic acids that may be reused many times, then destroyed and resynthesized in the nucleus. The rRNA is made in the dense portion of the nucleus called the nucleolus. The ribosomes last longer in the cell than the mRNA.

As tRNA molecules attach to mRNA while it is sliding over the ribosomes, they bring the correct activated amino acids into contact with each other so that peptide bonds are formed and a polypeptide is synthesized. As the polypeptide grows and becomes a protein, it folds into the unique shape determined by the amino acid sequence. This characteristic shape allows the protein to perform its specific function. If one DNA cistron base is out of sequence, the amino acid sequence will not be correct and the protein configuration will not allow the protein to function properly. For instance, some diabetics may not produce insulin properly because a mutation in the DNA of their chromosomes caused a rearrangement of the bases in the cistron that codes for insulin. DNA errors are the basis for most genetic and inherited diseases, such as phenylketonuria (PKU), sickle cell anemia, cerebral palsy, cystic fibrosis, cleft lip, clubfoot, extra fingers, albinism, and many other birth defects. Likewise, nonpathogenic microbes may mutate to become pathogens, and pathogens may lose the ability to cause disease by *mutation* or a change in the DNA sequence.

The relatively new science of genetic engineering attempts to repair the genetic damage in some diseases. As yet, the morality of manipulation of human genes has not been resolved by society. However, many genetically changed microbes are able to produce substances, such as human insulin, interferon, growth hormones, new pharmaceutical agents, and vaccines, that will have a substantial effect on the medical treatment of humans.

SUMMARY

To understand how a cell lives, metabolizes, and reproduces, one must comprehend some of the chemicals and chemical reactions involved. In this chapter, a brief introduction to atoms and molecules and how they combine by ionic, covalent, and polar bonds was given. The importance of water, hydrolytic reactions, and acid–base reactions was discussed. The structure, function, characteristics, and breakdown products of carbohydrates, lipids, proteins, and nucleic acids were emphasized so that the functions of enzymes and metabolic reactions may be better understood.

PROBLEMS AND QUESTIONS

1. Describe the relationships between electrolyte, solvent, solute, and solution.
2. Differentiate between acids and bases, and give some examples of each.
3. If a nutrient growth medium is too acidic or too basic, might a bacterial culture die? Why?
4. Why is it possible for humans to digest starch but not cellulose and chitin? Which microorganisms can digest cellulose?
5. Is the peptidoglycan layer of a bacterial cell wall a polymer? If so, provide reasons to support this claim.
6. Are the fat-soluble vitamins lipids? Why? Where would lipids be found in bacterial cells?
7. What are the differences between structural proteins and enzymes in a cell? Describe the primary, secondary, tertiary, and quaternary structure of proteins. How does an enzyme function?
8. Draw a hypothetical bacterium and indicate the types of macromolecules found in its various parts.
9. Explain the chemical characteristics of polymers, carbohydrates, polysaccharides, sugars, cellulose, starch, lipids, proteins, fats, waxes, enzymes, peptides, and nucleic acids.
10. Differentiate between DNA, mRNA, rRNA, and tRNA. Which four bases are found in DNA? In RNA?

Self Test

After you have read Chapter 3, examined the objectives, reviewed the chapter outline, studied the new terms, and answered the problems and questions above, complete the following self test.

Matching Exercises

Complete each statement from the list of words provided with each section.

BASIC CHEMISTRY

neutrons	ionic	dehydration synthesis
covalent	hydrolysis	ions
molecule	atoms	isotope
electrons	polar	metabolism

1. Matter is composed of fundamental units called ________________.
2. Charged atoms that have gained or lost electrons are ________________.
3. Isotopes of an element differ in the number of ________________.
4. When electrons are actually transferred from one atom to another, ________________ bonds form.

5. When atoms share electrons rather than transferring them, ____________ bonds form.
6. ________________ involves the breakdown of large organic molecules into their subunits through the process of ________________, and the synthesis of large organic molecules by dehydrolysis or ________________.

BIOCHEMISTRY

primary
monosaccharides
RNA
secondary
enzymes
glycerol
tertiary
substrate
glucose
polysaccharides
DNA
proteins
nucleotides

1. A fat molecule is composed of three fatty acid molecules and one ________________ molecule.
2. The order, or sequence, of amino acids in a protein molecule constitutes its ________________ structure.
3. ________________ is the genetic material of living systems, the subunits of which are called ______________.
4. Glycogen, starch, and cellulose are examples of ________________.
5. Most metabolic reactions are made possible through the action of ________________, or organic catalysts.
6. The substance acted on by an enzyme is called a/an ________________.

Match the chemicals in Column I with an appropriate group in Column II. An answer from Column II may be used more than once.

Column I

__ **1.** Cholesterol
__ **2.** Triglyceride
__ **3.** Hemoglobin
__ **4.** Glucose
__ **5.** DNA
__ **6.** Sucrose
__ **7.** Glycerol and fatty acids
__ **8.** RNA
__ **9.** Phospholipid
__ **10.** Cellulose

Column II

a. Lipid
b. Carbohydrate
c. Nucleic acid
d. Peptide
e. Protein

True or False (T or F)

__ **1.** In general, DNA molecules are double-stranded.

__ **2.** RNA contains the base cytosine instead of guanine.

___ 3. Transcription occurs in the nuclear area of eucaryotic cells, whereas translation occurs on ribosomes.

___ 4. In DNA, the molecule guanine normally pairs with cytosine.

___ 5. Lipids are generally insoluble inorganic solvents.

___ 6. Steroids are a form of lipid.

___ 7. The chief polysaccharide of animals is starch.

___ 8. Uracil is a common nitrogenous base of DNA.

___ 9. The chemical bond linking amino acids to one another is a disulfide bond.

Multiple Choice

1. The four major organic macromolecular groups include
  a. carbohydrates, sugars, starch, cellulose
  b. amino acids, proteins, enzymes, peptides
  c. lipids, proteins, nucleic acids, carbohydrates
  d. carbohydrates, lipids, vitamins, proteins

2. Proteins are polymers of
  a. sugars
  b. enzymes
  c. amino acids
  d. glycerol and fatty acids
  e. nucleotides

3. DNA is a polymer of
  a. deoxyribose
  b. phosphate
  c. organic bases
  d. nucleotides
  e. peptides

4. RNA is a polymer of
  a. triglycerides
  b. carbohydrates
  c. peptides
  d. nucleotides
  e. amino acids

5. Which of the following is *not* a true difference between DNA and RNA?
  a. DNA is a double-stranded molecule and RNA is a single-stranded molecule.
  b. DNA contains deoxyribose and RNA contains ribose.
  c. DNA contains thymine and RNA contains uracil.
  d. DNA is synthesized in the nuclear area of a eucaryotic cell, whereas RNA is synthesized on the ribosome.

6. Which of the following is an example of a lipid?
  a. Glycogen
  b. Amide
  c. Starch
  d. Cholesterol
  e. Deoxyribose

7. Which of the following is *not* an example of a complex polysaccharide?
  a. Cellulose
  b. Glycogen
  c. Starch
  d. Glucose

8. Which of the following components is *not* found in DNA?
   a. Thymine
   b. Cytosine
   c. Uracil
   d. Guanine
   e. Adenine
9. The term "peptide bonds" is associated with which compound?
   a. Carbohydrates
   b. Fats
   c. Phospholipids
   d. Proteins
   e. Nucleic acids
10. Digestion is equivalent to
   a. hydrolysis
   b. dehydration synthesis
   c. metabolism
   d. nutrition
   e. dehydrolysis
11. A charged atom that has achieved structural stability through the gain or loss of electrons is called a/an
   a. isotope
   b. isomer
   c. ion
   d. radical
   e. molecule
12. The notation C=O indicates that a carbon atom and an oxygen atom share how many electrons?
   a. One
   b. Two
   c. Three
   d. Four
   e. Five
13. Isotopes of an element differ in the number of
   a. protons
   b. neutrons
   c. electrons
14. The chloride ion carries a net charge of -1 because it has more
   a. protons than neutrons
   b. electrons than protons
   c. neutrons than electrons
   d. protons than electrons
   e. neutrons than protons
15. Lipids are commonly found, along with proteins, as part of the
   a. plasma membrane
   b. cell walls
   c. ribosomes
   d. nuclei
16. DNA:
   a. exists in the cell as a double helix
   b. carries messages for the synthesis of specific proteins
   c. carries messages that control the activities of the cell
   d. all of the above
   e. none of the above

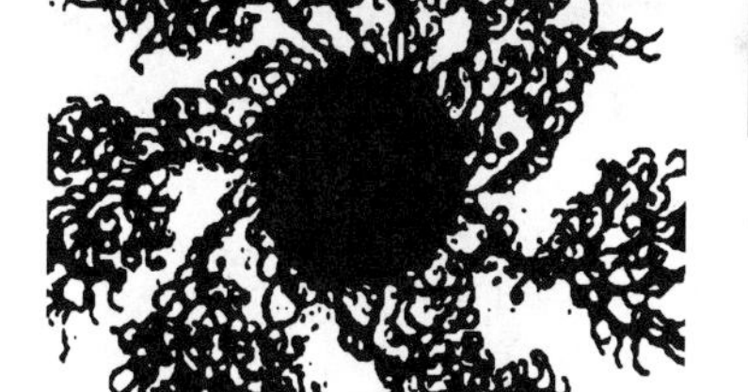

CHAPTER

4

Microbial Physiology

OBJECTIVES

After studying this chapter, you should be able to

- List the various nutritional types of bacteria
- Discuss how these nutritional types fit into the biosphere
- State the meaning of phototroph, autotroph, chemotroph, and heterotroph
- Define producers, consumers, and decomposers
- Describe and give an example of catabolism, anabolism, respiration, and photosynthesis
- List six uses for energy in a cell
- Draw and label the bacterial growth curve
- List the reasons bacteria die during the death phase
- Describe the bacterial chromosome
- List and describe five ways by which the genetic constitution of bacteria can be changed

CHAPTER OUTLINE

Nutrition
Nutritional Requirements
Nutritional Types
Enzymes, Metabolism, and Energy
Enzymes
Metabolic Enzymes
Inhibition of Enzymes
Cellular Metabolism
Energy Metabolism
Energy Production (Catabolism)
Cellular Respiration of Glucose
Anaerobic Fermentation
Aerobic Oxidation by Chemolithotrophs
Anaerobic Respiration by Chemotrophs
Metabolic Biosynthesis (Anabolism)
Energy Conversion
Energy Use
Photosynthesis
Chemosynthesis
Microbial Growth
Culture Media
Population Counts
Population Growth Curve
Bacterial Genetics
Changes in Bacterial Genetic Constitution
Lysogenic Conversion
Transduction
Transformation
Conjugation
Genetic Engineering

NEW TERMS

Anabolism
Autotroph
Catabolism
Catalyze
Chemoautotroph
Chemoheterotroph
Chemolithotroph
Chemoorganotroph
Chemostat
Chemosynthesis
Chemotroph
Citric acid cycle
Competence
Death phase
Dehydrogenation
Differential media
Ecosystem
Electron transport system
Endoenzyme
Enriched media
Episome
Exoenzyme
Fastidious bacterium
Fermentation
Generation time
Genetic engineering
Glycolysis
Growth curve
Heterotroph
In vitro
Lag phase
Lithotroph
Logarithmic scale
Logarithmic growth phase
Mutagen
Mutant
Organotroph
Oxidation
Oxidation-reduction reactions
Photoautotroph
Photoheterotroph
Photolithotroph
Photoorganotroph
Photosynthetic
Phototroph
Plasmid
Protoplast
Reduction
Selective media
Stationary phase
Transduction
Transformation
Viable plate count

Microorganisms are perhaps the best organisms to use in studies of the basic metabolic processes of life. Species of bacteria can be found that represent each of the nutritional types of organisms on earth. We can learn much about ourselves by studying the nutritional needs of bacteria, their metabolic cycles, and why they grow or die under certain conditions. The population growth cycles of bacteria illustrate the growth phases of any population of a species of organism, including humans.

Each tiny single-celled bacterium strives to produce more cells like itself, and it often does so at a rate that is alarming, as long as water and a nutrient supply are available. Under favorable conditions, in 24 hours, the offspring of a single *Escherichia coli* bacterium would outnumber the entire human population on the earth!

Bacteria are easy to find, grow, and maintain in the laboratory. Their morphology, nutritional needs, and some of their metabolic reactions are easily observable; thus, when these usual characteristics change in a pure culture, the resultant mutant (a genetically changed organism) can be quickly identified. Because some bacteria, molds, and viruses produce generation after generation so rapidly and easily, they have been used extensively in genetic studies. In fact, most of the genetic knowledge of today was and is being obtained from the study of these microorganisms.

NUTRITION

The study of bacterial nutrition and other aspects of microbial physiology helps us understand the vital chemical processes of every living cell, including those of the human body.

Nutritional Requirements

All living protoplasm consists of the following six major chemical elements: carbon, hydrogen, oxygen, nitrogen, phosphorus, and sulfur. Other elements usually necessary in lesser amounts include sodium, potassium, chlorine, magnesium, calcium, iron, iodine, and some trace elements. The combinations of all of these elements make up the vital macromolecules of life, including carbohydrates, fats, proteins, and nucleic acids (DNA and RNA).

Each organism must have a source of energy and nutrient chemicals to build the necessary cellular materials of life. Those materials that organisms cannot synthesize, but are required in building the macromolecules of protoplasm, are termed essential nutritional requirements. These are the nutrients that must be continually supplied to every organism for it to live. Essential nutrients vary from species to species.

Nutritional Types

Because microorganisms have been evolving since the beginning of life on earth, there are microbes representing each of the various nutritional types. Various terms are used to indicate the type of energy source and type of carbon source. As you will see, the various terms can be used in combination.

The terms *phototroph* and *chemotroph* are used to describe an organism's energy source. A phototroph uses light as an energy source. Organisms able to convert light energy into chemical energy are called *photosynthetic* organisms, and the process by which they do that is called photosynthesis. Chemotrophs use either inorganic or organic chemicals as an energy source.

The terms *autotroph, lithotroph,* and *heterotroph* (or *organotroph*) are used to describe an organism's carbon source. Autotrophs use carbon dioxide (CO_2) as their carbon source, lithotrophs use other inorganic compounds, and heterotrophs use organic compounds. Photosynthetic organisms such as plants, algae, and cyanobacteria are examples of autotrophs. All animals (including humans), protozoa, and fungi are examples of heterotrophs. Both saprophytic fungi, which live on dead and decaying organic matter, and parasitic fungi are heterotrophs.

Terms can be combined to indicate both an organism's energy source and its carbon source. For example, *photoautotrophs* are organisms (such as plants, algae, cyanobacteria, and purple and green sulfur bacteria) that use light as an energy source and CO_2 as a carbon source. *Photoheterotrophs* (or *photo-organotrophs*), like purple nonsulfur and green nonsulfur bacteria, use light as an energy source and organic compounds as a carbon source. *Chemoautotrophs* (such as nitrifying bacteria and, hydrogen, iron, and sulfur bacteria) use chemicals as an energy source and CO_2 as a carbon source. *Chemoheterotrophs* (or *chemoorganotrophs*), including most bacteria, fungi, protozoa, and animals, use chemicals as an energy source and organic compounds as a carbon source. All animals, protozoa, fungi, and most bacteria are chemoorganotrophs.

The term *ecosystem* refers to the interaction between living organisms and the nonliving environment. Interrelationships among the different nutritional types are of prime importance in the functioning of the ecosystem. *Photolithotrophs* (plants) are the producers of food and oxygen for the chemoorganotrophs (animals). Dead plants and animals would clutter the earth as debris if the chemoorganotrophic saprophytic decomposers (fungi and bacteria) did not break down the dead organic matter into inorganic compounds (carbon dioxide, nitrates, phosphates) of the soil and air so that they could be used and recycled by the photolithotrophs. Most plants, algae, and the photosynthetic bacteria are photolithotrophs. They contribute energy to the ecosystem by trapping energy from the sun and using it to build organic compounds (carbohydrates, fats, nucleic acids, and proteins) from inorganic materials in the soil, water, and air. In photosynthesis, oxygen also is released for respiration by animals.

ENZYMES, METABOLISM, AND ENERGY

Microorganisms are able to grow only if they obtain the proper raw materials for use as nutrients and for the manufacture of the enzymes necessary to promote metabolic reactions. These processes are similar to those in our own body cells. The term metabolism refers to all chemical reactions that occur within any cell. Metabolic reactions are enhanced and regulated by enzymes.

Enzymes

METABOLIC ENZYMES

Enzymes are organic catalysts, that is, they are organic molecules that accelerate (catalyze) the rate of biochemical reactions at certain temperatures without being used up in the process. These reactions might occur at the same temperature without enzymes, but at a much slower rate. A substrate is a compound on which an enzyme exerts its effect. The enzyme must fit the combining site of the substrate, as a key fits a lock. Usually an enzyme is specific, that is, it works on only one type of substrate, but occasionally it can attach to different substrates because the shapes of the combining sites on the substrate molecules are similar (Fig. 4 – 1).

All enzymes are protein molecules. The three-dimensional shape of the enzyme enables it to attach to one or more substrates to accelerate a very slow reaction, without causing the enzyme to change in the process. The enzyme continues to move from substrate molecule to substrate molecule at a rate of several hundred each second, producing a supply of the end product for as long as that particular end product is needed by the cell. However, enzymes do not last indefinitely; they finally degenerate and lose their activity. Therefore, the cell must synthesize and replace these important proteins. Because there are hundreds of metabolic reactions continually occurring in the cell, there must be thousands of enzymes available to control and direct the essential metabolic pathways. At any particular time, all of the required enzymes need not be present; this situation is controlled by genes on the chromosomes and the needs of the cell, which are determined by the internal and external environment. Those enzymes found inside the cell are called *endoenzymes,* and those secreted outside the cell to perform extracellular functions are *exoenzymes.* The digestive enzymes within phagocytes are good examples of endoenzymes. Examples of exoenzymes are the enzymes secreted by saprophytic fungi, which are then able to digest materials (*e.g.,* cellulose and pectin) outside the organism. Nutrients that are digested outside are then absorbed into the organism.

Some enzymes require cofactors to activate them to perform their intended functions because they normally exist in the cell in an inactive state. These cofactors are usually mineral ions such as magnesium, calcium, or iron. Other

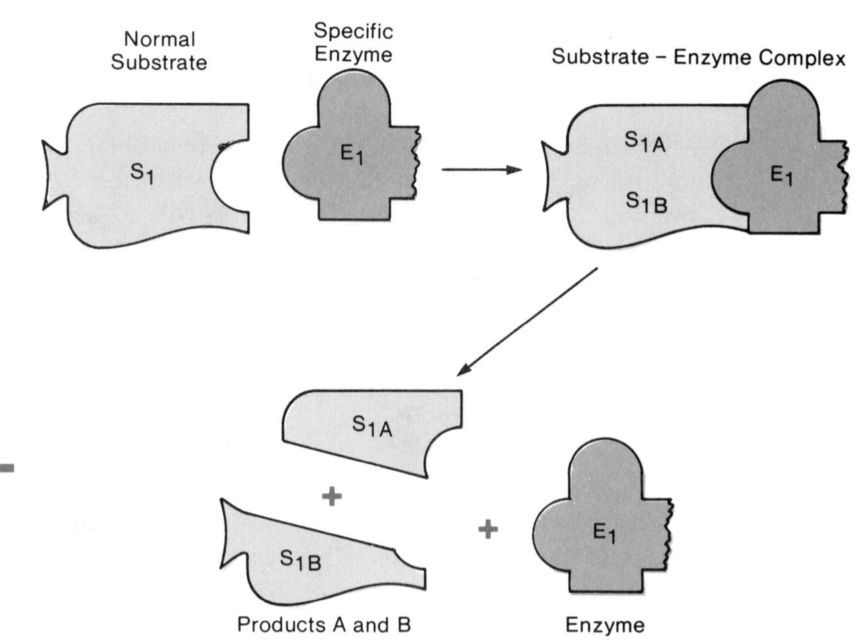

FIGURE 4 – 1
Action of a specific enzyme breaking down a substrate molecule.

enzymes function only in the presence of a coenzyme that acts as a carrier of small chemical groups (such as H_2) that are removed from the substrate. Coenzymes are small organic vitamin-type molecules such as flavin-adenine dinucleotide (FAD) and nicotinamide-adenine dinucleotide (NAD). These coenzymes participate in the citric acid cycle, which is discussed later in this chapter. Coenzymes, like enzymes, do not have to be present in large amounts because they are recycled through many reactions. However, the lack of certain vitamins from which the coenzymes are synthesized will halt all reactions involving that particular coenzyme–enzyme complex.

Some enzymes, called hydrolases, break down macromolecules by the addition of water in a process called hydrolysis. These hydrolytic processes enable saprophytes to break apart such complex materials as leather, wax, cork, wood, rubber, hair, some plastics, and even mechanical equipment. Some of the enzymes involved in the formation of large polymers, DNA, and RNA are called polymerases. These polymerases are active each time the DNA of a cell is replicated and during the synthesis of RNA molecules (Chapter 3).

INHIBITION OF ENZYMES

Many factors affect the activity of enzymes (Fig. 4 – 1). Any physical or chemical change may diminish or completely stop enzyme activity because these protein molecules function properly only under optimum conditions. Optimum conditions for enzyme activity include a relatively limited range of pH (acidity) and temperature as well as the appropriate concentration of enzyme and substrate. Extremes in heat and acidity can denature (or alter) enzymes by breaking the bonds responsible for their three-dimensional shape, resulting in the loss of enzymatic activity. This explains why certain bacteria grow best at certain temperatures and pH levels.

Although mineral ions, calcium, magnesium, and iron, enhance the activity of enzymes by serving as cofactors, other heavy metal ions such as lead, zinc, mercury, and arsenic usually act as poisons to the cell. These toxic ions inhibit enzyme activity by replacing the cofactors, or sometimes replacing only hydrogen, at the combining site of the enzyme, thus inhibiting normal metabolic processes. Some disinfectants containing mineral ions are effective in inhibiting the growth of bacteria by this means.

Sometimes, a similar substrate can be used as an inhibitor to deliberately interfere with a particular metabolic pathway. It binds with the enzyme; thus, the end product is not produced. For example, a chemotherapeutic agent, such as a sulfonamide drug, may bind with certain enzymes to prevent essential metabolites from being formed and thereby inhibits the growth of a pathogen.

Cellular Metabolism

Because metabolism is the total of all the chemical reactions in the cell, it includes the production of energy and intermediate products (metabolites) as well as end products. Furthermore, these reactions must proceed in many directions simultaneously, breaking down some materials to provide energy and raw materials for the synthesis of other compounds. Most metabolic reactions fall into two categories: *catabolism* and *anabolism.* Catabolism is the metabolic degradation (breakdown) of organic compounds that results in the production of energy and smaller molecules. Catabolic reactions involve the breaking of chemical bonds. Any time chemical bonds are broken, energy is released. Anabolism refers to those biosynthetic processes that use energy for the synthesis of protoplasmic materials needed for growth, maintenance, and other cellular functions. Anabolic reactions require energy because chemical bonds are being formed (it takes energy to create a chemical bond). The energy that is released during catabolic reactions is used to drive anabolic reactions. In this manner, the cell works much like a factory. It gathers and produces raw materials to be used in the production of macromolecules for building, maintenance, and repair. It must also have fuel or energy available to run this metabolic machinery. This energy may be trapped from the rays of the sun (as in photosynthesis) or it may be produced by certain catabolic reactions. Then the energy is bound into high-energy bonds in special molecules, usually adenosine triphosphate (ATP). These high-energy molecules serve as the fuel, just as coal is used to fire the furnaces in the production of steel and other important alloys.

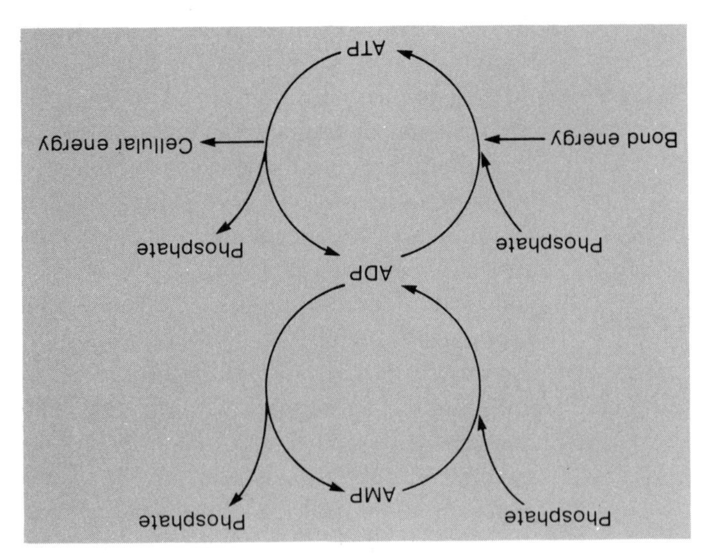

FIGURE 4 – 2
Conversion of bond energy to cellular energy.

The most important high-energy compound found within the cell is ATP, but it is not the only one. It is found in all cells because it is used to transfer energy from energy-yielding molecules, like glucose, to an energy-using reaction. Thus, ATP is a temporary, intermediate molecule. If ATP is not used shortly after it is formed, it is soon hydrolyzed to adenosine diphosphate (ADP), a more stable molecule, and adenosine monophosphate (AMP) in catabolic reactions. ADP can also be used as an emergency energy source by the removal of another phosphate group to produce AMP (a catabolic reaction; Fig. 4 – 2). Both AMP and ADP bind with high-energy phosphate groups to produce ATP to store energy released by energy-yielding reactions.

In addition to the energy required for metabolic pathways, energy must also be available to the organism for growth, reproduction, sporulation, and movement. Some organisms even use energy for bioluminescence, such as the plankton that glow in the darkness of the ocean. Much of the energy of any system is lost in the form of heat.

Energy Metabolism

Chemical reactions are essentially energy transformation processes during which the energy that is stored in chemical bonds is transferred to other newly formed chemical bonds. The cellular mechanisms that release small amounts of energy as the cell needs it usually involve a sequence of catabolic and anabolic reactions, many of which are *oxidation-reduction reactions.*

In oxidation-reduction reactions, electrons are transferred from one chemical to another. Whenever an atom, ion, or molecule loses one or more electrons

(e^-) in a reaction, the process is called *oxidation* and the molecule is said to be oxidized. The electrons lost do not float about at random but, since they are very reactive, attach immediately to another molecule. The resulting gain of one or more electrons by a molecule is called *reduction* and the molecule is said to be reduced. Within the cell, an oxidation reaction is always paired (or coupled) with a reduction reaction; thus the term oxidation-reduction or "redox" reaction.

Many biological oxidations are referred to as *dehydrogenations* because hydrogen ions (H^+) as well as electrons are removed. Concurrently, those hydrogen ions must be picked up in a reduction reaction. Many good illustrations are found in the metabolism of glucose to form pyruvic acid and the concurrent synthesis of ATP and water (see the discussion of the citric acid cycle that follows).

Energy Production (Catabolism)

CELLULAR RESPIRATION OF GLUCOSE

The complete catabolism of glucose takes place in three phases: (1) *glycolysis,* (2) the *citric acid cycle,* and (3) the *electron-transport system.* The first phase is anaerobic, whereas the last two require aerobic conditions (Fig. 4 – 3).

Glycolysis Glycolysis is the term for the stepwise breakdown of glucose into pyruvic acid in the absence of oxygen. It is the basic set of reactions in the anaerobic phase of cellular respiration. Somewhere in the metabolism of almost all cells, glycolysis, or the degradation of glucose, takes place to produce small amounts of ATP. Heterotrophs can degrade starch and glycogen to provide glucose for these glycolytic reactions. Other sugars, such as fructose, can also be used in these reactions. Autotrophs synthesize glucose during photosynthesis so that they can then derive energy from the glucose to drive other metabolic reactions. The amount of energy (ATP) derived from a glucose molecule depends on how much oxygen is available to bond with the hydrogen atoms released during the aerobic phase of cellular respiration. It should be noted that aerobes and facultative anaerobes are much more efficient in energy production than obligate anaerobes because they have oxygen available to aid in the production of many more ATP molecules.

Citric Acid Cycle and Electron-Transport System In the aerobic phases of cellular respiration, oxygen is used as the final hydrogen acceptor following a long series of molecular reactions controlled by specific enzymes. Aerobic microorganisms and facultative anaerobes use pyruvic acid to produce about nine times more energy than obligate anaerobes can produce by the fermentation of glucose and other sugars. Even in the presence of oxygen, obligate anaerobes

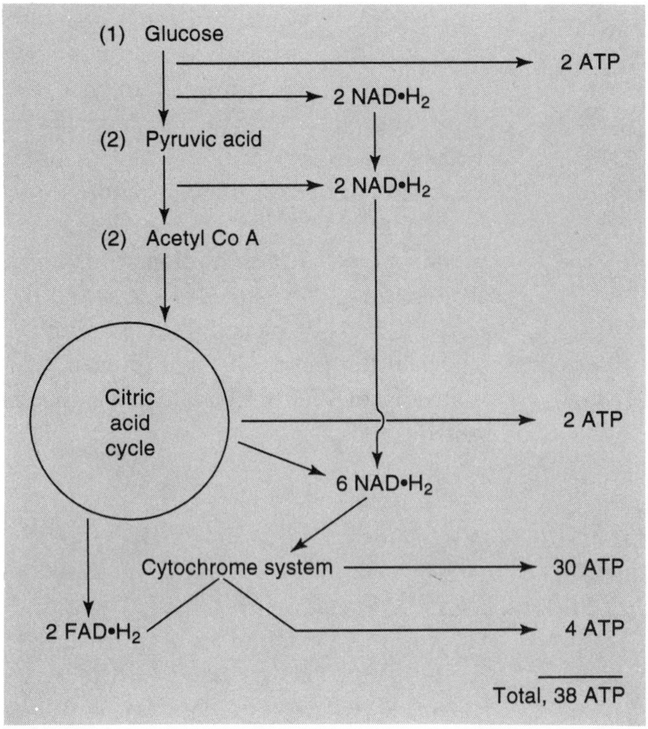

FIGURE 4 – 3
Summary tabulation of high-energy molecules produced by aerobic cellular respiration of one molecule of glucose.

do not have the appropriate enzymes and coenzymes to catalyze this metabolic pathway.

The citric acid cycle is also known as the tricarboxylic acid cycle (TCA) and is often called the Krebs cycle after the scientist who defined this phase of aerobic cellular respiration. The citric acid cycle involves a series of reactions, which are controlled by specific enzymes, that yield the end products carbon dioxide and hydrogen atoms. These activated hydrogen atoms are temporarily bound to NAD and FAD and are transferred to the electron-transport system, where cytochromes aid in the oxidative phosphorylation of ADP to ATP, and hydrogen bonds with oxygen to form water (cellular water).

The complete process of cellular respiration of one molecule of glucose yields 38 ATP molecules. Two ATP molecules are gained from the glycolysis phase and two from the citric acid cycle; the other 34 result from oxidative phosphorylation in the electron transport system. The chemical equation representing this highly efficient catabolic reaction is

$$C_6H_{12}O_6 + 6\ O_2 + 38\ ADP + 38\ Ⓟ \rightarrow 6\ H_2O + 6\ CO_2 + 38\ ATP$$

where Ⓟ indicates activated phosphate groups.

Although the metabolic pathway and amount of energy that can be produced from cellular respiration of glucose has been shown as an illustration,

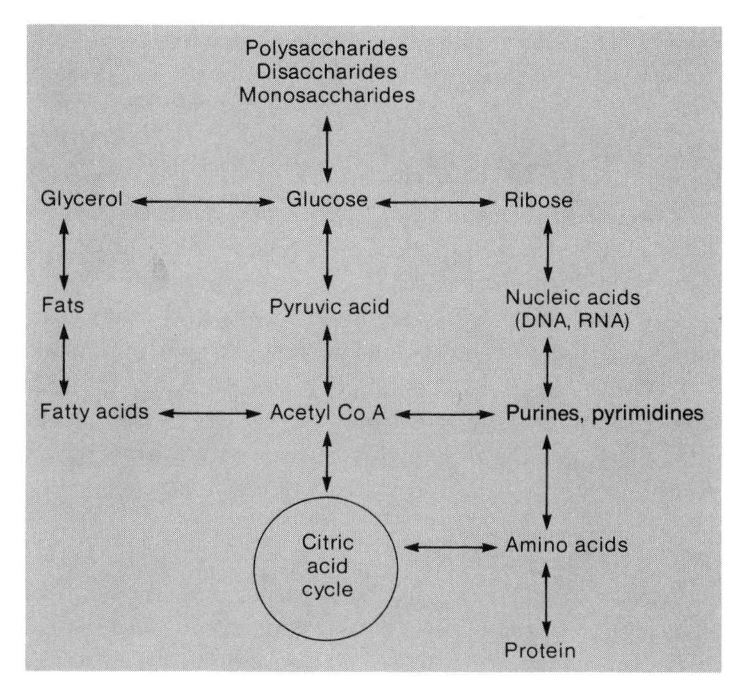

FIGURE 4 – 4
Other nutrients entering and exiting from the metabolism of glucose.

one must be aware that there are many variations to this pathway, depending on the individual organism and its available nutrient and energy resources. Some bacteria degrade glucose to pyruvic acid by other metabolic pathways. Also, glycerol, fatty acids from lipids, and amino acids from protein digestion may be fed into the citric acid cycle to produce energy for the cell when necessary, that is, when there are insufficient carbohydrates available (Fig. 4 – 4).

ANAEROBIC FERMENTATION

When the hydrogen atoms that are released from the breakdown of sugars bind to organic molecules instead of oxygen, the glycolytic process is referred to as *fermentation.* Pyruvic acid usually accepts the hydrogen atoms to produce lactic acid or ethanol (ethyl alcohol), but other end products may be formed. The specific end products depend on the species of organism and on the sugar used as the source of carbohydrate. In human muscle cells, the lack of oxygen during extreme exertion results in pyruvic acid being converted to lactic acid. The presence of lactic acid in muscle tissue is the cause of the soreness that develops in exhausted muscles. Some bacteria (*Lactobacillus* and *Streptococcus*) also produce lactic acid during the fermentation process. These organisms are found in the mouth, where the presence of lactic acid can promote tooth decay, and their presence in milk causes the normal souring of milk into curd and whey. The yeast, *Saccharomyces,* can ferment grain and fruit sugars into ethanol, the alcohol found in beer, wines, and liquors. Acetic acid bacteria (*Acetobacter*) are able to oxidize ethanol to acetic acid and spoil beer and wine by changing them to vinegar. These and other various end products of fermentation have many industrial applications.

AEROBIC OXIDATION BY CHEMOLITHOTROPHS

Chemolithotrophs are able to perform respiration reactions by oxidizing hydrogen (H_2) to water (H_2O), carbon (C) and carbon monoxide (CO) to carbon dioxide (CO_2), ammonia (NH_3) and nitrite ions (NO_2^-) to nitrate ions (NO_3^-), hydrogen sulfide (H_2S) and free sulfur (S) to sulfates (SO_4^{2-}), and iron to iron oxides. As one can imagine, these microorganisms (usually soil and water bacteria) are very important in recycling the elements and some compounds into forms more usable by plants and other microorganisms. They also must be acknowledged as major factors in the destruction of iron parts of machinery because they also promote rust.

ANAEROBIC RESPIRATION BY CHEMOTROPHS

Oxidation-reduction reactions occur, biologically, in certain anaerobes in the absence of oxygen. The electron donor usually is an organic compound such as glucose, but it may be inorganic iron or sulfur compounds (as used by *Thiobacillus*). Inorganic compounds (nitrates, sulfates, and carbonates) serve as the final electron acceptors in place of oxygen. Some facultative anaerobes convert nitrates (NO_3^-) to nitrites (NO_2^-) and free atmospheric nitrogen gas (N_2). Others reduce carbonates (CO_3^{2-}) to carbon dioxide (CO_2) and methane (CH_4). Some obligate anaerobes convert sulfate (SO_4^{2-}) to free sulfur (S) or to hydrogen sulfide gas (H_2S). These bacteria are found in the soil, sea water, marine mud, fresh water, acid mine waters, sewage, and sulfur springs. Obviously, they are unable to produce as much ATP from their energy sources as heterotrophic organisms because their nutrients do not contain as much bound chemical energy.

Metabolic Biosynthesis (Anabolism)

Energy Conversion In general, chemoheterotrophs produce energy from organic compounds by fermentation, anaerobic respiration, and aerobic respiration, as previously discussed. However, all of the phototrophs (algae, cyanobacteria, other photosynthetic bacteria, and plants) must derive their energy from light, usually the sun, by photosynthesis. This process provides energy to the greatest mass of organisms, not only all of the photosynthetic organisms but heterotrophs as well, because heterotrophs consume phototrophs.

Energy Use The biosynthesis of organic compounds requiring the use of energy is called anabolism, or an anabolic reaction. In living cells this biosynthetic metabolism may be one of two types: photosynthesis by photolithotrophs or photoorganotrophs, or *chemosynthesis* by chemolithotrophs or chemoorganotrophs.

In photosynthesis, light energy is converted to chemical bond energy to be used to synthesize organic biochemicals. Those phototrophic organisms using inorganic raw materials (CO_2, H_2O, H_2S, S) for biosynthesis are the photolithotrophs (photoautotrophs). Those phototrophs using small organic molecules, such as acids and alcohols, to build carbohydrates, fats, proteins, nucleic acids, and other important biochemicals are the photoorganotrophs (photoheterotrophs).

Photosynthesis The goal of photosynthetic processes is to trap the radiant energy of light and convert it into chemical bond energy in ATP and carbohydrates, particularly glucose, which can then be converted into more molecules of ATP via the respiratory pathways. The general overall photosynthesis reaction is

$$6\ CO_2 + 12\ H_2O \xrightarrow[\text{ATP}]{\text{light}} C_6\ H_{12}\ O_6 + 6\ O_2 + 6\ H_2O + \text{ADP} + \text{Ⓟ}$$

Notice that this reaction is almost the reverse of the oxidative respiration reaction and it is nature's way of balancing substrates in the environment.

Photosynthesis does not necessarily require the presence of oxygen; in other words, it can take place anaerobically, as occurs in the purple and green bacteria (obligate anaerobic photoautotrophs). These bacteria do not use H_2O but instead use H_2 or H_2S gas or other hydrogen donor molecules produced by other bacteria in soil or mud. The overall reaction of anaerobic bacterial photosynthesis then becomes

$$6\ CO_2 + 12\ H_2S \xrightarrow{\text{light}} C_6\ H_{12}\ O_6 + 6\ H_2O + 12\ S$$

or

$$6\ CO_2 + 12\ H_2 \xrightarrow{\text{light}} C_6\ H_{12}\ O_6 + 6\ H_2O$$

The bacterial photosynthetic pigments use shorter wavelengths of light, which penetrate deep within a pond or into mud where it appears to be dark.

In the absence of light, some photolithotrophic organisms may survive anaerobically by the fermentation process alone. Other phototrophic bacteria also have a limited ability to use simple organic molecules in photosynthetic reactions; thus, they become photoorganotrophic organisms under certain conditions. A few species of cyanobacteria have also been found to exist as facultative phototrophs and facultative autotrophs, meaning that in certain environments they become photoheterotrophs. In other words, they have back-up metabolic systems.

Chemosynthesis The chemosynthetic process involves a chemical source of energy and raw materials to synthesize the necessary metabolites and macromolecules for growth and function of the organisms. These chemotrophic organisms may be either autotrophs or heterotrophs.

The chemoautotrophs are the same previously discussed chemolithotrophic bacteria that obtain energy by aerobic oxidation of inorganic compounds or by anaerobic respiration of inorganic substances. These are the only organisms that do not depend on radiant energy from the sun. They are considered among the most primitive bacteria and are frequently found near volcanoes deep within the ocean.

The chemoorganotrophs have been defined as those organisms that derive both their energy and nutrients from organic materials. Most bacteria, as well as protozoa, fungi, and all animals, belong to this group. Although they vary greatly in the details of their metabolism, they all use carbohydrates, lipids, and proteins to synthesize their own carbohydrates, lipids, proteins, nucleic acids, and high-energy molecules such as ATP. These metabolic pathways may be carried on with or without oxygen by aerobic or anaerobic respiration, fermentation, and other biodegradation and biosynthetic reactions.

MICROBIAL GROWTH

Bacterial growth refers to an increase in the number of organisms rather than in their size. When each bacterial cell reaches its optimum size, it divides by binary fission ("bi" means "two") into two daughter cells, *i.e.*, each bacterium simply splits into two similar cells. These in turn divide, and as a result, a viable, healthy colony of cells is maintained as long as the nutrient supply, water, and space allow. This process continues until the waste products from cells build up to a toxic level or until the nutrients are depleted. The actual division of staphylococci by binary fission is shown in the electron micrograph in Figure 4 – 5.

The growth of microorganisms in nature as well as in the laboratory (referred to as *in vitro*) is greatly influenced by temperature, pH (acidity), moisture content, available nutrients, and the character of other organisms present. Therefore, the number of bacteria in nature fluctuates unpredictably because these factors vary with the seasons, rainfall, temperature, and time of day.

In the laboratory, however, a pure culture of a single species of bacteria can usually be grown if the appropriate growth medium and environmental conditions are provided. The temperature, pH, and amount of oxygen are quite easily controlled to provide optimum conditions for growth. Then the appropriate nutrients must be provided in the growth medium. Some bacteria are so

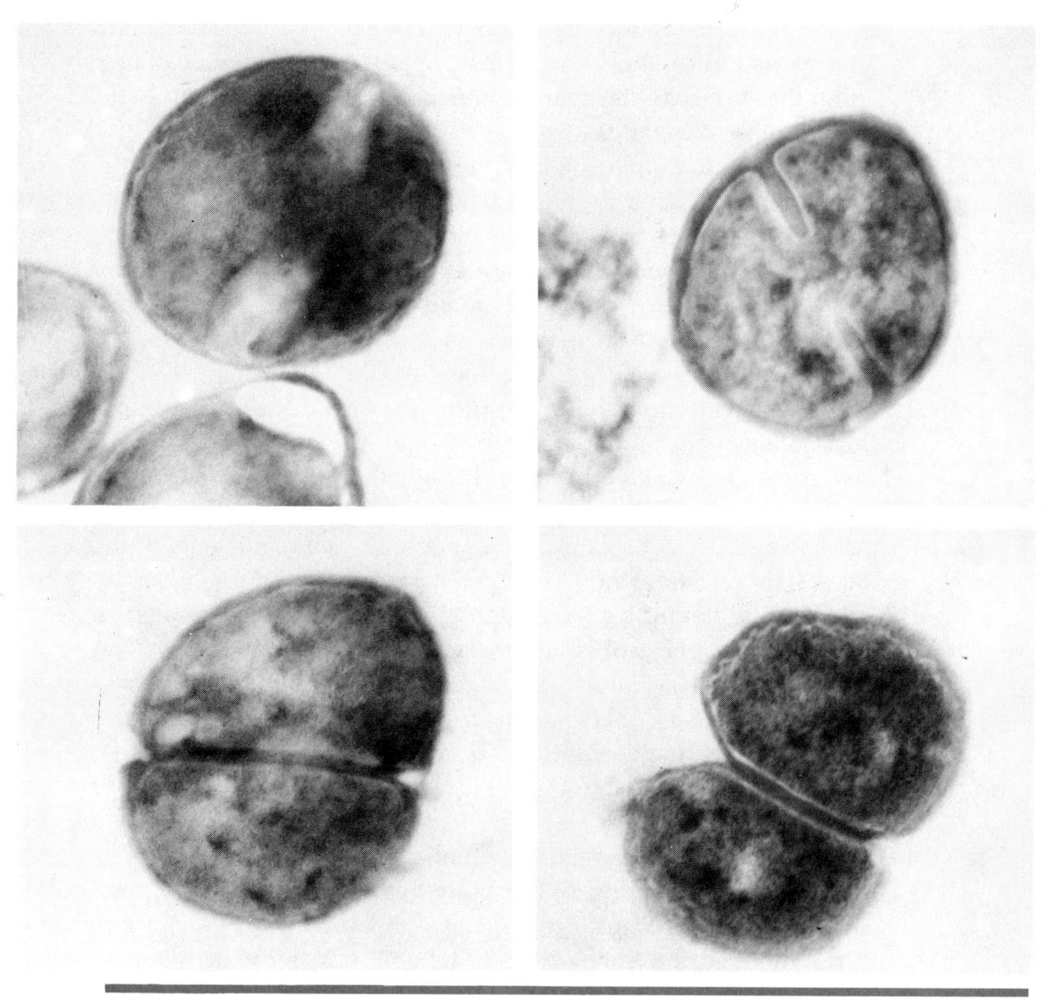

FIGURE 4 – 5
Binary fission of staphylococci (original magnification ×30,000). (Photograph courtesy of Ray Rupel)

fastidious (having complex nutritional requirements) that they will not grow outside of living cells; thus, they must be cultured in living animals, embryonated chicken eggs, or cell cultures.

Culture Media

Basically, there are two types of media for culturing bacteria: (1) a chemically defined synthetic medium and (2) a rich, natural, complex medium containing digested extracts from meats, fish, and plants providing the necessary nutrients, vitamins, and minerals. These media can be used in liquid (broth) form, which is available in tubes, or they may be solidified by the addition of agar and poured into tubes or petri dishes so that the bacteria can be grown within or on the surface of the agar. Agar is a complex polysaccharide that is obtained from a red marine alga.

An *enriched medium* is a broth or solid medium containing a rich supply of special nutrients that promotes the growth of fastidious organisms. Blood agar (nutrient agar plus 5% sheep red blood cells) and chocolate agar (nutrient agar plus powdered hemoglobin) are examples of solid enriched media that are used routinely in the clinical microbiology laboratory. Blood agar is bright red, whereas chocolate agar is brown.

A *selective medium* has added inhibitors that discourage the growth of certain organisms without inhibiting the growth of the one that is sought. For example, MacConkey agar inhibits growth of gram-positive bacteria and is, thus, selective for gram-negative bacteria. Phenylethyl alcohol (PEA) agar and colistin-nalidixic acid (CNA) agar are selective for gram-positive bacteria. Thayer-Martin agar (a chocolate agar containing extra nutrients and antimicrobial agents) is selective for *Neisseria gonorrhoeae.* Mannitol salt agar (MSA) only allows salt-tolerant bacteria to grow. Sabouraud dextrose agar is selective for fungi; its low pH (5.6) is inhibitory to most bacteria.

A *differential medium* permits the differentiation of organisms that grow on the medium. MacConkey agar is frequently used to differentiate between various gram-negative bacilli isolated from fecal specimens. Gram-negative bacteria able to ferment lactose (an ingredient of MacConkey agar) produce pink colonies, whereas those unable to ferment lactose produce clear, colorless colonies. Thus, MacConkey agar differentiates between lactose-fermenting (LF) and non–lactose-fermenting (NLF) gram-negative bacteria. Mannitol salt agar is used to screen for *Staphylococcus aureus;* not only will *S. aureus* grow on MSA, it turns the medium yellow due to its ability to ferment mannitol. In a sense, blood agar is also a differential medium because it is used to determine the type of hemolysis (alteration or destruction of red blood cells) that the bacterial isolate produces (see Color Figures 14, 15, 16).

The various categories of media (enriched, selective, differential) are not mutually exclusive. For example, as just seen, blood agar is both enriched and differential. MacConkey agar and MSA are selective and differential. Because they start as blood agar to which selective inhibitory substances are added, PEA and CNA are enriched and selective. Thayer-Martin agar is highly enriched and highly selective.

Population Counts

Once the desired species of bacteria has been separated from other organisms in a specimen, it can be grown as a pure culture under the best possible conditions. The changes in a bacterial population over an extended period

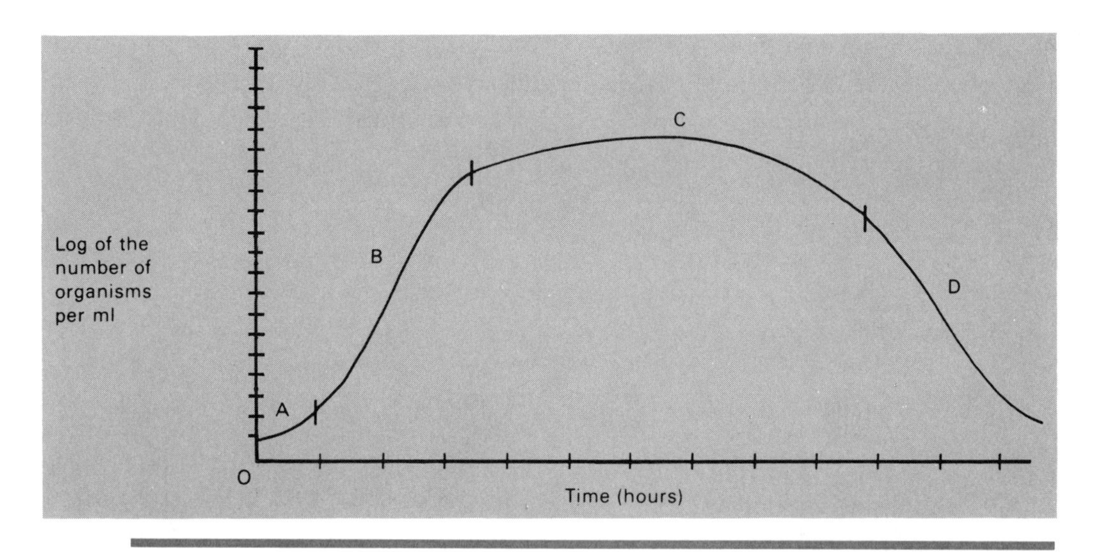

FIGURE 4 – 6
Population growth curve of living organisms. The logarithm of the number of bacteria per milliliter of medium is plotted against time. (*A*) Lag phase. (*B*) Logarithmic growth phase. (*C*) Stationary phase. (*D*) Death phase.

follows a definite predictable pattern that can be shown by plotting the population growth curve on a graph (Fig. 4 – 6; described later).

Microbiologists often need to know the rate of bacterial growth and how many bacteria are present at any given time. This information is particularly important in determining the degree of bacterial contamination in drinking water, milk, and other foods. The number can be determined by counting the total number of bacterial cells, living and dead, in 1 ml of solution, or by counting only the viable (or living) bacteria present. A total count is easier and faster but it differs from a viable cell count because it includes both living and dead cells; however, if the 1-ml sample is taken when the microorganisms are growing and dividing rapidly in the growth phase, few dead cells are found. In many hospitals, electronic cell counters are incorporated into the instruments used in blood, urine, and spinal fluid analyses. Many research laboratories use spectrophotometers, which determine the number of organisms present by measuring the turbidity (cloudiness) of the solution. The turbidity varies as a result of the number of organisms that are present. The turbidity increases (*i.e.,* the solution becomes more cloudy) as the number of organisms increases. Chemical analyses of nitrogen or carbon content also can be used to determine the number of bacteria present.

The *viable plate count* is usually the most accurate method for determining the number of living bacteria in a milliliter of liquid, which may be milk, water, diluted food, or broth. In this procedure, as shown in Figure 4 – 7, 1 ml of solution is diluted to 100 ml three times in sequence, and samples are taken

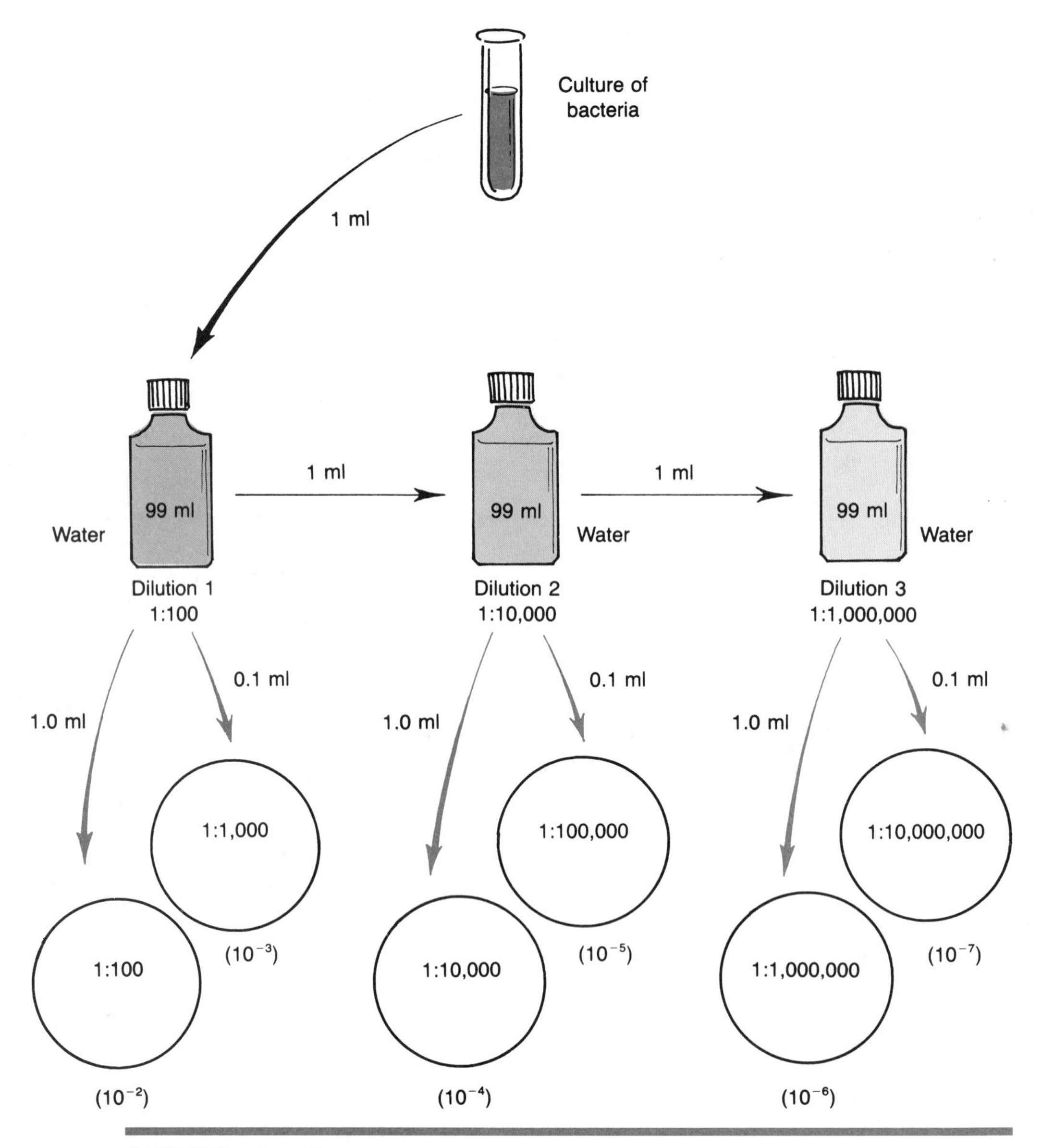

FIGURE 4 – 7
The viable plate count technique. One milliliter of the original culture is successively diluted and cultured on agar plates.

from each dilution. Then 0.1-ml and/or 1-ml samples are grown on nutrient agar. The number of colonies observed growing on the nutrient agar plates the following day indicates the number of viable bacteria present at that particular dilution. This number multiplied by the dilution factor indicates the number of living bacteria in the original culture at the time the sample was taken. For example, if 220 colonies were counted on an agar plate grown from a 1-ml sample from the second dilution bottle (1:10,000 dilution), there were 220 × 10,000 = 2,200,000 bacteria in 1 ml of the original material at the time the dilutions were made and cultured. Practically, it is easier to culture only 0.1 ml of each dilution, while increasing the dilution factor by 10, as shown in Figure 4 – 7. For the count to be statistically significant and most accurately representative of the number of living microorganisms in the solution, the number used in the calculations should be taken from the agar plate that has between 30 and 300 colonies.

A similar technique has been developed to count viruses. One ml of the diluted viruses is inoculated onto a "lawn" of bacteria or a layer of culture cells. Each cell lysed by a virus causes a clear zone (plaque) in the culture of cells following incubation. Thus, the number of plaques represents the number of viruses in 1 ml of the diluted solution. This number must then be multiplied by the dilution factor.

A method of approximating the number and type of bacteria in urine involves streaking a known volume (either 0.1 ml or 0.01 ml) of urine from a calibrated inoculation loop onto various appropriate differential media. The color and number of colonies enables the microbiologist to determine the presence of large numbers of certain bacteria that often cause urinary tract infections (UTIs), including species of *Staphylococcus, Streptococcus,* and particularly *E. coli* and other coliforms (bacteria closely related to *E. coli*).

Population Growth Curve

The population *growth curve* for any particular species of bacteria may be determined by growing the organism in pure culture (a culture of only one organism) at a constant temperature. The graph in Figure 4 – 6 is constructed by plotting the logarithm (refer to your math book) of the number of bacteria against the incubation time.

The first stage of the growth curve is the *lag phase* (see *A* in Fig. 4 – 6), during which the bacteria absorb nutrients, synthesize enzymes, and prepare for reproduction. In the *logarithmic growth phase* (or exponential growth phase; *B* in Fig. 4 – 6), the bacteria multiply so rapidly that the population number doubles with each *generation time.* The generation time, which is the time between the formation of a new bacterium and its division into two daughter cells, varies with the species of bacteria. The growth rate is the greatest during the logarithmic growth phase. In the laboratory, under ideal growth conditions, *E. coli* from the intestine, *Vibrio cholerae* (which causes cholera), *Staphylococcus,* and *Streptococcus* all have a generation time of about 20 minutes, whereas *Pseudomonas* from the soil may divide every 10 minutes, and *Mycobacterium tuberculosis* may divide only every 18 hours. The logarithmic growth phase is always brief unless the rapidly dividing culture is maintained by constant addition of nutrients and oxygen and frequent removal of the waste products and excess microorganisms. Many industrial and research procedures depend on the maintenance of an essential species of microorganism. These are continuously cultured in a controlled environment called a *chemostat* (Fig. 4 – 8), which regulates the supply of nutrients and the removal of waste products and excess microorganisms. Chemostats are used in industries where yeast is grown to produce beer and wine, where fungi and bacteria are cultivated to produce antibiotics, where *E. coli* is grown for genetic research, and in any other process needing a constant source of microorganisms.

As the nutrients and oxygen in the culture tube are used up and waste products from the metabolizing bacteria build up and change the pH of the culture medium, the rate of division slows, so that the number of bacteria dividing equals the number dying. The result is the *stationary phase* (C in Fig. 4 – 6). It is during this phase that the culture maintains its greatest population density.

As overcrowding occurs, the toxic waste products increase and the nutrient and oxygen supply decreases. The microorganisms then die at a rapid rate; this is the *death phase* or decline phase (*D* in Fig. 4 – 6). The culture may die completely or a few microorganisms may continue to survive for months. If the bacterial species is a spore-former, it will form spores to survive beyond this phase. When cells are observed in old cultures of bacteria in the death phase,

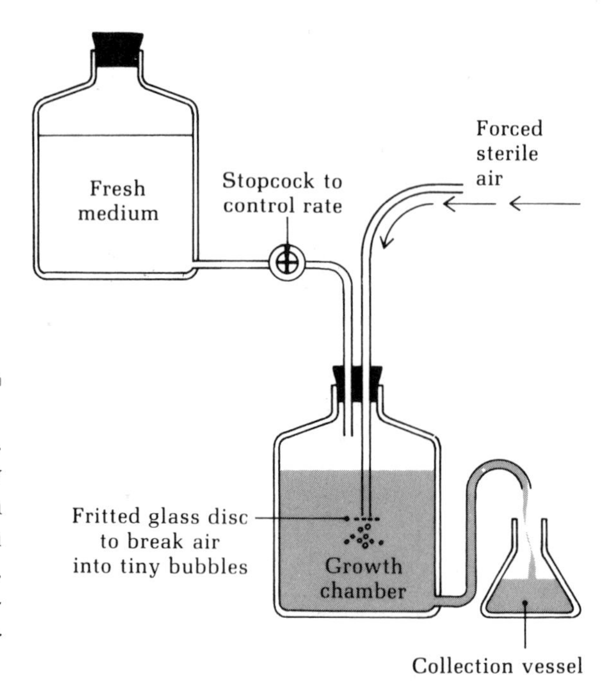

FIGURE 4 – 8
Chemostat used for continuous cultures. Rate of growth can be controlled either by controlling the rate at which new medium enters the growth chamber or by limiting a required growth factor in the medium. (Volk WA, et al.: Essentials of Medical Microbiology, 4th ed. Philadelphia, JB Lippincott, 1991)

some of them look different from healthy organisms seen in the growth phase. As a result of unfavorable conditions, morphological changes in the cells may appear. Some cells undergo involution and assume a variety of shapes, becoming long, multinucleated, filamentous rods, or branching or globular forms that are difficult to identify. Some develop without a cell wall (*protoplasts*) and others take an "L" shape because they have very little cell wall (L-forms). When these involuted forms are inoculated into a fresh nutrient medium, they usually revert to the original shape of the healthy bacteria.

A population growth curve may be plotted for all organisms, including humans. Currently, the human population is in the logarithmic growth phase with a generation time of 35 years. No population of organisms is known to continue in this phase forever without careful control of the food supply, numbers of individuals, and proper disposal of waste products. Eventually, the forces of nature (food shortages, epidemics, wars, toxic waste products) will probably cause the world population to stabilize in the stationary growth phase and, it is hoped, not proceed to the terminal death phase.

BACTERIAL GENETICS

The bacterial chromosome usually consists of only one long, continuous DNA molecule, with no protein on the outside as is found in eucaryotic chromosomes. The chromosome is a circular strand of genes, all linked together. Genes are the fundamental units of heredity that carry the information needed for the special characteristics of each different species of bacteria. Genes direct all functions of the cell, providing it with its own particular traits and individuality. Because there is only one chromosome that replicates just before cell division, identical traits of a species are passed from the parent bacterium to the daughter cells after binary fission has occurred.

The DNA of any gene on the chromosome is subject to accidental alteration, which changes the trait controlled by that gene. If the change in the gene alters or deletes (eliminates) a trait in such a way that the cell does not die or become incapable of division, the altered trait is transmitted to the daughter cells of each succeeding generation. A change in the characteristics of a cell caused by a change in the DNA molecule (genetic alteration) that is transmissible to the offspring is called a mutation. A lethal mutation is one that causes the cell to die because an essential functional gene is missing. It may perhaps be a gene that codes for an essential enzyme. Spontaneous mutations usually occur about once in every 10 million cell divisions. The mutation rate can be increased by exposing cells to physical or chemical agents that affect the DNA molecule. These agents are called *mutagens.* In the research laboratory, x-rays, ultraviolet light, radioactive substances, and certain chemical agents are used to induce more frequent mutations (Fig. 4 – 9). Bacterial *mutants* are used in genetic and

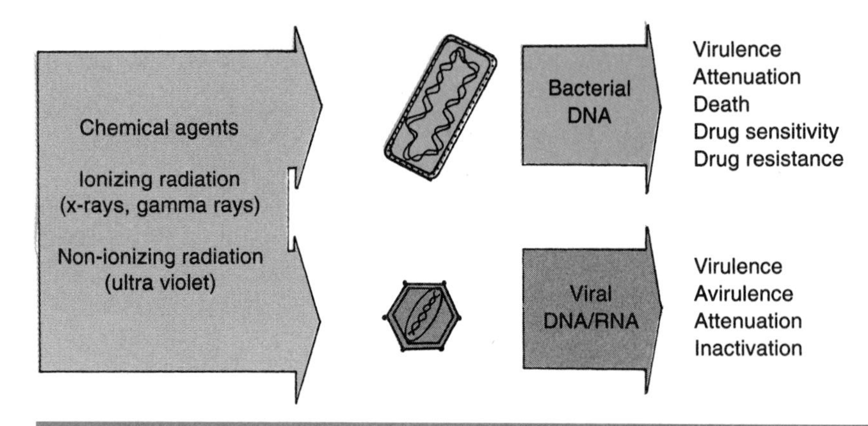

FIGURE 4 – 9
Agents that may cause mutagenic changes.

medical research and in the development of vaccines. The types of mutagenic changes frequently observed in bacteria involve color, colony appearance, cell shape, biochemical activities, nutritional needs, antigenic sites, virulence, pathogenicity, and drug resistance. Nonpathogenic attenuated virus vaccines, such as the Sabin vaccine for polio, are examples of laboratory-induced mutations of pathogenic microorganisms.

Changes in Bacterial Genetic Constitution

There are at least four ways, other than mutation, that the genetic composition of bacteria can be changed: lysogenic conversion, *transduction, transformation,* and conjugation. These types of gene transfers result in extra genetic material in the recipient cell. If this extra bit of DNA remains in the cytoplasm of the cell, it is called a *plasmid* (Fig. 4 – 10). Because they are not part of the chromosome, plasmids are referred to as extrachromosomal DNA. Some plasmids contain many genes, others only a few, but the cell is changed by the addition of these genetic components. Plasmids can replicate themselves simultaneously with chromosomal DNA replication or at various other times. When a plasmid becomes incorporated into the chromosome, it is referred to as an *episome.* When the gene product (usually a protein) has been produced, the gene is said to have been expressed. Some plasmid genes can be expressed as extrachromosomal genes, but others must become chromosomal episomes before the genes become functional.

LYSOGENIC CONVERSION

Lysogenic conversion occurs when a temperate bacteriophage (bacterial virus) infects a bacterium changing it to a lysogenic bacterium (*i.e.,* a bacterium that

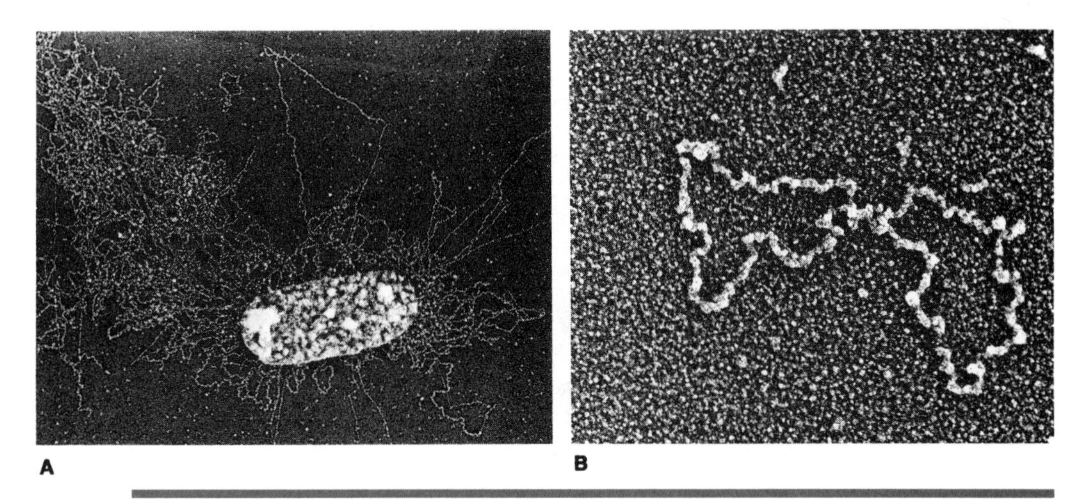

FIGURE 4 – 10
(*A*) Disrupted cell of *Escherichia coli;* the DNA has spilled out and a plasmid can be found slightly to the left of top center. (*B*) Enlargement of a plasmid (about 1 μm from side to side). (Volk WA, et al.: Essentials of Medical Microbiology, 4th ed. Philadelphia, JB Lippincott, 1991)

has the potential to be lysed by viral gene products) (Fig. 4 – 11). The bacteriophage injects its DNA into the cytoplasm of the bacterium, and then the phage DNA incorporates into the bacterial chromosome. When the phage DNA is integrated into the host cell's chromosome, the phage is referred to as a prophage (see the discussion of bacteria and bacteriophages in Chapter 2). The number of genes in the bacterium is increased by the number of genes injected by the phage. The diphtheria bacterium (*Corynebacterium diphtheriae*) is pathogenic only when it has been infected by a bacteriophage. It is actually a viral gene (called the tox gene) that codes for the toxin that causes diphtheria.

TRANSDUCTION

Transduction means "to carry across." Some bacterial genetic material may be "carried across" from one bacterial cell to another by a bacterial virus. This phenomenon may occur following infection of a bacterial cell by a temperate bacteriophage. The viral DNA (prophage) combines with the bacterial chromosome (Fig. 4 – 11). If a stimulating chemical, heat, or ultraviolet light activates the prophage, it begins to produce new viruses by the production of phage DNA and proteins. As the chromosome disintegrates, small pieces of bacterial DNA may remain attached to the maturing phage DNA. During the assembly of the virus particles, one or more bacterial genes may be incorporated into some of the mature bacteriophages. When all the phages are freed by cell lysis, they proceed to infect other cells, injecting bacterial genetic material as well as viral genes. Thus, bacterial genes that are attached to the phage DNA are carried to new cells by the virus.

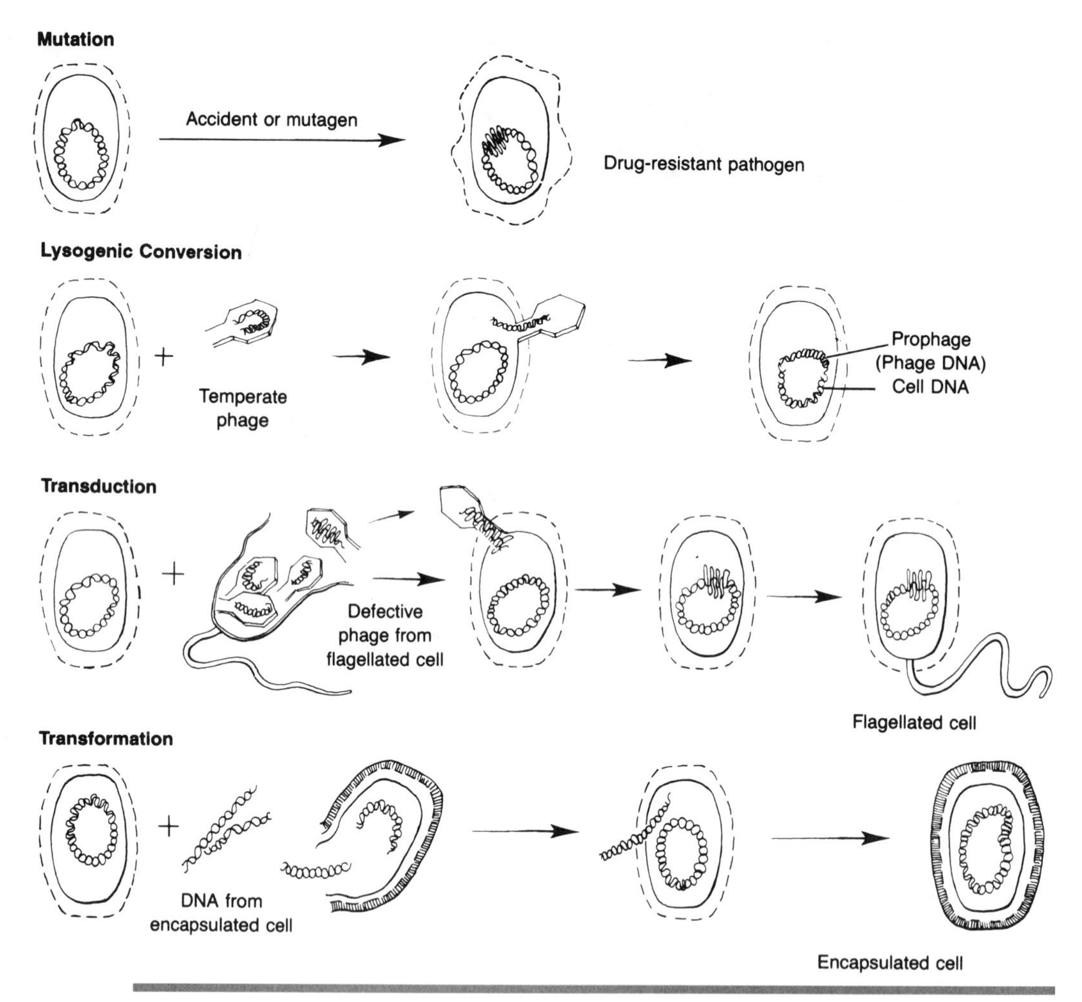

FIGURE 4 – 11
Four ways in which the genetic constitution of bacteria may be changed: mutation, lysogenic conversion, transduction, and transformation. See text for details

There are two types of transduction: specialized and generalized. The explanation in the previous paragraph describes specialized transduction, in which the infecting phage integrates into the bacterial chromosome or a plasmid. As the virus genome breaks away to replicate and produce more viruses, it carries a few identifiable bacterial genes with it to the newly infected cell. In this way genetic capabilities involving the fermentation of certain sugars, antibiotic resistance, and other phenotypic characteristics can be transduced to other bacteria. This process has been shown in the laboratory to occur among species of *Bacillus, Pseudomonas, Haemophilus, Salmonella,* and *Escherichia,* and it is assumed to occur in nature.

In generalized transduction, the bacteriophage is a virulent lytic phage that does not incorporate into the bacterial genome or plasmid. Rather, it picks up fragments of bacterial DNA during the assembly of new virus particles and carries these genes to other cells that the new viruses infect. This generalized transduction has been observed in species of *Streptococcus, Staphylococcus,* and *Salmonella* and in *Vibrio cholerae.*

Only small segments of DNA are transferred from cell to cell by transduction compared with the amount that can be transferred by transformation and conjugation.

TRANSFORMATION

In the transformation process, a recipient bacterial cell is genetically transformed following the uptake of DNA fragments from another strain of bacteria with at least one different observable characteristic (Fig. 4 – 11). It was transformation experiments that proved that DNA is, indeed, the genetic material. When a DNA extract from encapsulated, pathogenic *Streptococcus pneumoniae* type 1 was added to a growing culture of nonencapsulated, nonpathogenic *S. pneumoniae* type 2, the resulting culture showed the presence of live, encapsulated type 2 because some of the type 1 DNA was incorporated into the dividing type 2 DNA. Thus, the nonencapsulated streptococci must have been transformed by the genes coding for capsules that were incorporated into the cells. Although this type of genetic recombination is not widespread, it has been demonstrated in several genera including *Bacillus, Escherichia, Haemophilus, Pseudomonas,* and *Neisseria.* Transformations have even been shown to occur between two different species (*e.g., Staphylococcus* and *Streptococcus*).

Large DNA molecules from a donor cell can only penetrate the cell wall and cell membrane of certain bacteria; bacteria capable of taking up "naked" DNA molecules are said to be *competent.* Recipient bacteria usually become competent during the late logarithmic growth phase when the cell secretes a protein competence factor that increases its permeability to DNA.

Some competent bacterial cells have incorporated DNA fragments from certain animal viruses (*e.g.,* cowpox), retaining the latent virus genes for long periods. This knowledge may have some importance in the study of viruses that remain latent in humans for many years before they finally cause disease, as may be the case in Parkinson's disease. These human virus genes may hide in the bacteria of the indigenous microflora until they are released to cause disease.

CONJUGATION

Conjugation occurs when two bacteria attach to each other, usually by a pilus bridge, and some genetic material is transferred from the donor cell to the recipient. Many biologists compare this transfer of genetic material to the sexual process in animals, labeling the donor "male" and the recipient "female." This type of genetic recombination occurs mostly among species of enteric

INSIGHT
Genetically Engineered Bacteria

The term "genetic engineering" refers to the manufacture and manipulation of genetic material *in vitro* (in the laboratory). Genetic engineering has been possible only since the late 1960s, when a scientist named Paul Berg demonstrated that fragments of human or animal DNA can be attached to bacterial DNA. Such a hybrid DNA molecule is referred to as recombinant DNA. When a molecule of recombinant DNA is inserted into a bacterial cell, the bacterium is able to produce the gene product, usually a protein. Thus, microorganisms (primarily bacteria) can be genetically engineered to produce substances (gene products) that they would not normally manufacture.

Molecules of self-replicating, extrachromosomal DNA, called plasmids, are frequently used in genetic engineering and are referred to as vectors. A particular gene of interest is first inserted into the vector DNA, forming a molecule of recombinant DNA. The recombinant DNA is then inserted into or taken up by a bacterial cell. The cell is next allowed to multiply, creating many genetically identical bacteria (clones), each of which is capable of producing the gene product. From the clone culture, a genetic engineer may then remove (harvest) the gene product. The gram-negative bacillus *Escherichia coli* has often been used because it can be easily grown in the laboratory, has a relatively short generation time (about 20 minutes under ideal conditions), and its genetics are well understood by researchers.

An example of a product produced by genetic engineering is insulin, a hormone produced in *E. coli* cells and used to treat diabetic patients. Human growth hormone (somatotropin), somatostatin (a hormone used to limit growth), and interferon are also produced by genetically engineered *E. coli.* The hepatitis B vaccine that is administered to health-care workers is produced by a genetically engineered yeast, called *Saccharomyces cerevisiae.* Bovine growth hormone (BGH) and porcine growth hormone (PGH) are produced in genetically engineered *E. coli.*

New uses for recombinant DNA and genetic engineering are being discovered every day, causing profound changes in medicine, agriculture, and other areas of science.

bacteria, but it has been reported within species of *Pseudomonas* and *Streptococcus* as well. In electron micrographs, microbiologists have observed that sex pili cus as well. In electron micrographs, microbiologists have observed that sex pili are larger than other pili. The donor sex pilus attaches to the recipient cell, which is usually nonpiliated, to form the pilus bridge. Genetic material from a plasmid or from the chromosome is then transferred across the pilus bridge from donor to recipient cell (Fig. 4 – 12 and 4 – 13). Although many genes may be transferred by conjugation, the ones most frequently noted include those coding for antibiotic resistance, colicin (a protein that kills certain enteric bacteria), and the fertility factor (F^+, HFr^+). Bacteria with the F^+ or HFr^+ genes have the ability to produce pili, including sex pili; therefore, those bacteria may become donor cells. If the fertility factor is on a plasmid, it is an F^+ gene; if it is incorporated within the chromosome, it is referred to as the HFr^+ gene. A complete copy of the plasmid (with F^+) usually moves to the recipient (F^-) cell; thus, the recipient usually becomes F^+. However, the recipient cell usually receives only a portion of the chromosome from an HFr^+ cell, not including the HFr^+ gene; thus, the recipient remains HFr^- in that circumstance, does not

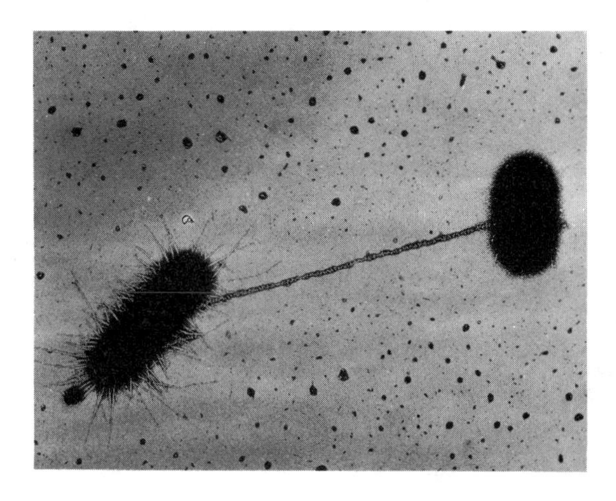

FIGURE 4 – 12
Initiation of conjugation in *Escherichia coli*. The donor cell (with numerous short pili) is connected to the recipient bacterium by an F^+ pilus, as described in the text (original magnification ×3,000). (Anderson TF: Cold Spring Harbor Symp Quant Biol 18:1970)

produce pili, and cannot become a donor cell (F = fertility; HFr = high frequency of recombination).

Transduction, transformation, and conjugation are excellent tools for mapping bacterial chromosomes and for studying bacterial and viral genetics. Although each of them is frequently used in the laboratory, it is believed that they occur in natural environments under certain circumstances.

Genetic Engineering

An array of techniques has been developed to transfer eucaryotic genes, particularly human genes, into other easily cultured cells to facilitate the large-scale production of important proteins. Bacteria, yeasts, human leukocytes, macrophages, and fibroblasts have been used as manufacturing plants for proteins, such as human growth hormone, insulin, and interferon. In the near future, perhaps the genes themselves will be implanted into people with genetic deficiencies. This work is not conducted without some risk to humanity. The greatest fear is that genes introduced into bacteria may produce yet unknown toxins or diseases that cause harm to or even destroy the human race. With this in mind, extreme care and caution is exercised during genetic manipulation experiments.

Many industrial and medical benefits may be derived from *genetic engineering* (or genetic recombination) research. There is a potential for incorporating nitrogen-fixing capabilities into more of the soil microorganisms, for making animal proteins for food and medicines, for increasing the production of antibiotics, for synthesizing important enzymes and hormones for treatment of inherited diseases, and for making portions of viruses and bacterial pathogens to be used as vaccines. Such vaccines would contain only part of the pathogen (for

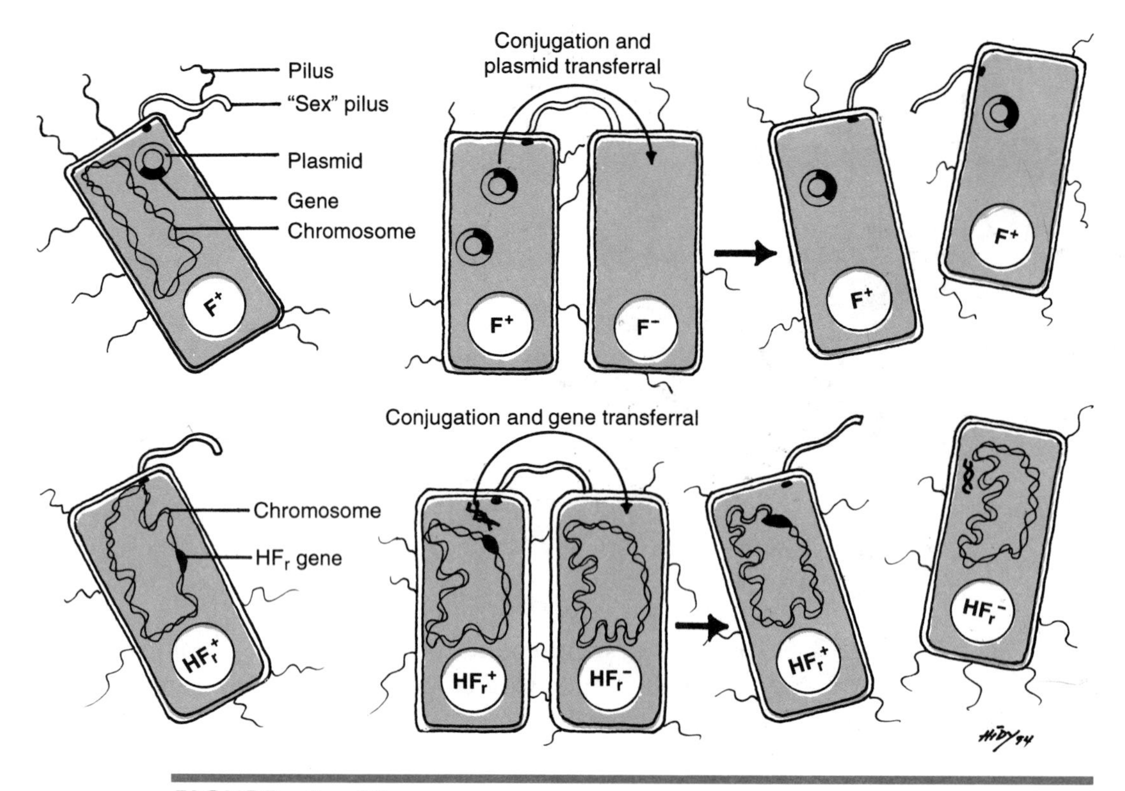

FIGURE 4 – 13
Bacterial conjugation. See text for details.

instance, the capsid of the smallpox virus) to which the person would form protective antibodies in the absence of complete virulent pathogens.

SUMMARY

This discussion of the physiology of microorganisms included information about the nutrient and energy sources that enable microbes to maintain the metabolic cycles of living organisms. These metabolic reactions include those involved in respiration, photosynthesis, catabolism, and anabolism.

When an organism has all the nutrients and energy it needs to maintain itself, it can then reproduce sexually or asexually. The growth curve represents the growth and decline of any population of organisms.

As you learn about bacterial genetics and the ways that bacteria can be changed by the addition of genetic material, try to envision how many ways human body cells could be changed by similar procedures. These techniques opened up a new field called genetic engineering.

PROBLEMS AND QUESTIONS

1. List five factors that influence the growth of microorganisms in nature.
2. What four factors are kept constant to produce a population growth curve in the laboratory?
3. List three factors that contribute to the death of a pure culture of bacteria in a tube of nutrient broth.
4. What are the six major elements found in living cells? List seven other elements that are also necessary for the metabolic functions of the cell.
5. Compare (a) heterotrophs and autotrophs; (b) chemotrophs and phototrophs; (c) chemoorganotrophs and chemolithotrophs; (d) photoorganotrophs and photolithotrophs.
6. Why are saprophytic decomposers necessary for ecological balance?
7. Describe catabolism, anabolism, respiration, fermentation, and photosynthesis.
8. Describe five ways by which the genetic constitution of bacteria may be altered.
9. What is a plasmid? an episome?
10. Describe the process of genetic engineering and give an example.

Self Test

After you have read Chapter 4, examined the objectives, reviewed the chapter outline; studied the new terms and answered the problems and questions above, complete the following self test.

Matching Exercises

Complete each statement from the list of words provided with each section.

NUTRITIONAL TYPES

heterotrophs	autotrophs	chemotrophs
phototrophs	chemoorganotrophs	photoorganotrophs
chemolithotrophs	photolithotrophs	

1. Organisms that use light as a source of energy are ________.
2. Organisms that get their energy from a chemical source are ________.
3. Organisms that use an organic carbon source are ________.
4. Organisms that use an inorganic source of carbon are ________.
5. Organisms that use a chemical source of energy and an organic source of carbon are ________.
6. Organisms that use a chemical source of energy and an inorganic source of carbon are ________.

7. Those organisms that use light as a source of energy and use an organic source of carbon are ______________.
8. Organisms that use light as a source of energy and use an inorganic source of carbon are ______________.
9. Which three terms describe algae? ______________, ______________, ______________.
10. Which three terms describe plants? ______________, ______________, ______________.
11. Which three terms describe animals? ______________, ______________, ______________.
12. Which three terms describe fungi and protozoa? ______________, ______________, ______________.

METABOLIC REACTIONS

catabolism
photosynthesis
fermentation
anabolism
aerobic respiration

1. The metabolic process by which plants and algae use light, carbon dioxide, and water to build carbohydrates is called ______________.
2. The process in which simple molecules are used to build complex ones is ______________.
3. The process in which complex macromolecules are broken down into simple molecules is ______________.
4. A chemical reaction in which oxygen participates with carbohydrates to yield energy and carbon dioxide is ______________.
5. An anaerobic respiration reaction that yields energy is ______________.
6. The breakdown and recycling of red blood cells in the liver is an example of ______________.
7. The synthesis of enzymes within a cell is an example of ______________.
8. The production of alcoholic beverages from grain is an example of ______________.

GROWTH CURVE

lag phase
logarithmic growth phase
death phase
stationary phase

1. The organisms absorb nutrients, synthesize enzymes, and prepare to reproduce in the ______________.
2. More organisms are dying than are reproducing in the ______________.
3. The organisms are all alive and reproducing rapidly in the ______________.
4. The number of living bacteria remain about the same in the ______________.
5. The healthiest stage of growth is the ______________.
6. A chemostat keeps the organisms in the ______________.
7. An industry that produces products from microorganisms maintains the microbes in the ______________.
8. Sporulation of certain genera of bacteria occurs during the ______________.

BACTERIAL GENETICS

mutation
mutagens
conjugation
transformation
transduction
lysogenic conversion

1. When living cells absorb and incorporate DNA from the surrounding medium, they are genetically changed; this process is called ______________.
2. The spontaneous change in the arrangement of the DNA molecule within a living cell is called ________.
3. The process by which a nonpathogenic bacterium could be produced for use in a vaccine is called ________________.
4. When a bacteriophage carries a bit of bacterial DNA from the cell it came from to another cell, the process is called ________________.
5. When a temperate bacteriophage injects its own DNA into a bacterial cell, the process is called ____________.
6. When two bacteria join by a pilus bridge and genetic material is transferred from one to another, the process is called ________________.
7. Ultraviolet light, x-rays, and some chemicals are used to increase the rate of mutation; these agents are called ________________.

True or False (T or F)

___ 1. Inorganic ions are basic nutrients required by living cells.
___ 2. A mutant is an organism that has survived mutation.
___ 3. Carbon, hydrogen, oxygen, and phosphorus are among the most necessary elements in living protoplasm.
___ 4. The macromolecules of living cells include carbohydrates, fats, proteins, and nucleic acids.
___ 5. Saprophytic fungi can get energy from the sun by photosynthesis.
___ 6. Autotrophs use carbon dioxide as a carbon source.
___ 7. Only a few groups of bacteria are chemoorganotrophs.
___ 8. The process of photosynthesis releases oxygen into the air for use by animals.
___ 9. Enzymes control metabolism in all living cells.
___ 10. All enzymes are proteins.
___ 11. All photosynthetic organisms must contain some form of photosynthetic pigment.

Multiple Choice

1. A characteristic that humans, fungi, and saprophytic bacteria have in common is that they
   a. can be facultative anaerobes
   b. obtain carbon atoms from organic materials
   c. use carbon dioxide as a basic carbon source
   d. obtain their energy from light
2. An *in vitro* culture that contains only one organism is known as
   a. singular
   b. specific
   c. a pure culture
   d. chemostatic
3. Viable plate counts are used to determine
   a. the total number of bacterial cells present
   b. the turbidity of the solution
   c. the toxic levels in the original solution
   d. the number of living bacteria present
4. The greatest number of living organisms are present during
   a. the lag phase
   b. the logarithmic growth phase
   c. the death phase
   d. the stationary phase
5. The largest number of adenosine triphosphate (ATP) molecules are produced in what part of cellular respiration?
   a. Citric acid cycle
   b. Glycolysis
   c. Electron transport system
   d. Krebs' cycle
6. In conjugation, the DNA that is transferred is
   a. viral
   b. defective
   c. bacterial
   d. piliated
7. Saprophytes digest organic materials outside the organism by means of
   a. endoenzymes
   b. cofactors
   c. coenzymes
   d. exoenzymes
8. The process of converting light energy to chemical bond energy is
   a. oxidation
   b. photosynthesis
   c. catabolism
   d. phosphorylation
9. Bacterial DNA is transferred by viruses in
   a. transduction
   b. conjugation
   c. lysogenic conversion
   d. transformation
10. The soreness in muscles after extreme exertion is due to
    a. the conversion of lactic acid to pyruvic acid
    b. excess oxidation
    c. the conversion of pyruvic acid to lactic acid
    d. the accumulation of phosphates
11. Transduction and transformation differ in that transduction
    a. involves DNA
    b. is restricted to pneumococci
    c. requires a phage
    d. is a sexual process

CHAPTER

5

Control of Microbial Growth

OBJECTIVES

After studying this chapter, you should be able to

- List three reasons why microbial growth must be controlled
- Define sterilization, disinfection, bactericidal agents, and bacteriostatic agents
- Differentiate between sterilization, pasteurization, and lyophilization
- Describe aseptic, antiseptic, and sterile techniques
- List the factors that influence the growth of microbial life
- Describe the following types of microorganisms: psychrophilic, mesophilic, thermophilic, halophilic, haloduric, alkalinophilic, acidophilic, and barophilic
- List several factors that influence the effectiveness of antimicrobial methods
- List common physical antimicrobial methods
- List common chemical antimicrobial compounds
- Describe the mode of action of sulfonamide drugs on bacteria

- Describe the action of penicillin on bacteria
- List four reasons why antibiotics should be used with caution

CHAPTER OUTLINE

Definition of Terms
Sterilization
Disinfection
Microbicidal Agents
Microbistatic Agents
Asepsis
Sterile Technique

Factors Influencing Microbial Growth
Temperature
Moisture
Osmotic Pressure
pH
Barometric Pressure
Gases

Antimicrobial Methods
Physical Antimicrobial Methods
Heat
Dry heat
Moist heat
Pressurized steam
Cold
Drying
Radiation
Ultrasonic Waves
Filtration
Chemical Antimicrobial Methods
Antisepsis
How Antimicrobial Chemicals Work

Chemotherapy
Major Discoveries
Characteristics of Antimicrobial Agents
How Antimicrobial Agents Work
Side Effects of Antimicrobial Agents

NEW TERMS

Acidophile
Alkaliphile
Antibiotic
Antimicrobial agent
Antisepsis
Antiseptic
Antiseptic technique
Asepsis
Autoclave
Bactericidal agent
Bacteriostatic agent
Barophile
Chemotherapeutic agent
Chemotherapy
Contamination
Crenated
Crenation
Desiccation
Disinfect
Disinfectant
Disinfection
Fungicidal agent
Germicidal agent
Haloduric
Halophile
Hemolysis
Hypertonic solution
Hypotonic solution
In vivo
Isotonic solution
Lyophilization
Mesophile
Microbicidal agent
Microbistatic agent
Osmosis
Osmotic pressure
Phenol coefficient test
Plasmolysis
Psychroduric
Psychrophile
Sanitization
Sporicidal agent
Sterile
Sterilization
Superinfection
Thermal death point (TDP)
Thermal death time (TDT)
Thermoduric
Thermophile
Tuberculocidal agent
Virucidal agent

The factors or agents that influence the growth of microorganisms are subject to continual study. On the basis of these studies, researchers learn how beneficial microbes can be encouraged to grow while growth of pathogenic microbes can be controlled, inhibited, or destroyed. Control of certain microorganisms is important (1) to prevent and control infectious diseases in humans, animals, and plants; (2) to preserve food; (3) to prevent contaminating microbes from interfering with certain industrial processes; and (4) to prevent contamination of pure culture research. Preventing the spread of infectious diseases and controlling infections require many different procedures. The source of infection can be controlled by (1) destroying or inhibiting disease-causing microbes; (2) blocking the sources, routes, and vectors of transmission of disease agents; and (3) protecting an infected person from the consequences of disease by building up the body's defenses and administering appropriate chemotherapeutic drugs.

An infectious disease is any disease caused by the invasion and multiplication of pathogenic microorganisms in the body. Infection indicates the presence of pathogens in living tissues. An infected person may or may not exhibit signs and symptoms of disease. *Contamination* means that pathogenic microorganisms are present on or in nonliving materials, such as bed linens, discharges from human or animal sources, or food and water.

Long before people were aware of the existence of microorganisms, they tried to prevent the spoilage of food and wines, the transmission of diseases, and the infection of wounds. They developed many primitive procedures and "cures" that were often more harmful than helpful to the recipient. Today we know that the avoidance, inhibition, and destruction of potentially pathogenic microorganisms is imperative to control diseases. It is essential that those who work in the health fields appreciate the importance of controlling microbes in patients' rooms, operating rooms, treatment rooms, and emergency rooms. They must be aware of proper aseptic procedures to be followed for dressing wounds, giving injections and respiratory treatments, and assisting physicians, dentists, and other personnel who have the responsibility for patient care. Health-care workers must be able to properly handle contaminated linen and wound dressings, bedside equipment, and laboratory specimens to avoid infecting themselves, the patients for whom they are providing care, or other patients.

DEFINITION OF TERMS

Before we can discuss the various methods used to destroy or inhibit microbes, a number of terms should be understood as they apply to microbiology.

Sterilization The complete destruction of all living organisms, including cells, viable spores, and viruses, is called *sterilization.* Sterilization of objects can be accomplished by heat, autoclaving (heat and steam under pressure), gas (ethylene oxide), various chemicals (such as formaldehyde), and certain levels of radiation with ultraviolet or gamma rays. These procedures are discussed later in this chapter. When something is *sterile,* it is devoid of microbial life.

Disinfection *Disinfection* is the destruction or removal of infectious or harmful microorganisms from nonliving objects by physical or chemical methods. The heating process developed by Pasteur to disinfect beer and wines is called pasteurization. It is still used to eliminate pathogenic microorganisms from milk and beer. It should be remembered that pasteurization is not a sterilization procedure, because not all the microbes are destroyed. Chemical agents are also used to eliminate pathogens. Chemicals used to *disinfect* inanimate objects, such as bedside equipment and operating rooms, are called *disinfectants.* An *antiseptic* is a solution used to disinfect the skin or other living tissues. Disinfectants are strong chemical substances and are more destructive to living tissues than antiseptics. *Sanitization* reduces microbial populations to levels considered safe by public health standards, such as those applied to restaurants.

Microbicidal Agents The suffix "-cide" or "-cidal" refers to "killing." Thus, a *microbicidal agent* (*microbicide*) is one that kills microbes. A *bactericidal agent* (*bactericide*) kills bacteria, but not necessarily endospores of bacteria. A disinfectant that kills fungi is a *fungicide,* and similarly, an agent that destroys viruses is a *virucide.* The general term *germicide* refers to any agent that destroys germs or harmful microorganisms; such an agent might be used in sanitization procedures.

Microbistatic Agents A *microbistatic agent* is a drug or chemical that inhibits growth and reproduction of microorganisms, whereas a *bacteriostatic agent* is one that inhibits the metabolism and reproduction of bacteria. Important microbistatic agents and processes include desiccation (drying), freezing temperatures, concentrated sugar and salt solutions, and some antimicrobial agents (including certain antibiotics).

Asepsis Because sepsis refers to the growth of infectious microbes on living tissues, *asepsis* means the absence of pathogens on living tissues. Thus, aseptic technique is designed to eliminate and exclude all infectious microbes by sterilization of equipment, disinfection of the environment, and cleansing of body tissues with antiseptics. *Antiseptic technique* was developed by Lister in 1867. He used dilute carbolic acid to cleanse surgical wounds and equipment and a carbolic acid aerosol to prevent harmful microorganisms from entering the surgical field or contaminating the patient. *Antisepsis* is the prevention of infection.

Sterile Technique When it is necessary to prevent *all* microorganisms from gaining entrance into a laboratory or onto a surgical field, sterile technique is followed.

FACTORS INFLUENCING MICROBIAL GROWTH

There are many environmental factors that enhance or inhibit the growth of microorganisms, including temperature, moisture, osmotic pressure, pH, barometric pressure, gases, radiations, chemicals, and the presence of neighboring microbes. Many concepts involving these factors may be applied to our everyday lives as well as to laboratory and hospital situations.

Temperature

For every microorganism, there is an optimum temperature at which the organism grows best, a minimum temperature below which it ceases to grow, and a maximum temperature above which it is destroyed. These temperature ranges differ greatly among organisms. Their rate of growth and metabolism is generally slower at lower temperatures and faster at higher temperatures. The effect of temperature changes varies from one species to another. Microbes that thrive at 20° to 40° C (68° to 104° F) are called *mesophiles,* a group that includes most of the species that grow on plants and animals and in soil and water. *Psychrophiles* are capable of growth at temperatures near the freezing point; they thrive in oceans, soil, and refrigerated foods at temperatures between 0° and 20° C (30° to 68° F). Microorganisms that grow at temperatures above 45° C (113° F) are called *thermophiles.* These heat-loving microbes may be found in hot springs, compost pits, and silage; because they thrive at high temperatures, boiling is not an effective means of killing them. These microbes cause much of the color observed in hot springs. Refer to Table 5 – 1 for the minimum, optimum, maximum temperature ranges of psychrophilic, mesophilic, and thermophilic bacteria.

Most pathogens and members of the indigenous microflora are mesophilic because they grow best at normal body temperature, 37° C (98.6° F). Thus, most

TABLE 5 – 1.
Categories of Bacteria on the Basis of Temperature Tolerance

	Temperature Range (°C)		
Group	Minimum	Optimum	Maximum
Psychrophiles	−20 to 5	0 to 20	19 to 35
Mesophiles	10 to 15	20 to 40	35 to 47
Thermophiles	40 to 45	55 to 75	60 to 90

pathogens are easily destroyed by boiling. Exceptions are the spores produced by spore-forming bacteria, mycobacteria having resistant cell walls, and those microbes that are encased in a protective coating of organic material, such as mucus, vomitus, pus, or feces.

Microorganisms that can survive or endure very cold temperatures and can be preserved in the frozen state are known as *psychroduric* organisms. Fecal material left by early Arctic explorers contained psychroduric *Escherichia coli* that survived the Arctic temperatures. Also, many microbes and endospores can survive boiling and are therefore called *thermoduric* organisms.

Moisture

Living organisms require water to carry out their normal metabolic processes. However, some microorganisms can survive the complete drying process (*desiccation*). Such organisms are in a dormant or resting state after they have been dried. As soon as they are placed in a moist nutrient environment, they grow and reproduce normally.

Another method of inhibiting growth of microbes is by dehydration (or drying) of frozen organisms, a process often called *lyophilization.* Lyophilized materials are frozen in a vacuum; the container is then sealed to maintain the inactive state. This freeze-drying method is widely used in industry to preserve foods, antibiotics, antisera, microorganisms, and other biological materials. It should be remembered that lyophilization cannot be used to sterilize or kill microorganisms, but rather, it is used to prevent them from reproducing.

Osmotic Pressure

Osmotic pressure is that pressure exerted on the cell membrane by solutions inside and outside the cell (Fig. 5 – 1). When cells are suspended in a solution, the ideal situation is that the osmotic pressure inside the cell equals the pressure of the solution outside the cell. Substances dissolved in liquids are referred to as solutes. When the concentration of solutes in the environment outside of a cell is greater than the concentration of solutes inside the cell, the solution in which the cell is suspended is said to *hypertonic.* In such a situation, whenever possible, water leaves the cell by *osmosis* in an attempt to equalize the two concentrations. If the cell is a human cell, such as a red blood cell (erythrocyte), the loss of water causes the cell to shrink; this shrinkage is called *crenation* and the cell is said to be *crenated.* If the cell is a bacterial cell, having a rigid cell wall, the cell does not shrink. Instead, the cell membrane and cytoplasm shrink away from the cell wall; this condition, known as *plasmolysis,* inhibits bacterial cell growth and multiplication. Salts and sugars are sometimes added to certain

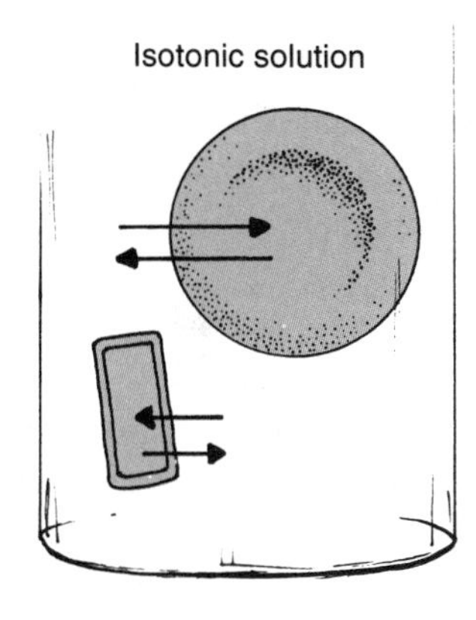

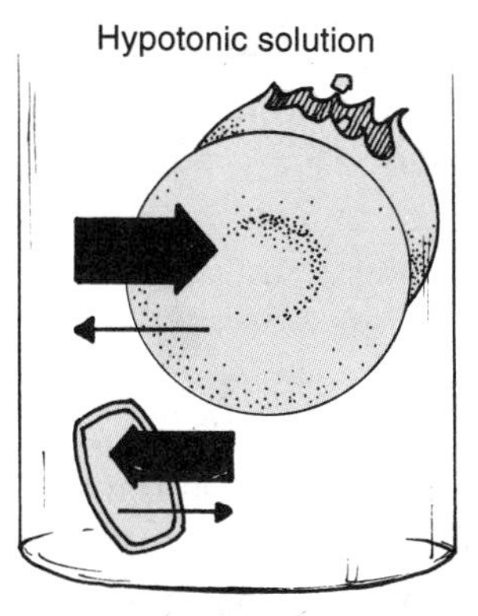

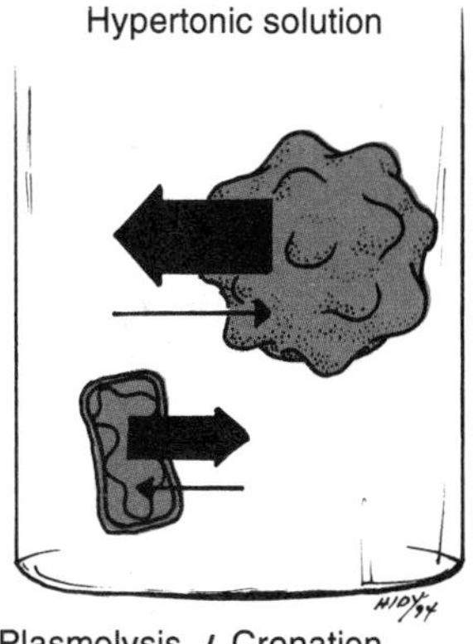

FIGURE 5 – 1
Changes in osmotic pressure. No change in pressure occurs inside the cell in an isotonic solution; pressure is increased in a hypotonic solution; and pressure is decreased in a hypertonic solution. Arrows indicate direction of water flow. The larger the arrow, the greater the amount of water flowing in that direction.

foods as a way of preserving them. Bacteria entering such hypertonic environments will die.

When the concentration of solutes outside a cell is less than the concentration of solutes inside the cell, the solution in which the cell is suspended is said to *hypotonic.* In such a situation, whenever possible, water enters the cell in an attempt to equalize the two concentrations. If the cell is a human cell, such as an erythrocyte, the increased water within the cell causes the cell to swell. If sufficient water enters, the cell will burst (or lyse). In the case of erythrocytes, this bursting is called *hemolysis.* If a bacterial cell is placed in a hypotonic solution (such as distilled water), the cell does not burst (due to the rigid cell wall), but the fluid pressure within the cell increases greatly. This increased pressure (plasmoptysis) occurs in cells having rigid cell walls such as plant cells and bacteria.

When the concentration of solutes outside a cell equals the concentration of solutes inside the cell, the solution is said to be *isotonic.* In an isotonic environment, excess water neither leaves nor enters the cell; thus, no plasmolysis or plasmoptysis occurs. The cell is said to have normal turgor (fullness). Refer to Figure 5 – 1 for a comparison of the differences of various solution concentrations on bacteria and human body cells.

Sugar solutions for jellies and pickling brines (salt solutions) for meats preserve these foods by inhibiting the growth of microorganisms. However, many types of molds and some types of bacteria can survive and even grow in a salty environment. Organisms capable of surviving salty environments are said to be *haloduric*. Those that prefer salty environments (such as the concentrated salt water found in the Great Salt Lake) are called *halophilic;* "halo" for "salt" and "philic" meaning "to love."

pH

The term pH refers to the acidity or alkalinity of a solution (see Chapter 3). Most microorganisms prefer a neutral growth medium (about pH 7), but *acidophilic* microbes, such as those that can live in the stomach and in pickled foods, prefer a pH of 2 to 5. *Alkaliphiles* prefer an alkaline environment (above pH 8.5) such as is found inside the intestine (about pH 9).

Barometric Pressure

Most bacteria are not affected by minor changes in barometric pressure. Some thrive at normal atmospheric pressure, and some, known as *barophiles* ("baro", referring to "pressure"), thrive deep in the ocean and in oil wells, where the atmospheric pressure is very high.

Autoclaves and home pressure cookers kill microbes by a combination of high pressure and high temperature. The increase in pressure raises the temperature above the temperature of boiling water. (At a pressure of 15 pounds per square inch [p.s.i.], the temperature of boiling water is 121° C.) However, home canning done without the use of a pressure cooker does not destroy the endospores of bacteria, notably the anaerobe, *Clostridium botulinum.* Occasionally, the local newspapers report cases of food poisoning resulting from the ingestion of C. *botulinum* toxins (poisons) in improperly canned vegetables and meats.

Gases

The types of gases present in a particular environment and their concentrations determine which species of microbes are able to live there. Most microbes grow best in an atmosphere containing oxygen, but obligate anaerobes die in its presence. For this reason, oxygen is sometimes forced into wound infections caused by anaerobes. For instance, wounds that may contain tetanus bacteria are lanced (opened) to expose them to the air. Another example is gas gangrene; this is a deep wound infection that is often treated by placing the patient in a hyperbaric (increased pressure) oxygen chamber or in a room with high oxygen pressure, because the causative bacteria, most often *Clostridium perfringens,* are anaerobic and cannot live in the presence of oxygen.

ANTIMICROBIAL METHODS

The methods used to destroy or inhibit microbial life are either physical or chemical, and sometimes both types are used. The effectiveness of any antimicrobial procedure depends on (1) the length of time it is applied, (2) the temperature, (3) its concentration, (4) the nature and number of microbes and spores present (bioburden), and (5) the presence of organic matter, such as proteins in feces, blood, vomitus, and pus, on the materials being treated (Fig. 5 – 2).

Physical Antimicrobial Methods

The physical methods commonly used in hospitals, clinics, and laboratories to destroy or control pathogens are heat, pressure, desiccation, radiation, sonic disruption, and filtration.

Heat Heat is the most practical, efficient, and inexpensive method of disinfection and sterilization of those inanimate objects and materials that can withstand high temperatures. Because of these advantages, it is the means most frequently used.

Two factors, *temperature* and *time,* determine the effectiveness of heat for sterilization. There is considerable variation from organism to organism in susceptibility to heat; pathogens usually are more susceptible than nonpathogens. Also, the higher the temperature, the shorter the time required to

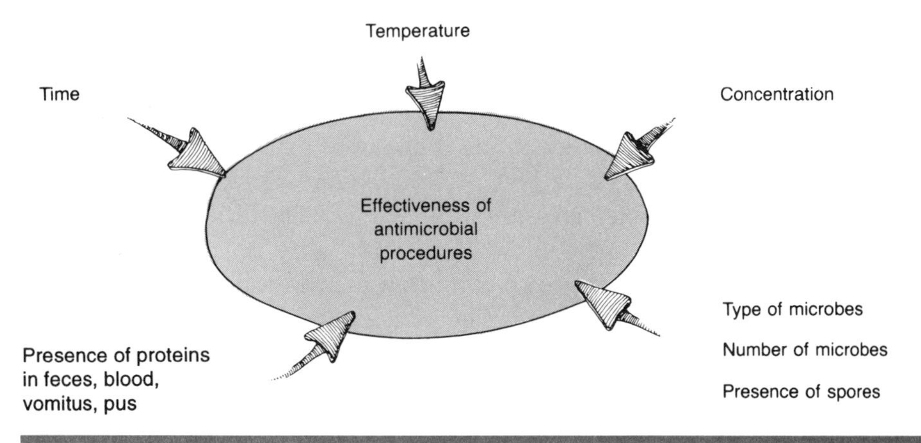

FIGURE 5 – 2
Factors that determine the effectiveness of any antimicrobial procedure: time, temperature, concentration, presence of other microbes or spores, and presence of protein materials.

kill the organisms. The *thermal death point* of any specific species of microorganism is the lowest temperature that will kill all the organisms in a standardized pure culture within a specified period. The *thermal death time* is the length of time necessary to sterilize a pure culture at a specified temperature.

In practical applications of heat for sterilization, one must consider the material in which a mixture of organisms and their spores may be found. Pus, feces, vomitus, mucus, and blood contain proteins that serve as a protective coating to insulate the pathogens; when these substances are present on bedding, bandages, surgical instruments, and syringes, very high heat is required to destroy vegetative (growing) microorganisms and spores. In practice, the most effective procedure is to wash away the protein debris with strong soap, hot water, and a disinfectant and then sterilize the equipment with heat.

Heat applied in the presence of moisture, as boiling or steaming, is more effective than dry heat because moist heat causes proteins to coagulate. Because cellular enzymes are proteins, they are also inactivated. This is exactly what happens when an egg is hard-boiled: the combination of heat and moisture causes the proteins to coagulate. Moist heat sterilization is faster than dry heat sterilization and can be done at a lower temperature; thus, it is less destructive to many materials that otherwise would be damaged by higher temperatures.

The vegetative forms of most pathogens are quite easily destroyed by boiling; however, bacterial endospores are particularly resistant to heat and drying. The autoclave, which combines heat and pressure, offers the most effective yet inexpensive means of destroying the spores. Two examples of spore-formers are *Clostridium tetani,* the causative agent of tetanus, and *C. botulinum,* which causes a severe form of food poisoning; their spores are usually found in contaminating dirt and dust. Botulism food poisoning is preventable by properly washing and pressure cooking (autoclaving) the food.

Certain viruses are remarkably resistant to heat. A case in point is the hepatitis virus, which is frequently transferred from one person to another by the reuse of contaminated syringes and needles that have not been adequately sterilized. It is recommended that all equipment used in the transfer of blood be sterilized in an autoclave at 121° C (250° F) for 20 minutes or boiled for 30 minutes or baked in an oven at 180° C (356° F) for 1 hour.

Dry Heat Dry heat baking in a thermostatically controlled oven provides effective sterilization of metals, glassware, some powders, oils, and waxes. These items must be baked at 160° to 165° C (320° to 329° F) for 2 hours or at 170° to 180° C (338° to 356° F) for 1 hour. An ordinary oven of the type found in most homes may be used if the temperature remains constant. The effectiveness of dry heat sterilization depends on how deeply the heat penetrates throughout the material, and the items to be baked must be placed so that the hot air circulates freely among them.

Incineration, or burning, is an effective means of destroying contaminated

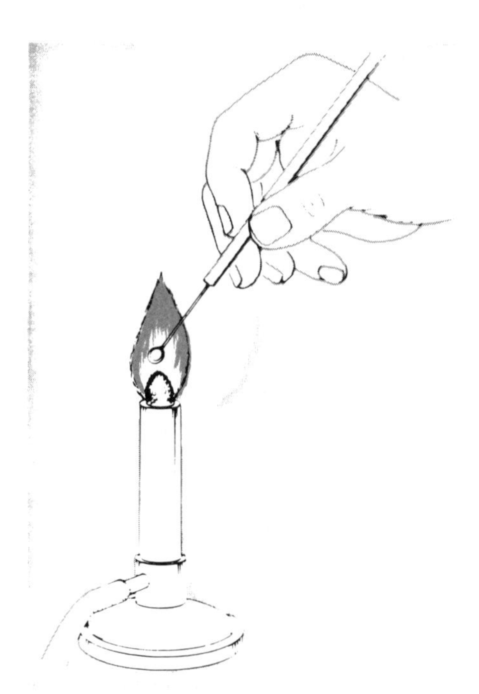

FIGURE 5 – 3
The technique for flaming a bacteriological loop.

disposable materials. An incinerator must never be overloaded with moist or protein-laden materials, such as feces, vomitus, or pus, because the contaminating microorganisms within these moist substances may not be destroyed if the heat does not readily penetrate and burn them. Flaming the surface of heat-resistant material is an effective way to kill microorganisms on forceps and bacteriological loops and is a common laboratory procedure. Flaming is accomplished by holding the end of the loop or forceps in the yellow portion of a gas flame (Fig. 5 – 3).

Moist Heat As previously stated, moist heat causes cellular proteins (including enzymes) in the microorganisms to become inactivated, and the cell dies. Boiling water and steam are favored in disinfection because no expensive equipment is necessary and the required time is short. Most pathogens die after 10 minutes of steaming at 70° C (158° F); also, boiling for 10 to 30 minutes at 90° to 100° C (194° to 212° F), depending on the altitude, destroys most viable bacteria, fungi, and viruses. Clean articles made of metal and glass, such as syringes, needles, and simple instruments, may be disinfected by boiling for 30 minutes. However, this technique is not always effective because heat-resistant bacterial endospores, mycobacteria, and viruses may be present. As mentioned in Chapter 2, the endospores of the bacteria that cause anthrax, tetanus, gas gangrene, and botulism as well as the hepatitis viruses are notably heat resistant and often survive normal disinfection procedures.

An effective way to disinfect clothing, bedding, and dishes is to use hot water, above 60° C (140° F) with detergent or soap and to agitate the solution around the items. This combination of heat, mechanical action, and chemical inhibition is deadly to most pathogens.

Pressurized Steam An *autoclave* is a large metal pressure cooker that uses steam under pressure to completely destroy all microbial life. Pressure raises the temperature of the steam and shortens the time necessary to sterilize materials that can tolerate the high temperature and moisture. Autoclaving at a pressure of 15 p.s.i. at a temperature of 121.5° C (250° F) for 20 minutes kills viable microorganisms, viruses, and exposed bacterial endospores if they are not protected by pus, feces, vomitus, blood, or other proteinaceous substances. Some types of equipment and certain materials, such as rubber, which may be damaged by high temperatures, can be autoclaved at lower temperatures for longer periods. The timing must be carefully determined based on the contents and compactness of the load. All articles must be properly packaged and arranged within the autoclave to allow steam to penetrate each package (Fig. 5 – 4). Cans should be open, bottles covered loosely with foil or cotton, and instruments wrapped in cloth. Sealed containers should not be autoclaved.

Cold Most microorganisms are not killed by cold temperatures and freezing, but their metabolic activities are slowed, greatly inhibiting their growth. Thus, freezing is a microbistatic method of preservation in which the microorganisms are in a state similar to suspended animation. When the temperature is

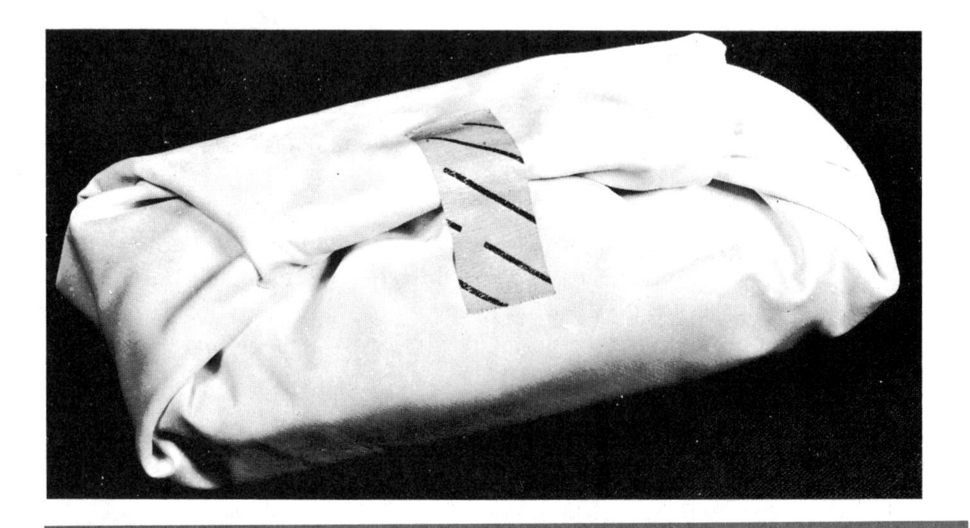

FIGURE 5 – 4
Pressure-sensitive autoclave tape shows dark stripes after sterilization. (Volk WA, Wheeler MF: Basic Microbiology, 5th ed. Philadelphia, JB Lippincott, 1984)

raised above the freezing point, the metabolic reactions speed up and the organisms slowly begin to reproduce again. Refrigeration merely slows the growth of microorganisms; it does not altogether inhibit them. Many foods, biologic specimens, and bacterial cultures are preserved by rapid freezing to very low temperatures. It should be noted that slow freezing causes ice crystals to form within cells and may rupture the cell walls of some bacteria; hence, if it is important to preserve a pure culture of bacteria, such slow freezing should be avoided and rapid freezing should be performed, usually using liquid nitrogen.

Persons who are involved in the preparation and preservation of foods must be aware that thawing to room temperature allows bacterial spores to germinate and microorganisms to resume growth. Consequently, refreezing of thawed foods is an unsafe practice, because it preserves the millions of microbes that might be present and the food deteriorates quickly when it is rethawed. Also, if bacterial endospores of *C. botulinum* or *C. perfringens* were present, the viable bacteria would begin to produce toxins that would cause food poisoning.

Drying For many centuries, foods have been preserved by drying. When moisture and nutrients are lacking, many dried microorganisms remain viable, although they cannot reproduce. Foods, serums, toxins, antitoxins, antibiotics, and pure cultures of microorganisms are often preserved by lyophilization (discussed previously).

In the hospital or clinical environment, health-care workers should keep in mind that dried viable pathogens may be lurking in dried matter including blood, pus, fecal material, and dust that are found on floors, in bedding, on clothing, and in wound dressings. Should these dried materials be disturbed, such as by dry dusting, the microbes would be easily transmitted through the air or by contact. They would then grow rapidly if they settled in a suitable moist, warm nutrient environment such as a wound or a burn. Therefore, important precautions that must be observed include the following: wet mop and damp dust floors and furniture, roll bed linens and towels carefully, and properly dispose of wound dressings.

Radiation The sun is not a particularly reliable disinfecting agent because it kills only those microorganisms that are exposed to direct sunlight. The rays of the sun include the long infrared (heat) rays, the visible light rays, and the shorter ultraviolet (UV) rays. The UV rays, which do not penetrate glass and building materials, are effective only in the air and on the surface of equipment. They do, however, penetrate cells and, thus, can cause damage to DNA. When this occurs, genes may be so severely damaged that the cell dies (especially unicellular microorganisms) or is drastically changed.

In practice, a UV lamp is useful for reducing the number of microorganisms in the air. A UV lamp is often called a germicidal lamp. Its main component is a low-pressure mercury vapor tube. Such lamps are found in newborn nurseries, operating rooms, elevators, entryways, cafeterias, and classrooms, where they are incorporated into louvered ceiling fixtures designed to radiate across the top of the room without striking persons in the room. The sterility of an area may also be maintained by having a UV lamp placed in a hood or cabinet containing instruments, paper and cloth equipment, liquid, and other inanimate articles. Many biological materials, such as sera, antisera, toxins, and vaccines, are sterilized with UV rays.

Those whose work involves the use of UV lamps must be particularly careful not to expose their eyes or skin to the rays because the rays can cause serious burns and cellular damage. Because UV rays do not penetrate cloth, metals, and glass, these materials may be used to protect persons working in a UV environment. It has been shown that skin cancer can be caused by excessive exposure to the UV rays of the sun; thus, extensive suntanning is harmful.

X-rays and gamma and beta rays of certain wavelengths from radioactive materials may be lethal or cause mutations in microorganisms and tissue cells because they damage DNA and proteins within those cells. Work done in radiation research laboratories has demonstrated that these radiations can be used for the prevention of food spoilage, sterilization of heat-sensitive surgical equipment, preparation of vaccines, and treatment of some chronic diseases such as cancer, all of which are very practical applications for laboratory research. When these radiations are used in the treatment of disease, care must be taken to focus the rays precisely on the specific area being treated to minimize damage to surrounding normal cells.

Ultrasonic Waves In hospitals and clinics, ultrasonic waves are a frequently used means of cleaning and sterilizing delicate equipment. Ultrasonic cleaners consist of tanks filled with liquid solvent (usually water); the short sound waves are then passed through the liquid. The sound waves mechanically dislodge organic debris on instruments and glassware.

Glassware and other articles that have been cleansed in ultrasonic equipment must be washed to remove the dislodged particles and solvent. They are then sterilized by another method before they are used.

Filtration Filters of various pore sizes are used to filter or separate cells, larger viruses, bacteria, and certain other microorganisms from the liquids or gases in which they are suspended. The filtered solution (filtrate) is not necessarily sterile because small viruses may not be filtered out. The variety of filters is large and includes sintered glass (in which uniform particles of glass are fused), plastic films, unglazed porcelain, asbestos, diatomaceous earth, and cellulose membrane filters. Small quantities of liquid can be filtered through a syringe; large quantities require larger apparatuses.

A cotton plug in a test tube, flask, or pipette is a good filter for preventing the entry of microorganisms. Dry gauze and paper masks prevent the outward passage of microbes from the mouth and nose, at the same time protecting the wearer from inhaling airborne pathogens and foreign particles that could damage the lungs. Biological safety cabinets and laminar flow hoods contain high-efficiency particulate air (HEPA) filters to protect workers from contamination.

Chemical Antimicrobial Methods

Chemical disinfection means the use of chemical agents to inhibit the growth of microorganisms, either temporarily or permanently. The effectiveness of a chemical disinfectant depends on many factors: the concentration of the chemical; the time allowed for the chemical to work; the pH or acidity of the solution; the temperature; and the presence of proteins, blood, pus, feces, mucous secretions, and vomitus. Directions for the preparation and dilution of the disinfectant must be carefully followed, and the proper concentration, pH, and temperature must be maintained for the specified period to ensure the best results. The items to be disinfected must first be washed to remove any material in which pathogens may be hidden. Although the washed article may then be clean, it is not safe to use until it has been properly disinfected. Health personnel need to understand an important limitation of chemical disinfection: many disinfectants that are effective against pathogens in the controlled conditions of the laboratory become ineffective in the actual hospital or clinical environment. Furthermore, the stronger and more effective antimicrobial chemical agents are of limited usefulness because of their destructiveness to human tissues and certain other substances.

Almost all bacteria in the vegetative state as well as fungi, protozoa, and most viruses are susceptible to many disinfectants, although the mycobacteria that cause tuberculosis and leprosy, bacterial endospores, pseudomonads, fungal spores, and hepatitis viruses are notably resistant. Therefore, chemical disinfection should never be attempted when it is possible to use proper physical sterilization techniques.

The disinfectant most effective for each situation must be carefully chosen. Chemical agents used to disinfect respiratory therapy equipment and thermometers must destroy all pathogenic bacteria, fungi, and viruses that may be found in sputum and saliva. One must be particularly aware of the oral and respiratory pathogens, including *Mycobacterium tuberculosis, Pseudomonas* spp., *Staphylococcus, Streptococcus,* the various fungi that cause candidiasis, blastomycosis, coccidioidomycosis, and histoplasmosis, and all of the respiratory viruses.

Because most disinfection methods do not destroy all bacterial endospores that are present, any instrument or dressing used in the treatment of an infected wound or a disease caused by spore-formers must be autoclaved or incinerated. Gas gangrene, tetanus, and anthrax are examples of diseases caused by spore-formers that require the health worker to take such precautions. Formaldehyde and ethylene oxide, when properly used, are highly destructive to spores, mycobacteria, and viruses. Certain articles are heat sensitive and cannot be autoclaved or safely washed before disinfection; such articles are soaked for 24 hours in a strong detergent and disinfectant solution, washed, and then sterilized in an ethylene oxide autoclave. The use of disposable equipment whenever possible in these situations helps to protect patients and health-care team members.

The effectiveness of a chemical agent depends to some extent on the physical characteristics of the article on which it is used. A smooth, hard surface is readily disinfected, whereas a rough, porous, or grooved surface is not. Thought must be given to selection of the most suitable germicide for cleaning patient rooms and all other areas where patients are treated.

The most effective antiseptic or disinfectant should be chosen for the specific purpose, environment, and pathogen or pathogens likely to be present. The following are characteristics of a good chemical antimicrobial agent.

1. It must kill pathogens within a reasonable period and in specified concentrations.
2. It must be nontoxic to human tissues and noncorrosive and nondestructive to materials on which it is used.
3. It must be soluble in water and easy to apply. If a tincture (*e.g.*, alcohol-water solution) is used, the proper concentration must be used. Evaporation of the alcohol solvent can cause a 1% solution to increase to a 10% solution; at this concentration, it may cause tissue damage.
4. It should be inexpensive and easy to prepare for use with simple, specific directions.
5. It must be stable in the dissolved or solid form so that it can be shipped and stored for a reasonable period.
6. It should be stable to pH and temperature changes within reasonable limits.

ANTISEPSIS

Most antimicrobial chemical agents are too irritating and destructive to be applied to mucous membranes and skin. Those that may be safely used on human tissues are called *antiseptics.* An antiseptic merely reduces the number of organisms on a surface but does not penetrate the pores and hair follicles to destroy microorganisms residing there. To remove organisms lodged in pores and folds of the skin, health personnel use an antiseptic soap and scrub with a brush. To prevent resident indigenous microflora from contaminating the surgical field, the surgeon wears sterile gloves on freshly scrubbed hands and a mask and hood to cover face and hair. Also, an antiseptic is applied at the site of the surgical incision to destroy local microorganisms.

HOW ANTIMICROBIAL CHEMICALS WORK

Injury of Cell Membranes Soap and detergents are referred to as surfactants; this means that they are surface-active agents that help to disperse the bacteria, allowing them to more readily be rinsed away. These agents concentrate on the surface and thus reduce the surface tension; this characteristic makes them good wetting and dispersing agents. Some agents, such as Dial® and Safeguard® soaps, contain disinfectants, which also aid in killing bacteria. Certain concentrations of weak acids such as acetic and benzoic acids may also be used in disinfectant soaps.

Inactivation of Enzymes Alcohols, such as ethyl and isopropyl, are good skin antiseptics at 70% solution. Ethyl alcohol has a low toxicity for humans; hence, it is frequently used to disinfect clinical thermometers and other instruments. However, when taken internally, isopropyl alcohol causes severe gastrointestinal upset and methanol causes brain damage. Alcohols are *tuberculocidal* (destructive to tuberculosis-causing organisms), but not *sporicidal* (destructive to spores).

The phenolics including phenol, carbolic acid, xylenols, orthophenylphenol, and cresol are used as disinfectants in hospitals and laboratories. However, they are too irritating and toxic to be used on skin. The commercial mixture of phenolics, Lysol®, is an effective germicide because it works in the presence of organic material and remains active on hard surfaces for extended periods. These chemicals are tuberculocidal but not sporicidal.

The effectiveness of phenol was demonstrated by Joseph Lister in 1867, when it was used to reduce the incidence of infections following surgical procedures. The effectiveness of other disinfectants is compared with that of phenol using the *phenol coefficient test.* To perform this test, a series of dilutions of phenol and the experimental disinfectant are inoculated with the test bacteria, *Salmonella typhi* and *Staphylococcus aureus,* at 37° C. The highest dilutions (lowest concentrations) that kill the bacteria after 10 minutes are used to calculate the phenol coefficient.

Salts of heavy metals such as mercury chloride (Merthiolate®, Mercurochrome®, Metaphen®—generic names: thimerosal, merbromin, nitromersol, respectively) and silver nitrate (Argyrol®, Protargol®) are bacteriostatic antiseptics, but they are not sporicidal and are ineffective against many pathogens. Silver nitrate in low concentrations has been used in the eyes of newborns to kill *Neisseria gonorrhoeae;* this prevents gonococcal infections, which could cause blindness.

Chemical oxidizing agents are useful disinfectants. Two of these are hydrogen peroxide and sodium perborate, which destroy bacteria and tissue debris and prevent anaerobic growth in damaged tissues. A third, potassium permanganate, is used in weak solutions to treat urethral infections and fungal infections of the skin. Another agent in this group is ethylene oxide (Carboxide®, Cryoxide®, Oxygume®), which is used as a sterilant in gas autoclaves to sterilize heat-sensitive materials. Although it is a good microbicide and sporicide, this gas must be used with great care because it is flammable and toxic to humans.

The elements and many compounds of chlorine, iodine, bromine, and fluorine are also useful disinfectants. For example, chlorine compounds (Clorox®, Halozone®, hypochlorites, Warexin®) are used to disinfect water and sewage and for sanitization of dishes, floors, and plumbing fixtures. It has been found that the human immunodeficiency virus (HIV) can be destroyed on syringes and needles by soaking them in a Clorox® (chlorine bleach) solution for 10 minutes. The iodine compounds, such as Wescodyne®, Betadine®, Isodine®, and tincture of iodine, are effective skin antiseptics and disinfectants. However, these compounds can be dangerous. If an alcohol solution of iodine is left open to the air, allowing the alcohol to evaporate, the solution may become too concentrated, and an iodine burn may result if it is used on the skin. Most compounds of bromine and fluorine are too toxic at the effective concentrations to be used as antiseptics. All of these compounds are viricidal, bactericidal, and tuberculocidal; however, none is sporicidal.

Damage to Genetic Material The DNA of cells is inactivated by caustic compounds such as formalin. Formalin is a 37% aqueous solution of gaseous formaldehyde that inactivates proteins and nucleic acids. It is one of the few antimicrobial agents that are also sporicidal; however, it is so irritating to skin and mucous membranes that it cannot be used on living tissues. Frequently, it is used in the laboratory to preserve tissue specimens.

Basic aniline dyes also inactivate nucleic acids. This group includes gentian violet and crystal violet, which are useful in treatment of fungal skin infections (ringworm) and vaginal infections caused by yeasts (*Candida*) and gram-positive bacteria, as well as intestinal roundworm infections. Pyridium is a dye classified in the same group. It is sometimes prescribed for urinary tract infections caused by gram-negative enteric organisms.

CHEMOTHERAPY

Chemotherapeutic agents are substances (drugs) used to treat diseases, including infectious diseases. For thousands of years, people have been finding and using herbs and chemicals to cure diseases. Native witch doctors in Central and South America long ago discovered that the herb, ipecac, aided in the treatment of dysentery and that a quinine extract of cinchona bark was effective in treating malaria. During the 16th and 17th centuries, the alchemists of Europe searched for ways to cure smallpox, syphilis, and the many other diseases that were rampant during that period of history. Many of the mercury and arsenic chemicals that were used frequently caused more damage to the patient than to the pathogen. Chemotherapeutic agents used to treat infectious diseases are called *antimicrobial agents.*

Major Discoveries

The true beginning of modern *chemotherapy* was in the late 1800s when Paul Ehrlich began his search for chemicals that would destroy bacteria, yet would not damage normal body cells. By 1909, he had tested and discarded more than 600 chemicals. Finally, in that year, he discovered an arsenic compound that proved effective against syphilis. Because this was the 606th compound Ehrlich had tried, he called it "compound 606." The technical name for it is arsphenamine and the trade name was Salvarsan®. Until the purification of penicillin in 1938, arsphenamine was used to treat syphilis. In 1928, Alexander Fleming noted that a substance produced by a mold, *Penicillium notatum,* inhibited the growth of staphylococci on an agar plate (Fig. 5 – 5). He also found that broth cultures of the mold were nontoxic to his laboratory animals and destroyed staphylococci and other bacteria. During World War II, two biochemists, Sir Howard Walter Florey and Ernst Boris Chain, purified penicillin and demonstrated its effectiveness in the treatment of various bacterial infections. By 1942, the U.S. drug industry was able to produce sufficient penicillin for human use, and the search for other antibiotics began. In 1935, a chemist

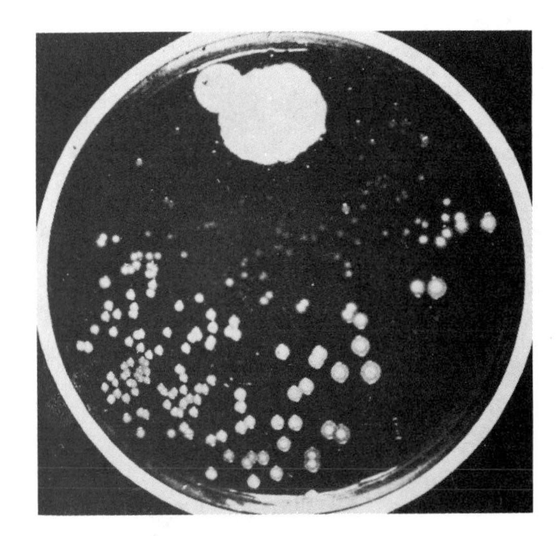

FIGURE 5 – 5
The discovery of penicillin by Fleming. Colonies of *Staphylococcus aureus* are not growing near the contaminant colony of *Penicillium notatum* because the mold is producing an antibiotic (penicillin) that kills the bacteria. (Davis BD: Microbiology, 4th Ed. Philadelphia, JB Lippincott, 1990; this photograph originally appeared in the *British Journal of Experimental Pathology* in 1929).

INSIGHT
"Superbugs," Part 1: What Are They?

Undoubtedly, you have heard much about drug-resistant bacteria, or "superbugs," as they've been labeled by the press. "Superbugs" are microorganisms, mainly bacteria, that have become resistant to one or more antimicrobial agents. The worst of them are multiresistant, that is, resistant to a variety of antimicrobial agents. Microorganisms become resistant by chromosomal mutations or by inheriting genes, most frequently via conjugation and the transfer of plasmids. An organism may become resistant in several ways: (1) by some surface alteration that prevents the drug from binding to the cell, (2) by some membrane alteration that prevents the drug from entering the cell, or (3) by developing the ability to produce an enzyme that destroys the drug. Let's examine some examples of "superbugs."

Mycobacterium tuberculosis—the etiologic agent of tuberculosis (TB)—is the organism that has received the most publicity in the past few years. Tuberculosis remains one of the biggest killers worldwide, with about 8 million new cases of active TB every year and approximately 3 million deaths annually. In the United States, there are about 27,000 new cases and approximately 2,000 deaths each year. Outbreaks of multi–drug-resistant TB have been reported in 35 states. One particular strain, designated "strain W," is resistant to most of the antitubercular drugs such as isoniazid, rifampin, streptomycin, and ethambutol; some strains cannot be treated with any drug or combination of drugs. Patients infected with these strains may have to have a lung or section of lung removed, as in preantibiotic days, and many will die.

MRSA and MRSE: Another ever-growing problem concerns the multi–drug-resistant *Staphylococcus aureus* (MRSA) and *Staphylococcus epidermidis* (MRSE). Today, more than 90% of *S. aureus* strains are penicillin-resistant, and many strains (the MRSA) have developed resistance to methicillin, nafcillin, oxicillin, and cloxicillin, as well as other antimicrobial agents. *S. aureus* is the second most common cause of nosocomial infections, including skin and wound infections, bacteremia, and lower respiratory infections. Today, 40% of nosocomial *S. aureus* infections are due to MRSA. Coagulase-negative staphylococci, like *S. epidermidis,* are the most frequent cause of infections related to intravenous catheters and prosthetic devices and also cause urinary tract infections and endocarditis. Depending on the particular hospital, anywhere from 60% to 90% of coagulase-negative staphylococci are methicillin-resistant. Vancomycin, a very expensive and potentially toxic agent, must be used to treat infections with MRSA and MRSE. Scientists worry that the frequency with which vancomycin is used will ultimately lead to the emergence of vancomycin-resistant MRSA and MRSE. Should that occur, there will be no drugs left to treat the numerous types of infections caused by these staphylococci.

Other "superbugs" include **vancomycin-resistant *Enterococcus* spp., penicillin-resistant strains of *Neisseria gonorrhoeae, Haemophilus influenzae,* and *Streptococcus pneumonia.*** *Haemophilus influenzae* is one of the three major causes of bacterial meningitis, causes about one-third of the cases of ear infections, and is a cause of pneumonia. *S. pneumoniae* is the most common cause of bacterial pneumonia (about 500,000 cases in the United States per year), is one of the three major causes of bacterial meningitis (about 6,000 cases in the United States per year), causes about one-third of the cases of ear infection (about 6 million cases in the United States per year), and causes about 55,000 cases of bacteremia in the United States per year. *S. pneumoniae* causes about 40,000 deaths in the United States every year. [Other multi–drug-resistant organisms that cause especially severe and sometimes untreatable infections are *Pseudomonas aeruginosa* and *Pseudomonas cepacia; P. aeruginosa* is the fourth most common cause of nosocomial infections, and both of these organisms are frequently isolated from the lungs of cystic fibrosis patients. Additional "superbugs" include certain strains of *Escherichia coli, Salmonella,* and *Shigella.*]

named Gerhard Domagk discovered that the red dye, Prontosil®, was effective against streptococcal infections in mice. Further research demonstrated that Prontosil® was degraded or broken down in the body into sulfanilamide, and that sulfanilamide was the effective agent. For their outstanding contributions to scientific progress, these investigators, Ehrlich, Fleming, Florey, Chain, and Domagk, were all Nobel Prize recipients at various times.

Characteristics of Antimicrobial Agents

An *antimicrobial agent* is any chemical used to treat infectious disease by inhibiting or killing pathogens *in vivo* (in the living animal). An antimicrobial substance that is derived from a living organism and that in small amounts kills or inhibits the growth of microorganisms is called an *antibiotic;* these substances are produced by molds, bacteria, and some plants. The term antibiotic was intended to distinguish between chemical antimicrobial agents, such as sulfonamide drugs, and those that are extracted from secretions of living organisms, such as penicillin, streptomycin, and erythromycin; however, many of these drugs and their derivatives are now synthesized or manufactured in the laboratory.

The ideal antimicrobial agent should (1) kill or inhibit the growth of pathogens, (2) cause no damage to the host, (3) cause no allergic reaction in the host, (4) be stable when stored in solid or liquid form, (5) remain in specific tissues in the body long enough to be effective, and (6) kill the pathogens before they mutate and become resistant to it. However, almost all antimicrobial agents have some side effects, produce allergic reactions, or permit development of resistant mutant pathogens.

How Antimicrobial Agents Work

To be acceptable, an antimicrobial drug must inhibit or destroy the pathogen without damaging the host. The agent does this by disrupting the pathogen's metabolism in an area that is slightly different from normal human metabolism. The following examples illustrate this action.

The sulfonamide drugs inhibit production of folic acid in those bacteria that require para-aminobenzoic acid (usually abbreviated PABA) to synthesize folic acid. The folic acid (a vitamin) is essential to these bacteria. Because the sulfonamide molecule is similar in shape to the PABA molecule, bacteria attempt to metabolize sulfonamide to produce folic acid (Fig. 5 – 6). However, the enzymes that convert PABA to folic acid cannot produce folic acid from the sulfonamide molecule. Without folic acid, bacteria cannot produce some essential proteins and finally die. Sulfa drugs, therefore, are called competitive inhib-

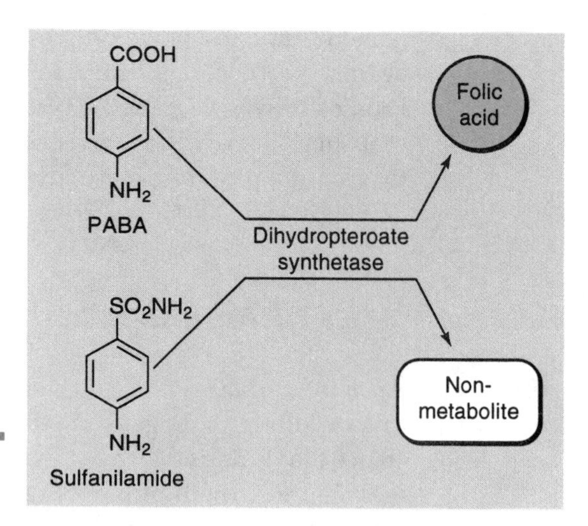

FIGURE 5 – 6
The effect of sulfonamide drugs.

itors, that is, by competing with an enzyme that metabolizes an essential nutrient, they inhibit growth of microorganisms. They are, therefore, bacteriostatic. Cells of humans and animals do not synthesize folic acid from PABA—they get it from the food they eat; consequently, they are unaffected by sulfa drugs.

In most gram-positive bacteria, including streptococci and staphylococci, penicillin interferes with the synthesis of the peptidoglycan that is required in bacterial cell walls. Thus, by inhibiting cell wall synthesis, penicillin destroys the bacteria. Why does penicillin not also destroy human cells? Human cells do not have a cell wall.

There are other antimicrobial agents that have similar action; they inhibit a specific step that is essential to the microorganism's metabolism and thereby cause its destruction. Antibiotics are used in this way against bacteria and they are highly effective. Some specifically destroy gram-positive bacteria; others specifically destroy gram-negative bacteria; those that are destructive to both gram-positive and gram-negative bacteria are called broad-spectrum antibiotics. Examples of broad-spectrum antibiotics are tetracycline, streptomycin, and ampicillin. Table 5 – 2 lists some of the antimicrobial drugs most frequently used against many common pathogens.

Antimicrobial drugs work well against bacterial pathogens because the bacteria (being procaryotic) have different cellular structures and metabolic pathways that can be disrupted by drugs that do not damage the host's (eucaryotic) cells. Antibacterial agents work in one of several ways.

1. To inhibit cell wall synthesis; *e.g.*, penicillins, vancomycin, and cephalosporins.
2. To act as competitive enzyme inhibitors to block the formation of essential metabolites; these "antimetabolites" include sulfa drugs, trimethoprim, and ethambutol.

TABLE 5 – 2.
Some Antimicrobial Agents Used Against Important Pathogens

Pathogen	Disease	Antimicrobial Agent or Treatment
Bacteria		
Bacillus anthracis	Anthrax	Penicillin, tetracyclines, erythromycin
Bordetella pertussis	Whooping cough	Erythromycin, ampicillin
Brucella abortus and *B. melitensis*	Brucellosis, undulant fever	Tetracyclines, streptomycin
Chlamydia trachomatis	Lymphogranuloma venereum	Sulfonamides, tetracyclines
Clostridium botulinum	Botulism (food poisoning)	Antitoxin, penicillin, kanamycin
Clostridium perfringens	Gas gangrene, wound infections	Antitoxin, penicillin, kanamycin
Clostridium tetani	Tetanus (lockjaw)	Antitoxin, penicillin, kanamycin
Corynebacterium diphtheriae	Diphtheria	Antitoxin, penicillin, erythromycin, cephalosporin
Escherichia coli	Urinary tract infections	Sulfonamides, gentamicin, cefazolin, nalidixic acid, norfloxacin, ampicillin
Francisella tularensis	Tularemia	Streptomycin, tetracyclines
Haemophilus ducreyi	Chancroid	Streptomycin, tetracyclines
Haemophilus influenzae	Meningitis, pneumonia	Ampicillin, streptomycin, tetracyclines, cefamandole
Klebsiella pneumoniae	Pneumonia	Colistin, cefazolin, gentamicin
Legionella pneumophilia	Legionellosis	Erythromycin, tetracycline
Mycobacterium leprae	Leprosy	Dapsone, tetracyclines, rifamide
Mycobacterium tuberculosis	Tuberculosis	Isoniazid, streptomycin, PAS, rifampin, ethambutol
Mycoplasma pneumoniae	Atypical pneumonia	Tetracyclines, erythromycin
Neisseria gonorrhoeae	Gonorrhea	Penicillin, tetracyclines, spectinomycin
Neisseria meningitidis	Nasopharyngitis, meningitis	Penicillin, sulfonamides, ampicillin, tetracyclines, rifamide
Proteus vulgaris and *P. morgani*	Gastroenteritis, urinary tract infections	Kanamycin, streptomycin, nalidixic acid
Pseudomonas aeruginosa	Respiratory and urogenital infections	Gentamicin, sulfonamides, mezlocillin, polymyxin
Rickettsia rickettsii	Rocky Mountain spotted fever	Tetracyclines, chloramphenicol

(continued)

TABLE 5 – 2. (Continued)

Pathogen	Disease	Chemotherapy or Treatment
Salmonella typhi	Typhoid fever	Chloramphenicol, ampicillin, tetacyclines
Salmonella spp.	Gastroenteritis	Chloramphenicol, ampicillin, tetracyclines, co-trimoxazole
Shigella spp.	Shigellosis (bacillary dysentery)	Ampicillin, tetracyclines, sulfonamides, nalidixic acid
Staphylococcus aureus	Boils, carbuncles, pneumonia, septicemia	Bacitracin, erythromycin, gentamicin, vancomycin, cephalosporins
Streptococcus pyogenes	"Strep" throat, scarlet fever, rheumatic fever, septicemia	Penicillin, cephalosporin, erythromycin
Streptococcus pneumoniae	Pneumonia	Penicillin, cephalosporin, erythromycin
Treponemia pallidum	Syphilis	Penicillin, erythromycin, tetracyclines
Vibrio cholerae	Cholera	Trimethoprim plus sulfamethoxazole
Yersinia pestis	Plague	Streptomycin, tetracyclines
Fungi		
Dermatophytes	Tinea ("ringworm") infections	Nystatin, amphotericin B, griseofulvin
Candida	Mucosal candidiasis	Clotrimazole, miconazole, nystatin
Blastomyces, *Histoplasma*, *Cryptococcus*, *Coccidioides*	Systemic mycosis	5-Flucytosine, amphotericin B
Candida		Amphotericin B
Protozoa		
Trichomonas	Trichomoniasis	Metronidazole
Giardia lamblia	Giardiasis	Metronidazole, quinacrine
Entamoeba histolytica	Amebiasis	Emetine, metronidazole
Toxoplasma	Toxoplasmosis	Clindamycin
Plasmodium spp.	Malaria	Clindamycin, chloroquine, pyrimethamine, sulfadiazine
Pneumocystis carinii	Pneumonia	Pentamidine, co-trimoxazole

(*continued*)

TABLE 5 – 2. (Continued)

Pathogen	Disease	Chemotherapy or Treatment
Viruses		
Herpes spp.	Eye and lip herpes	Idoxuridine, acyclovir
	Genital herpes	Acyclovir
	Encephalitis	Vidarabine, acyclovir
Influenza A	Influenza	Amantidine, rimantadine
Human immunodeficiency virus (HIV)	AIDS	Azidothymidine (AZT), dideoxycytidine, ribavirin, suramin

3. To inhibit protein synthesis by acting on 70S ribosomes; *e.g.*, tetracyclines, erythromycin, streptomycin, neomycin, and chloramphenicol.
4. To damage plasma membranes, such as polymyxin B.
5. To inhibit nucleic acid synthesis; *e.g.*, rifamycin, nalidixic acid, and norfloxacin.

Frequently, a single antimicrobial agent is not sufficient to destroy all the pathogens that develop during the course of a disease; thus, two or more drugs may be used simultaneously to kill all of the pathogens and to prevent resistant mutant pathogens from emerging. In tuberculosis, for example, three drugs are routinely prescribed, and as many as 12 drugs may be required in resistant cases. Many urinary, respiratory, and gastrointestinal infections respond particularly well to a combination of trimethoprim and sulfamethoxazole; this combination is called co-trimoxazole.

It is much more difficult to use antimicrobial drugs against fungal and protozoal pathogens because they are eucaryotic cells; thus, the drugs are much more toxic to the host. Antifungal agents work by (1) binding with cell membrane sterols; *e.g.*, nystatin and amphotericin B; (2) by interfering with sterol synthesis; *e.g.*, clotrimazole and miconazole; and (3) by blocking mitosis or nucleic acid synthesis; *e.g.*, griseofulvin and 5-flucytosine. Antiprotozoal drugs are usually quite toxic and work (1) by interfering with DNA and RNA synthesis; *e.g.*, chloroquine, pentamidine, and quinacrine; or (2) by interfering with protozoal metabolism; *e.g.*, metronidazole.

Antiviral chemotherapeutic agents are particularly difficult to find and use because viruses are produced within the host's cells. A few have been found to be effective in certain conditions, and these work by interfering with the action of certain enzymes necessary for viral replication. These include amantadine and rimantadine for influenza, acyclovir for herpes, and azidothymidine (AZT) for acquired immunodeficiency syndrome (AIDS). Some may interfere with DNA replication and RNA transcription; the action of others is unknown.

In cancer, in which malignant cells are dividing more rapidly than normal cells, chemical agents that inhibit DNA and RNA synthesis can be used, provided the dosage and total period of administration are carefully controlled. Cancer drugs interfere with normal DNA function in rapidly dividing cells, regardless of whether the cells are normal or malignant. Thus, normal cells that are rapidly dividing, including skin cells, erythroblasts that later become red blood cells, and sperm cells, are damaged along with the malignant cells. This is why blood cell counts are done frequently in cancer patients; the physician must be able to determine at what point the chemotherapy must be discontinued to avoid critical damage to the patient's normal cells.

Side Effects of Antimicrobial Agents

There are many reasons why antimicrobial agents should not be used indiscriminately.

1. The microorganisms may mutate and become resistant to the agent. Their metabolism may change and they may produce an enzyme that can destroy the agent or produce one that uses it as a nutrient; they may also become impermeable to the drug. To prevent these developments, several drugs, each of which has a different mode of action, often are administered simultaneously. If not properly treated, the drug-resistant pathogens may continue to flourish, causing an overgrowth or *superinfection* with these organisms.
2. The patient may become allergic to the agent. For example, penicillin G in low doses often sensitizes those who are prone to allergies; when these persons receive a second dose of penicillin at some later date, they may have a severe reaction known as anaphylactic shock, or they may break out in hives. This reaction is described in more detail in Chapter 9.
3. Many antimicrobial agents are toxic to humans, and some are so toxic they are administered only for serious diseases for which no other agents are available. One such drug is chloramphenicol (Chloromycetin®), which, if given in high doses for a long period, may cause a very severe type of anemia called aplastic anemia. Another is streptomycin, which can damage the auditory nerve and cause deafness.
4. With prolonged use, broad-spectrum antibiotics may destroy the normal flora of the mouth, intestine, or vagina. The person no longer has the protection of the indigenous microflora and, thus, becomes much more susceptible to infections caused by opportunists or secondary invaders. The result is a superinfection. An example of such an infection is diarrhea, which can result from prolonged antibiotic therapy owing to the loss of the normal protective microbes. A vaginal yeast infection often follows

antibacterial therapy because many bacteria of the vaginal flora were destroyed, allowing the indigenous yeast to overgrow. This topic is discussed more fully in Chapter 6.

Therefore, antimicrobial agents, including antibiotics, should be taken only when prescribed and only under a physician's supervision. Also, the proper dosage must be administered for the recommended period to prevent resistant organisms from gaining a foothold.

In recent years, microorganisms have developed resistance at such a rapid pace that many people, including some scientists, are beginning to fear that pathogens are becoming so resistant to antibiotics that the human race may be destroyed. Some problems already have arisen—a case in point is pneumonia. In the more-developed countries of the world, the types of pneumonia currently occurring are different from those that were seen 30 years ago and sometimes are highly resistant to treatment. It is hoped that antibiotics will be used with greater care and that effective vaccines may become available. *Haemophilus influenzae* and *Streptococcus pneumoniae* vaccines are available to protect against some types of pneumonia, but not all. Also, resistant strains of gonococci, penicillinase-producing *Neisseria gonorrhoeae* (PPNG), have developed. Scientists hope to someday see vaccines against gonorrhea, syphilis, and AIDS. The difficulty then will be to convince the U.S. population to take the vaccines to protect themselves against these pathogens.

SUMMARY

It is important to learn how the growth of microorganisms can be inhibited to prevent and control infectious diseases, prevent contamination of industrial processes, prevent spoilage of food and crops, and prevent contamination of pure culture research. People working in the health field must take special care not to transfer potentially pathogenic microbes from patient to patient, from themselves to patients, or from patients to themselves. Health-care workers need to know the physical and chemical methods used to control pathogenic microorganisms and the conditions for the best use of these methods.

PROBLEMS AND QUESTIONS

1. Discuss why it is necessary to control microbial growth.
2. Define sterilization, disinfection, pasteurization, and lyophilization.
3. What are the differences among sterile, aseptic, and antiseptic techniques? In what circumstances might each be used?
4. What are the characteristics of bacteria indicated by the following terms: psychrophilic, mesophilic, thermophilic, halophilic, haloduric, alkalinophilic, acidophilic?
5. List six characteristics of a good antimicrobial agent.
6. List some effective physical and chemical means of controlling microbial growth.
7. How do antimicrobial agents destroy microbes without also harming the patient?
8. Discuss why antibiotics should be used discriminately.

Self Test

After you have read Chapter 5, examined the objectives, reviewed the chapter outline, studied the new terms, and answered the problems and questions above, complete the following self test.

Matching Exercises

Complete each statement from the list of words provided with each section.

TERMS

disinfection	fungistatic agent	antiseptic technique
sterilization	sepsis	sterile technique
pasteurization	asepsis	aseptic technique
fungicidal agent		

1. A chemical that kills fungi is a ________________.
2. A chemical that inhibits growth of the fungus that causes athlete's foot is a ________________.
3. The growth of infectious microorganisms in living tissues is ________.
4. The surgical technique of using disinfectants and antiseptics to cleanse skin, instruments, and so on is ________________.
5. The lack of infectious microorganisms on living tissues is ________________.
6. The surgical technique that eliminates and avoids pathogens is ___________.
7. The process of opening a sterile packet without exposing the sterile equipment to any microorganisms is ________________.

8. A process during which all microorganisms are killed is called ________.
9. The process of destroying the harmful microorganisms that are on or in nonliving objects is called ________.
10. Heating milk to destroy pathogens is called ____________.
11. To spray the base of the shower stall to kill the fungi growing there, one could use Lysol®, a ____________.
12. When you apply iodine or Merthiolate® to a cut or abrasion, you are using ____________ technique.
13. The microbes that spoil beer and wine can be destroyed by ________.
14. Washing tabletops with an antimicrobial agent is an example of ____________.

MICROBIAL TYPES

thermophiles
mesophiles
psychrophiles
halophiles
aerobes
acidophiles
alkalinophiles
barophiles
obligate anaerobes
facultative anaerobes

1. Organisms that can thrive deep within the ocean, in a high barometric pressure, are ____________.
2. Organisms that can live in the digestive tract or in the air are ________.
3. Microbes that die in the presence of oxygen are ____________.
4. Microbes that thrive best in the air are considered ____________.
5. Organisms that are not inhibited by chlorine or iodine disinfectants and tolerate a high salt concentration are called ____________.
6. Microbes that can live in the acid environment of the stomach are ______.
7. Microbes that prefer the alkaline environment of the intestine are ______.
8. Microbes found living in an iceberg are ____________.
9. Organisms that can live in hot springs are ____________.
10. Most pathogens are ____________ because they grow best at body temperature.
11. When you gargle with salt water most microbes are inhibited except the staphylococci, which are ____________.
12. *Escherichia coli* and other enteric bacteria would be considered ________.
13. Pathogens of the genus *Clostridium* are all spore-formers and grow best in a closed wound or jar where there is no oxygen; they are therefore called ____________.
14. The pathogen that causes botulism is one of the ____________.

PHYSICAL ANTIMICROBIAL METHODS

ultraviolet rays
x-rays
filtration
desiccation
autoclaving
sonic waves
osmotic pressure

1. Microorganisms of various sizes may be removed from a solution by a process called ________, so that the remaining solution may be sterile.
2. The use of concentrated salt and sugar solutions inhibits the growth of bacteria by changing the ________.
3. An excellent method of cleaning and sterilizing delicate instruments is the use of ________.
4. The type of radiation used to keep certain areas sterile, such as cabinets containing sterile instruments and equipment, is ________.
5. When pressurized steam is used to kill microorganisms and spores, the process is called ________.
6. When all moisture is removed, microbes are inhibited by ________.
7. The rays of the sun that cause suntans and may cause skin cancer are called ________.

CHEMICAL ANTIMICROBIAL METHODS

ethyl alcohol
isopropyl alcohol
detergent or soap
phenolics
hydrogen peroxide
formalin
ethylene oxide
mercury salts

1. A type of alcohol that is toxic when taken internally, but is good for rubbing on the skin is ________.
2. The type of alcohol that is the best antiseptic is ________.
3. An oxidizing agent that is frequently used to cleanse wounds and remove pus is ________.
4. A solution that is sporicidal but is too caustic to use on living tissues is ________.
5. Lysol® and carbolic acid are ________.
6. Merthiolate® is one of the ________.
7. Before surgery, the doctor scrubs well with a ________ to destroy the surface bacteria on the skin.
8. A toxic, flammable gas that is sporicidal and is used in gas autoclaves is ________.

CHEMOTHERAPY

antibiotic
penicillin
sulfanilamide
Salvarsan®
broad-spectrum
Prontosil®
amphotericin B
antibiotics

1. The antibiotic that was discovered by Fleming, purified by Florey and Chain, and isolated from a mold is ________.
2. Antibiotics that are effective against many gram-positive and gram-negative pathogens are known as ________.
3. The drug that worked against *in vivo* streptococcal infections but was not effective *in vitro* was ________.
4. A drug that is effective against fungal infections is ________.
5. When Prontosil® was broken apart, one half of the molecule was effective against gram-positive infections *in vitro*. This drug is ________.

6. A chemotherapeutic drug that is derived from a living organism (fungi, bacteria) is called an ________________.

7. The antimicrobial agent that Ehrlich discovered after 605 unsuccessful attempts was ________________.

8. The first antimicrobial agent found to be effective against syphilis was ________________.

9. The antimicrobial agent used against syphilis today is ____________.

True or False (T or F)

___ **1.** Sulfonamide drugs inhibit production of the essential vitamin, folic acid, in all fungi.

___ **2.** Penicillin interferes with the synthesis of nuclei in bacteria.

___ **3.** Antimicrobial agents work by inhibiting a specific essential step in the metabolism of a pathogen, which is different from the metabolism of a human cell.

___ **4.** Drugs that destroy viral infections generally also destroy host cells.

___ **5.** Drugs that destroy cancer cells affect all rapidly dividing cells.

___ **6.** Many pathogens can mutate to become resistant to antimicrobial agents by changing their metabolic pathways.

___ **7.** The concentration of a disinfectant is not important; it will be effective at any concentration.

___ **8.** The mycobacteria that cause leprosy and tuberculosis are among the most resistant pathogens.

___ **9.** A rough, porous surface is more easily disinfected than a smooth, hard surface.

___ **10.** Formaldehyde and ethylene oxide are not effective against spores.

___ **11.** Pasteurization kills all the bacteria present in milk.

___ **12.** All bacteria must be destroyed because they all cause disease.

___ **13.** An infectious disease is any disease caused by the growth of microorganisms.

___ **14.** An antiseptic is a mild disinfectant used on the skin.

___ **15.** The antiseptic technique was developed by Lister in the late 1800s.

Multiple Choice

1. A bactericide would be effective against
- **a.** viruses
- **b.** endospores
- **c.** bacteria
- **d.** all of the above

2. Pasteurization is a good example of
- **a.** sterilization
- **b.** disinfection
- **c.** tyndallization
- **d.** antisepsis

3. A combination of freezing and drying of microbes is called
   a. desiccation
   b. disinfection
   c. lyophilization
   d. hemolysis
4. The organisms found in the Great Salt Lake are
   a. thermophilic
   b. barophilic
   c. mesoduric
   d. halophilic
5. When placed in a hypertonic solution the osmotic pressure of a cell
   a. increases
   b. stays the same
   c. diffuses
   d. decreases
6. In a hospital setting, an effective disinfectant must be effective against
   a. fungal spores
   b. *Mycobacterium tuberculosis*
   c. *Pseudomonas* species
   d. hepatitis viruses
   e. all of the above
7. *Clostridium perfringens* is a causative agent of
   a. tetanus
   b. leprosy
   c. gas gangrene
   d. ringworm
8. The most effective method of sterilization is
   a. moist heat combined with pressure
   b. chemical
   c. boiling
   d. dry heat combined with pressure
9. Sulfonamide drugs are considered
   a. bactericidal
   b. sporicidal
   c. bacteriostatic
   d. virucidal
10. Modern chemotherapy began with the development of
   a. penicillin
   b. Prontosil®
   c. Salvarsan®
   d. sulfanilamide
11. Thermal death time is the
   a. time required to kill all cells at a given temperature
   b. temperature that kills all cells in a given time
   c. time and temperature needed to kill all cells
   d. all of the above
   e. none of the above
12. Asepsis refers to
   a. no organisms are present
   b. procedures that reduce spread of microorganisms
   c. a barrier to infection is maintained
   d. sterility of materials is maintained
   e. all of the above
13. The usual autoclaving temperature is
   a. 100° C
   b. 62.8° C
   c. 50° C
   d. 15° C
   e. 121° C

CHAPTER

6

Human/Microbe Interactions

OBJECTIVES

After studying this chapter, you should be able to

- Discuss the importance of indigenous microflora and where it is found
- List four types of symbiotic relationships
- Differentiate between mutualism and commensalism and give examples of each
- Describe one parasitic relationship
- Discuss factors related to the pathogenicity of microbes
- Describe ecological interrelationships of plants, animals, and microorganisms

CHAPTER OUTLINE

NEW TERMS

Antagonism
Antibiosis
Asymptomatic infection
Candidiasis
Carrier
Commensalism
Ecology
Ectoparasite
Endoparasite
Endosymbiont
Host
Infestation
Mutualism
Neutralism
Parasitism
Resident microflora
Symbiont
Symbiosis
Synergism
Transient microflora

INDIGENOUS MICROFLORA

The indigenous microflora or indigenous microbiota (referred to in the past as "normal flora") of a person includes all the microbes usually found on or within the human body. These microorganisms include bacteria, fungi, protozoa, and viruses. A fetus has no indigenous microflora. During and after delivery, a newborn is exposed to many microorganisms from its mother, food, air, and everything that touches the infant. Both harmless and helpful microbes take up residence on the skin, at all body openings, and in mucous membranes that line the digestive tract (mouth to anus) and the urogenital tract. In these areas, the moist, warm environment provides excellent conditions for growth. Conditions for proper growth (moisture, pH, temperature, oxygen supply, nutrients) vary throughout the body; thus, the types of resident flora differ from one anatomical site to another. Blood, lymph, spinal fluid, and most internal organs should be free of microorganisms. See Table 6 – 1 for a list of the microorganisms frequently found on and within the human body.

Relatively few types of microbes establish themselves as indigenous microflora because most organisms in our external environment do not find the body to be a suitable *host.* In addition to the *resident microflora, transient microflora* take up temporary residence in humans. The body is constantly exposed to the flow of microorganisms from the external environment, and these transient mi-

TABLE 6 – 1.
Locations of Microorgainisms Usually Found on and in Humans

Organism	Skin	Eye	Ear	Mouth	Nose	Respiratory Tract	Intestine	Urogenital Tract
Bacteria								
Bacillus spp.	+	−	−	+	−	−	+	−
Bacteriodes	+	−	−	+	+	+	+	+
Borrelia spp.	−	−	−	+	−	−	−	+
Clostridium spp.	+	−	−	−	−	−	+	+
Coliforms	+	−	−	−	−	+	+	+
Escherichia coli	−	−	−	−	−	−	+	+
Corynebacterium spp.	+	+	+	+	+	+	+	+
Fusobacterium spp.	−	−	−	+	−	−	+	−
Haemophilus influenzae	−	+	+	−	+	+	−	−
Klebsiella pneumoniae	−	−	−	−	+	−	+	−
Lactobacillus spp.	+	−	+	+	−	−	+	+
Leptotrichia	−	−	−	+	−	−	−	+
Micrococcus spp.	+	−	−	+	−	−	+	−
Mycobacterium spp.	−	−	+	−	−	−	−	+
Mycoplasmas	−	−	−	+	+	+	+	+
Neisseria spp.	−	+	−	+	+	+	−	+
Proteus spp.	−	−	−	−	−	−	+	+
Pseudomonas aeruginosa	−	−	+	−	−	−	+	−
Staphylococci	+	+	+	+	+	+	+	+
S. aureus	+	−	+	+	+	−	−	−
S. epidermidis	+	+	+	+	+	−	−	−
Streptococci	+	+	+	+	+	+	+	+
S. mitis	+	+	−	+	+	−	−	−
S. pneumoniae	−	+	+	+	+	−	−	−
S. pyogenes	+	+	+	−	−	−	−	−
Veillonella spp.	−	−	−	+	−	−	+	−
Fungi								
Actinomyces spp.	−	−	−	+	−	−	−	−
Candida albicans	+	+	−	+	−	−	+	+
Cryptococcus spp.	+	−	−	−	−	−	−	−
Protozoa	−	−	−	+	−	−	+	+
Viruses	+	−	−	+	+	+	+	−

+, present; −, absent

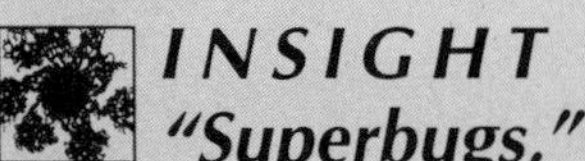

INSIGHT
"Superbugs," Part 2: What Can Be Done About Them?

"Superbugs" or multi–drug-resistant microorganisms (defined in Chapter 5) are an enormous public health problem. What can be done about them?

In the past, we merely relied on drug companies to come up with "bigger and better" drugs. However, due to a variety of factors, there are relatively few drugs ready for introduction today. If we can no longer rely on drug companies to solve the problems, what *can* be done? Education is crucial—education of health-care professionals and, in turn, education of patients. The following are some of the ways in which we can control "superbugs."

1. Patients must stop demanding antibiotics every time they are sick or have a sick child. Most sore throats and many respiratory infections are caused by viruses, and viruses are unaffected by antibiotics. Consequently, antibiotics should not be prescribed for viral infections. According to the *Journal of the American Medical Society (JAMA)*, instead of demanding antibiotics from the physician, consumers/patients should be asking why one *is* being prescribed.
2. Physicians must not let themselves be pressured by patients. They should prescribe antibiotics only when they are warranted—only when there is a demonstrated need for them. According to *JAMA*, by some estimates, at least half of current antibiotic use in the United States is inappropriate—antibiotics are either not indicated at all or they are incorrectly prescribed as the wrong drug, the wrong dosage, or the wrong duration.
3. Patients must take their antibiotics in the exact manner in which they are prescribed. Health professionals should emphasize this to patients and do a better job explaining exactly how medications are to be taken.
4. Patients must take *all* their pills—even after they are feeling better. As stated in *JAMA*, if treatment is cut short, there is selective killing of only the most susceptible members of a bacterial population. The more resistant variants are left behind to regrow into a new infection having a greater number of more resistant members. Shorter treatment schedules might lead to better patient compliance.
5. Never keep antibiotics in your medicine cabinet and never give them to anyone else. Unless prescribed by a physician, never use antibiotics in a prophylactic manner—such as to avoid "traveler's diarrhea" when traveling to a foreign country.
6. As always, practice good infection prevention and control procedures. Frequent and proper hand-washing is essential to prevent the transmission of pathogens from one patient to another.
7. Vaccines also help. They cut down on the number of infections with drug-resistant organisms such as ampicillin-resistant strains of *Haemophilus influenzae* and penicillin-resistant strains of *Streptococcus pneumoniae.*

crobes frequently are attracted to the moist body areas. These microbes are only temporary for many reasons: they may be washed from external areas by bathing; they may not be able to compete with the resident microflora; they may fail to survive in the acid or alkaline environment of the site; or they may be flushed out in excretions or secretions, such as urine, feces, tears, and perspiration.

Destruction of the resident microflora disturbs the delicate balance established between the host and its microorganisms. Prolonged therapy with certain antibiotics often destroys the intestinal microflora; diarrhea is usually the result of such an imbalance, which, in turn, leaves the body more susceptible to secondary invaders. When the number of normal resident microbes is greatly reduced, opportunistic invaders may establish themselves within those areas. One important opportunist usually found in small numbers near body openings is the yeast, *Candida albicans,* which, in the absence of sufficient numbers of other resident microflora, may grow unchecked in the mouth, vagina, or lower intestine, causing the disease *candidiasis* (also known as moniliasis; see Fig. 2 – 19).

Microflora of the Skin

The resident microflora of the skin consists primarily of bacteria and fungi. The number and species of microorganisms present depend on many factors. Moist, warm conditions in hairy areas where there are many sweat and oil glands, such as under the arms and in the groin, stimulate the growth of staphylococci, streptococci, diphtheroids, aerobic spore-forming bacilli, nonpathogenic mycobacteria, gram-negative enteric bacilli (*e.g., Escherichia coli*), as well as the fungi, *Candida albicans* and *Cryptococcus.* Dry, calloused areas of skin have few bacteria, whereas moist folds between the toes and fingers support many bacteria and fungi. The surface of the skin near mucosal openings of the body (the mouth, eyes, nose, anus, and genitalia) is inhabited by bacteria present in their secretions.

Frequent washing with soap and water removes most of the potentially harmful transient microorganisms harbored in sweat, oil, and other secretions from moist body parts. All persons involved in patient care must be particularly careful to keep their skin and clothing as free of transient microbes as possible, to help prevent personal infections, and to avoid transferring pathogens to patients. Such persons should always remember that most infections following burns, wounds, and surgery result from the growth of resident or transient skin microflora in these susceptible areas.

Microflora of the Mouth

The mouth and throat have an abundant and varied population of microorganisms. These areas provide moist, warm mucous membranes that furnish excellent conditions for microbial growth. Bacteria thrive especially well in particles of food and in the debris of dead epithelial cells around the teeth. The peculiar anatomy of the oral cavity and throat affords shelter for numerous anaerobic and aerobic bacteria. Anaerobic microorganisms flourish in gum margins, in cervices between the teeth, and in deep folds (crypts) on the surface of the tonsils.

The list of microbes that have been isolated from healthy human mouths reads like a manual of the main groups of microorganisms. It includes cocci, bacilli, and spiral-shaped bacteria, as well as yeasts, mold-like organisms, protozoa, and viruses. The first such list was made by Leeuwenhoek in 1690 (see Chapter 1).

Most microorganisms found in the healthy mouth and throat are beneficial (or at least harmless); these include diphtheroids, lactobacilli, and micrococci. Others, such as certain streptococci and staphylococci, are potentially pathogenic opportunists and are frequently associated with disease. Some people carry virulent pathogens in their nasal passages or throats but do not have the diseases associated with them, such as diphtheria, meningitis, pneumonia, and tuberculosis. These people are healthy *carriers* who are resistant to these pathogens but can transmit them to susceptible persons.

Food remaining on and between teeth provides a rich nutrient medium for growth of the many oral bacteria. Carelessness in dental hygiene allows growth of these bacteria, with development of dental caries (tooth decay), gingivitis (gum disease), and periodontitis. These bacteria include species of *Actinomyces, Lactobacillus, Streptococcus, Neisseria,* and *Veillonella.*

Many alpha-hemolytic (α-hemolytic) streptococci are indigenous inhabitants of the mouth and oropharynx. When large numbers of group A, beta-hemolytic (β-hemolytic) streptococci are present, however, the person should be treated with an antibiotic to destroy these pathogens that may cause "strep" throat and its complications (like scarlet fever, rheumatic fever, and glomerulonephritis).

Microflora of the Ears and Eyes

The middle and inner ear are normally sterile, whereas the outer ear and the auditory canal contain the same types of microorganisms as are found on moist areas, such as the mouth and nose. When a person coughs, sneezes, or blows his or her nose, these microbes are carried along the eustachian tube and into the middle ear where they can cause infection. Infection can also develop in the middle ear when the eustachian tube does not open and close properly to maintain correct air pressure within the ear.

Many microorganisms are found in the external opening of the eye, in the conjunctiva that lines the eyelid, and in tears. But these microbes are not a frequent cause of disease because the intact membranes serve as a barrier. These mucous membranes are constantly flushed by tears, which contain an enzyme called lysozyme that destroys bacteria. The indigenous microflora of the eye area includes species of *Staphylococcus, Streptococcus,* and *Corynebacterium,* as well as *Moraxella catarrhalis.*

Microflora of the Respiratory Tract

The respiratory tract consists of the nose, pharynx (throat), larynx (voice box), trachea, bronchi, bronchioles, and alveoli. The lower respiratory tract, below the larynx, is usually free of microbes because the mucous membranes and lungs have defense mechanisms that efficiently remove invaders. Thus, staphylococci, streptococci, *Pseudomonas* species, or yeasts found in sputum specimens would indicate either an infectious disease of the lungs or specimen contamination by indigenous microflora of the upper respiratory tract.

The membranes of the upper part of the tract, including the nasopharynx and the oropharynx, provide a suitable environment for the growth of many species of *Streptococcus, Staphylococcus, Neisseria, Corynebacterium,* yeasts, and other microorganisms. In susceptible persons, many of these opportunists can cause disease.

Microflora of the Urogenital Area

The healthy kidney, ureters, and bladder are sterile. However, the external opening of the urethra houses many microorganisms, such as nonpathogenic *Neisseria* species, staphylococci, enterococci, diphtheroids, mycobacteria, mycoplasmas, enteric (intestinal) gram-negative rods, yeasts, and viruses. As a rule, these organisms do not invade the bladder because the urethra is periodically flushed by acidic urine; however, persistent, recurring urinary infections can develop following obstruction or narrowing of the urethra and with infrequent urination, which allows the invasive organisms to multiply and cause urinary tract infections. Chlamydias and mycoplasmas are frequent causes of nongonococcal urethritis (NGU). All of these organisms are easily introduced into the urethra by sexual intercourse; however, they can be flushed out by urination following sexual activities.

The reproductive systems of both men and women are usually sterile, with the exception of the vagina; here the flora varies with the stage of sexual development. During puberty and following menopause, vaginal secretions are alkaline, supporting the growth of various diphtheroids, streptococci, staphylococci, and coliforms (enteric gram-negative rods); through the childbearing years, vaginal secretions are acidic, encouraging the growth mainly of lactobacilli, along with a few α-hemolytic streptococci, staphylococci, diphtheroids, and yeasts.

The *Neisseria* species, particularly *Neisseria gonorrhoeae* (also called gonococci), can survive the acidic environment of the vagina and the penis; hence, they may be harbored by infected persons who show no symptoms of gonorrhea. This disease, which is readily transmitted through sexual contact, is *asymptomatic* (causing no symptoms) in 80% of infected women and 20% of infected men.

Microflora of the Gastrointestinal Tract

The acidic environment of the stomach prevents growth of indigenous microflora. However, a few microbes, protected by foods, manage to pass through the stomach during periods of low acid concentration. Also, when the amount of acid is reduced in the course of diseases such as stomach cancer, certain bacteria may be found in this site.

Usually, in the upper portion of the small intestine (the duodenum) few microorganisms exist because bile inhibits their growth, but many are found in the lower part of the small intestine. The most abundant organisms include many species of *Staphylococcus, Lactobacillus, Streptobacillus, Veillonella,* and *Clostridium perfringens.*

The colon, or large intestine, contains large numbers of microorganisms growing on the food wastes collected there. Obligate anaerobes make up most of the colon population, including gram-positive *Clostridium* species and gram-negative *Bacteroides* and *Fusobacterium* species. Usually, less than 10% of the large intestine microflora are facultative anaerobes that are easily grown in the laboratory, including *Escherichia coli, Enterobacter aerogenes,* and species of *Proteus, Pseudomonas, Enterococcus, Lactobacillus,* and *Mycoplasma.* Also, many fungi, protozoa, and viruses are found in the intestine. Usually these microflora of the colon are opportunists, causing disease only when they lodge in the other areas of the body or when the balance among the microorganisms is upset.

BENEFICIAL ROLES OF INDIGENOUS MICROFLORA

Many benefits are derived by humans from the symbiotic relationship established with their indigenous microflora. Some nutrients, particularly vitamins K, B_{12}, pantothenic acid, pyridoxine, and biotin, are obtained from secretions of the coliform bacteria.

Evidence also indicates that these indigenous microbes provide a constant source of irritants and antigens to stimulate the immune system. This causes the immune system to respond more readily by producing antibodies to foreign invaders and substances, which in turn enhances the body's protection against disease-producing agents. It appears that by merely occupying a place and using the nutrients present, these resident microflora prevent other microorganisms that may be pathogenic from "gaining a foothold" and establishing a site of infection. This protection is maintained by competition for food, controlled pH and oxygen levels, and antibiotic production by certain resident microbes.

When the delicate balance of the various species in the population of indigenous microflora is upset by antibiotics or other chemotherapy, many complications may result. Certain microorganisms may flourish out of control, such as the yeast that is the cause of candidiasis. Also, diarrhea and pseudomembranous colitis may occur. Cultures of *Lactobacillus* species, which are present in yogurt and some medications, may be prescribed to reestablish and stabilize the microbial balance.

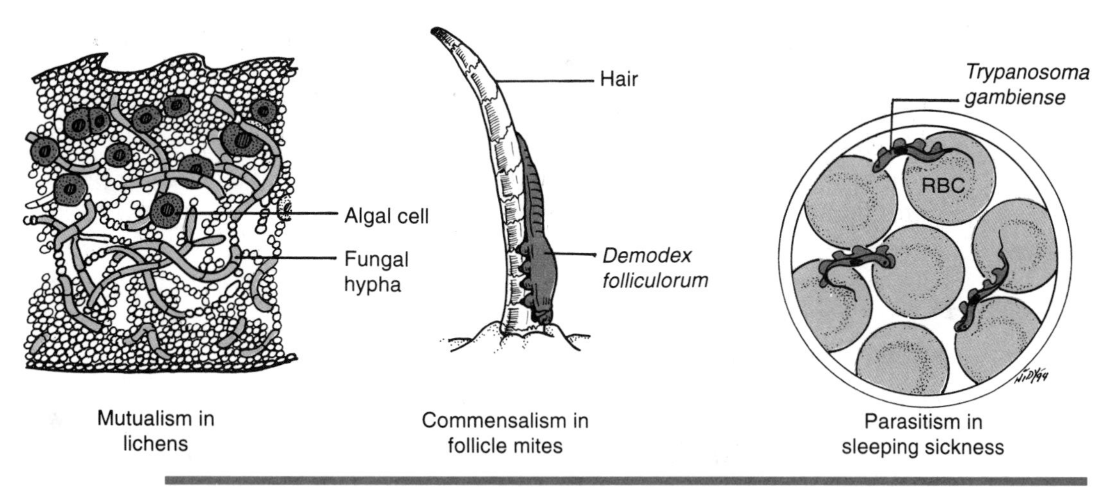

FIGURE 6 – 1
Various symbiotic relationships

Symbiotic Relationships

The relationships between the indigenous microflora and the human host are excellent examples of *symbiosis,* a term given to the general relationship between dissimilar organisms living together in close proximity. The *symbionts* (the organisms that live together) are two or more organisms of unlike species. The relationship may be beneficial, harmless, or harmful to one, several, or all of the symbionts. The types of relationships outlined in the following sections demonstrate that the pathogenicity of a microbe can be represented as a balance between the virulence of the pathogen and the resistance of the host. The dynamic symbiotic balance may shift toward the parasitic / pathogenic disease state if host defenses are reduced with an accompanying rise in host susceptibility. Recovery from disease occurs with a shift toward *mutualism* and *commensalism.* These factors are further discussed in later chapters. Various symbiotic relationships are illustrated in Figure 6 – 1.

Mutualism In the symbiotic relationship called mutualism, both organisms benefit and, in fact, depend on each other metabolically. An example is the intestinal bacterium *Escherichia coli,* which obtains nutrients from food materials ingested by the host and produces vitamin K to be used by the host. Vitamin K is a blood-clotting factor that is essential to humans. Also, some protozoa live symbiotically in the intestine of termites, enabling them to digest the wood they eat by breaking cellulose down into nutrients to be absorbed and used by the termites. In turn, the termite provides food and a place for the protozoa to live. Without these protozoa, the termites would die of starvation. Lichens that you see on rocks and tree trunks are further examples of mutualism. A lichen is composed of an alga and a fungus. The fungus uses some of the energy that the alga produces by photosynthesis, and the chitin in the fungal cell walls protects the alga from desiccation; thus, both symbionts benefit from the relationship.

In some mutualistic relationships, two organisms work together to produce a result that neither could accomplish alone. This is called *synergism,* or a synergistic relationship. Fusobacteria and spirochetes, which together cause the disease "trench mouth", represent such a relationship. Also, nitrogen-fixing bacteria and the roots of legumes where they exist have a true synergistic relationship because each depends on the other for nutrients.

Commensalism A relationship in which one symbiont is benefited but the other is neither benefited nor harmed is called commensalism. Most of the indigenous microflora of humans are considered to be commensals, in that the microbes are provided nutrients and "housing" with no effect on the welfare of the host. Indifference, or *neutralism,* exists when organisms occupy the same niche but do not affect each other. However, sometimes waste products of one microorganism can destroy certain neighboring bacteria. This situation is called *antagonism* or *antibiosis.* For example, *Penicillium* mold growing on a culture plate of certain strains of staphylococci can inhibit growth of the staphylococci by producing the antibiotic, penicillin (see Fig. 5 – 5).

Parasitism The relationship in which an organism benefits at the expense of the host organism is called *parasitism.* Depending on the parasite and the circumstances, the damage may be slight or it may be fatal. However, the "wise" parasite does not kill its host, but rather takes only the nutrients it needs to exist. Intestinal worms—such as pinworms and tapeworms—and external parasites—such as mites, lice, and ticks—usually cause only minor damage in humans. The presence of *ectoparasites* (parasites that live on the outside of the body) is referred to as an *infestation,* whereas the presence of *endoparasites* (parasites that live inside the body) is called an infection.

Pathogenic Relationships When microorganisms cause damage to the host during the infection process, a pathogenic relationship exists. The pathogen may be only a displaced commensal; for example, the staphylococci that normally inhabit the skin can cause an infection when the skin is wounded or burned. It may be a highly virulent airborne pathogen, such as the common cold virus, or it may be carried in food and water such as the dysentery pathogens. An opportunist (or opportunistic pathogen) causes disease in a host who is physically impaired or debilitated and the host's normal defenses against disease are weakened. Pneumonia developing in a bedridden patient is another example of a pathogenic relationship. In the usual course of events, the opportunist is harmless; it moves in to cause damage when an abnormal situation develops, such as a wound or a burn or the destruction of the indigenous microflora by antibiotic therapy. Opportunities can also cause disease in otherwise healthy persons, if they gain access to the blood, urinary bladder, lungs, or other organs or tissues.

Nonpathogenic Microbes Microbes that never cause disease are referred to as nonpathogenic microorganisms or nonpathogens. It is now known that many microorganisms that were at one time thought to be nonpathogens cause serious illness in immunosuppressed patients, such as patients with acquired immunodeficiency syndrome (AIDS).

MICROBIAL ECOLOGY

Ecology is the systematic study of the interrelationships of various organisms to other organisms and to their shared environment. The interactions of microorganisms with animals, plants, and other microbes have far-reaching effects on our lives. We are all aware of diseases caused by microbial infections; this is just one example of the effects of one type of organism on another. Most of the relationships with microbes in nature are beneficial rather than harmful.

Cycling of Nutrients

Bacteria are exceptionally adaptable and versatile and are found on the land, in all waters, in every animal and plant, and even inside of other microorganisms (in which case they are referred to as *endosymbionts*). Some bacteria and fungi serve valuable functions by recycling back into the soil the nutrients from dead, decaying animals and plants, as discussed in Chapters 1 and 4 (see Fig. 1 – 3). The free-living fungi and bacteria that decompose dead organic matter into inorganic materials are called saprophytes. The inorganic nutrients that are returned to the soil are used by autotrophic bacteria and plants for synthesis of biological molecules necessary for their growth; the plants are eaten by animals, which eventually die and are recycled again with the aid of saprophytes.

Good examples of the cycling of nutrients in nature are the nitrogen, carbon, oxygen, sulfur, and phosphorus cycles, in which microorganisms play very important roles. In the nitrogen cycle, free atmospheric nitrogen (N_2) is converted by nitrogen-fixing bacteria and cyanobacteria into ammonia (NH_3), nitrite (NO_2^-), and nitrate (NO_3^-) compounds in the soil (Fig. 6 – 2). The plants then use the nitrates to build plant proteins; these proteins are eaten by animals, which then use them to build animal proteins. The excreted waste products (such as urea in urine) are converted by certain bacteria to ammonia. Also, dead plant and animal debris and fecal material are transformed by saprophy-

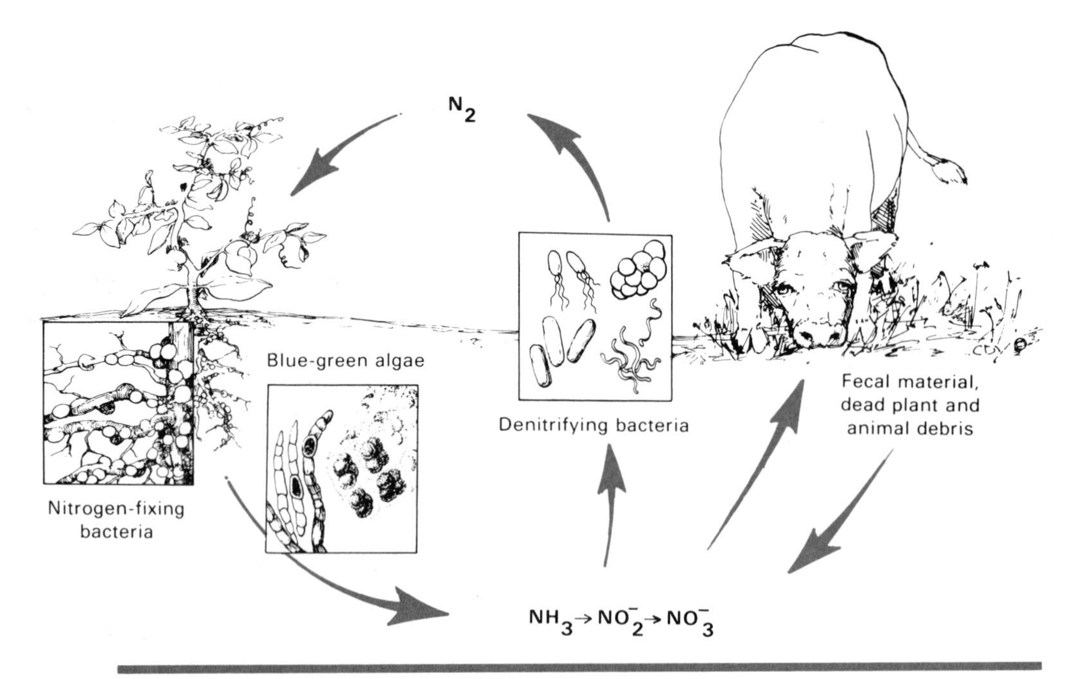

FIGURE 6 – 2
The nitrogen cycle. Nitrogen of the air is converted by nitrogen-fixing bacteria and algae into ammonia, nitrites, and nitrates. These inorganic nitrogen compounds are also derived from the breakdown of fecal material as well as from decaying plant and animal material. The nitrates are used as food nutrients by plants, which are eaten by animals. The cycle is completed by the denitrifying bacteria, which produce free nitrogen from inorganic nitrogen compounds

tic fungi and bacteria into ammonia, nitrites, and nitrates for recycling by plants. To replenish the free nitrogen in air, a group of bacteria called denitrifying bacteria convert nitrites and ammonia to atmospheric nitrogen (N_2). Thus, the cycle goes on and on.

Nitrogen-fixing bacteria are of two types, free-living and symbiotic. Symbiotic bacteria live in and near the root nodules of plants called legumes—alfalfa, clover, peas, soy beans, and peanuts (Fig. 6 – 3). These plants are often used in crop-rotation techniques by farmers to return nitrogen compounds to the soil and thus avoid the loss of nutrients.

The types and amounts of microorganisms living in soil depend on many factors: amount of decaying organic material, available nutrients, moisture content, amount of oxygen available, acidity, temperature, and the presence of waste products of other microbes. Likewise, the type and number of harmless microbes that live on and within the human body depend on pH (acidity), moisture, nutrients, antibacterial factors, and the presence of other microorganisms.

FIGURE 6 – 3
Bacterial nodules on the roots of a legume. (Lechavalier HA, Pramer D: The Microbes. Philadelphia, JB Lippincott, 1970)

SUMMARY

Various interrelationships among organisms—plants, animals, and microbes—were discussed in this chapter. You are now aware of how necessary microorganisms are for the survival of humans in this environment. Apparently by mutation, some of the previously harmless microorganisms have developed the ability to invade other organisms and cause disease. These are the opportunists and true pathogens.

PROBLEMS AND QUESTIONS

1. What types of relationships exist between humans and their indigenous microflora?
2. Where would you find symbiotic relationships in your environment?
3. What are the differences between mutualism, commensalism, neutralism, antagonism, and parasitism?
4. Why are the microbial decomposers so necessary for life on earth?
5. What factors control the number of microorganisms in the soil and on the human body?

Self Test

After you have read Chapter 6, examined the objectives, reviewed the chapter outline, studied the new terms, and answered the problems and questions above, complete the following self test.

Matching Exercises

Complete each statement from the list of words provided with each section.

SYMBIOTIC RELATIONSHIPS

symbiotic
mutualism
synergism
antibiosis
pathogen
opportunist
symbionts
commensalism
neutralism
parasite
infestation
infection

1. A parasitic microorganism that causes damage to its host is called a ________.
2. An organism that lives on or within a host organism is called a ____________.
3. If a parasite is an endoparasite, like pinworms, the host has an ________________.
4. If the parasite is an ectoparasite, the host has an ________________.
5. When two dissimilar organisms live together, they live in a ______________ relationship.
6. The two organisms that live together are called ________________.
7. If both organisms benefit from the relationship, they live in a state of ________________.
8. If the two organisms work together to produce an effect they live in a state of ________________.
9. If one organism secretes a material that damages or repels another organism, the two organisms live in a state of ________________.
10. The secretion of penicillin by a *Penicillium* mold in an area where bacteria are established would be an example of ________________.

11. When two organisms live together without harming or benefiting each other, the relationship is termed ________________.

12. If the relationship is beneficial to one party, but of no consequence to the other, the relationship is one of ________________.

True or False (T or F)

___ **1.** All indigenous microflora are nonpathogens.

___ **2.** Newborn infants acquire their first resident microflora organisms as they pass through the birth canal.

___ **3.** Saprophytes aid in the cycling of nutrients, which provides plants with proper carbon, oxygen, sulfur, phosphorus, and nitrogen sources.

___ **4.** Nitrogen-fixing protozoa change free nitrogen from the air into ammonia, nitrites, and nitrates.

___ **5.** Legumes, like alfalfa and clover, help fertilize the soil because of the bacteria that live on and in their roots.

___ **6.** The yeast *Candida albicans* is an indigenous microflora organism that is an opportunist.

___ **7.** The cool, dry areas of the skin support the growth of most indigenous microflora of the skin.

___ **8.** Wound and burn infections are frequently caused by resident microflora.

___ **9.** Careless dental hygiene encourages dental caries and gingivitis.

___ **10.** Beta-hemolytic streptococci are responsible for "strep" throat, scarlet fever, and rheumatic fever.

___ **11.** The lysozyme found in saliva and tears helps to destroy bacteria.

Multiple Choice

1. Symbionts might also be
- **a.** opportunists
- **b.** parasites
- **c.** mutualists
- **d.** all of the above

2. Indigenous microflora that are facultative anaerobes are most apt to be found
- **a.** in the mouth
- **b.** on the skin
- **c.** in the large intestine
- **d.** all of the above

3. The presence of the intestinal bacterium *Escherichia coli* as resident microflora is an example of
- **a.** mutualism
- **b.** neutralism
- **c.** parasitism
- **d.** commensalism

4. Indigenous microflora of the blood might be
- **a.** *Streptococcus*
- **b.** *Proteus*
- **c.** *Staphylococcus*
- **d.** none of the above

5. Nitrate compounds in the soil are utilized by plants to
   a. synthesize carbohydrates
   b. give off nitrogen
   c. build plant proteins
   d. neutralize toxins
6. *Escherichia coli* is a common inhabitant of the
   a. intestinal tract
   b. vagina
   c. urinary tract
   d. mouth
7. The synthesis of vitamins B and K by enteric bacteria is an example of
   a. commensalism
   b. mutualism
   c. opportunism
   d. infection
8. Indigenous microflora that are harmless are called
   a. aerobes
   b. commensals
   c. opportunists
   d. parasites
   e. cytopathogenic
9. The greatest microbial population on the skin is
   a. *Staphylococcus*
   b. *Streptococcus*
   c. *Klebsiella*
   d. *Neisseria*
10. A gram-negative facultative bacterium you would expect to see in fecal material is
    a. bacteriophage
    b. plasmid
    c. *Klebsiella*
    d. β-hemolytic streptococcus
    e. *Escherichia coli*
11. A microbe that lives on its host and gives no evidence of benefit or harm is known as a (an)
    a. commensal
    b. leech
    c. opportunist
    d. parasite

CHAPTER 7

Microbial Pathogenicity and Epidemiology

OBJECTIVES

After studying this chapter, you should be able to

- Differentiate between infectious, communicable, and contagious diseases
- List six reasons why an infection may not occur even though a pathogen is present
- Discuss the disease process
- Define acute and chronic diseases
- State the difference between primary and secondary diseases
- State the difference between local and generalized infections
- List three factors associated with the virulence of a pathogen
- List and discuss eight factors that affect the pathogenicity of bacteria
- Define the following terms: epidemiology, epidemic, endemic, and pandemic
- Describe the difference between sporadic and nonendemic diseases

- List three factors that contribute to an epidemic
- List six reservoirs of infection
- List five modes of disease transmission
- Discuss the procedure for stopping an epidemic

CHAPTER OUTLINE

Diseases and Infections
Why Infection Does Not Always Occur
The Development of Infection
The Disease Process
Mechanisms of Disease Causation
Virulence
Bacterial Morphological Characteristics Associated with Infection
Enzymes Associated with Invasiveness
Toxins and Enzymes Associated with Toxigenicity
Pathogenicity and Virulence
Epidemiology and Disease Transmission
Endemic Diseases
Epidemic Diseases
Pandemic Diseases
Sporadic and Nonendemic Diseases
Reservoirs of Infectious Agents
Modes of Disease Transmission
Control of Epidemic Diseases

NEW TERMS

Active carrier
Acute disease
Avirulent
Biological vector
Chronic disease
Coagulase
Collagen
Collagenase
Communicable disease
Contagious disease
Convalescent carrier
Endemic disease
Endotoxin
Enterotoxin
Epidemic disease
Epidemiology
Erythrocytes
Erythrogenic toxin
Exfoliative toxin
Exotoxin
Fibrinolysin
Fomite
Generalized infection
Hemolysin
Hyaluronic acid
Hyaluronidase
Incubatory carrier
Kinase
Latent infection
Lecithin
Lecithinase
Leukocidin
Leukocytes
Local infection
Mechanical vector
Neurotoxin
Nonendemic disease
Pandemic disease
Passive carrier
Parenteral route
Pathogenicity
Plasma
Primary disease
Pyogenic
Reservoirs of infection
Secondary disease
Sporadic disease
Staphylokinase
STD
Streptokinase
Systemic infection
Toxigenicity
Toxin
Virulence
Virulence factor
Virulent
Zoonosis (pl. *zoonoses*)

DISEASES AND INFECTIONS

There are many diseases that are not caused by pathogens. Included among them are those caused by malfunction of an organ, such as diabetes and hyperthyroidism; those caused by a vitamin deficiency, such as scurvy and rickets; those caused by an allergic response, such as asthma and hay fever; and those caused by uncontrolled cell growth, such as tumors and cancer. However, this chapter focuses on the many infectious diseases caused by the growth of pathogens on and in living tissues.

The ability of a pathogen to invade and infect the host, cause damage, and produce disease is termed its *virulence.* Those microbes that can cause disease with relative ease are referred to as *virulent* pathogens; microorganisms incapable of causing disease, are described as *avirulent.* Within a single genus or species, some strains may be virulent and others avirulent, as determined by the genetic characteristics of the microorganism. Whether a disease results when a person is exposed to a pathogen depends on many factors, including the host's resistance and the virulence of the microorganism.

The potential for infections always exists when organisms live in a parasitic relationship. When a pathogen finds the appropriate places and conditions in which it can grow and cause damage, it produces an infectious disease.

An infection occurs when a pathogenic microbe is able to multiply in the tissues where it lodges. Thus, an infectious disease results from the growth of a pathogenic microorganism. A *communicable disease* is an infectious disease that can be transmitted from one person to another, as with measles, gonorrhea, and diphtheria. A *contagious disease* is a communicable disease that is *easily* transmitted from person to person, for example via droplets in the air, such as occurs with the common cold and influenza.

The severity of an infectious disease and the amount of damage it causes are determined by the host's ability to resist invasion and to neutralize the damaging enzymes and *toxins* produced by the pathogen.

Why Infection Does Not Always Occur

Many people who are exposed to pathogenic microbes do not get sick for numerous reasons.

1. The microbe may land in the wrong place and be unable to multiply. For example, when a respiratory pathogen falls on the skin, it may be unable to grow because the skin lacks the necessary warmth, moisture, and nutrients required for growth of the particular microbe.
2. Many pathogens must be able to attach to specific host receptor sites before they can multiply and cause damage.

3. Antibacterial factors that destroy or inhibit the growth of microbes may be present; *e.g.*, the lysozyme found in tears, saliva, and perspiration.
4. The indigenous microflora may inhibit growth of the foreign microorganism in regions such as the mouth, vagina, and intestine by occupying space and using up the available nutrients.
5. The microbes already growing in the region may produce antibacterial factors (called bacteriocins) that have a local antibiotic effect, as when streptococci inhibit the diphtheria pathogen.
6. Antibodies may be present because the person had previously been infected with the organism or had been immunized against it. These antibodies attack and destroy the pathogen before it can multiply.
7. Phagocytes present in the blood and other tissues may engulf and destroy the invader.

The Development of Infection

Infection occurs when pathogens are able to enter the host, attach (adhere), multiply, and cause damage to the tissues of the host.

The body often responds to the infection via the inflammatory process (Fig. 7 – 1). The symptoms of inflammation are swelling (edema), redness, heat, and

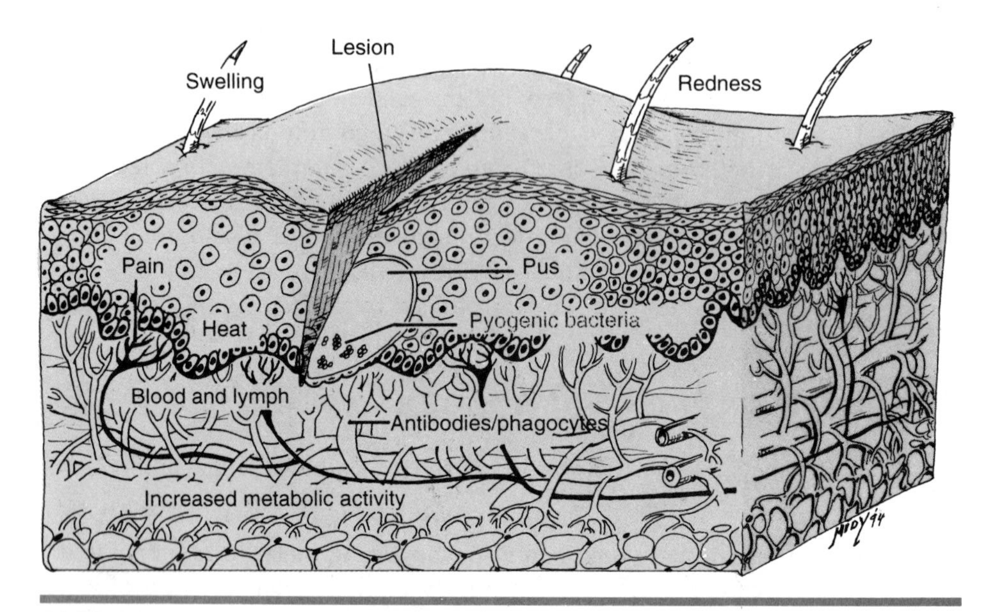

FIGURE 7 – 1
The development of infection and the inflammatory process. The symptoms of inflammation include swelling, redness, heat, and pain. Pus may also be present whether or not pyogenic bacteria are present.

pain. The swelling is due to the escape of large quantities of fluid from the highly permeable capillaries. The area becomes reddened because more blood is brought to it to fight the infection. The heat is caused by the increased blood flow and increased metabolic activity in the cells. Pain usually results from pressure on the nerve endings. Phagocytes rush in to help destroy the invaders and to rid the area of dead tissue. Pus is often found at the site of inflammation; it consists of lymph, plasma, dead tissue cells, leukocytes, and sometimes bacteria.

The production of antibodies begins slowly; the body first attempts to isolate the infection by forming a connective tissue wall that inhibits deeper penetration of the pathogens into surrounding tissue. If the pathogen is *pyogenic* (pus-forming), such as a staphylococcus or a streptococcus, the inflamed area contains much more pus than is normally present during the inflammatory process. Inflammation and antibody production are further discussed in Chapters 8 and 9, which are devoted to mechanisms of defense against disease.

The Disease Process

If the body wins the battle against the invading pathogens at the site of inflammation, the local infection is stopped. The person usually has some antibodies to protect the body against a later similar infection, but the individual may become a carrier if all of the pathogens have not been destroyed. A person may recover from a sore throat without complications, but may harbor a few resistant pathogens and transmit them to others.

Clinical disease occurs when the body's primary defenses lose the battle with the pathogen. The disease may be a *local infection* in which the pathogen is confined to a single area; this is often the situation in diphtheria, "staph" boils or carbuncles, "strep" throat, tuberculosis, primary syphilis, and gonorrhea. But if the pathogen is not stopped at the local level, it may emerge from this focal site to invade the tissues or be carried by phagocytes to other organs or by the bloodstream as a *generalized* or *systemic infection;* such is the case in systemic streptococcal infection, staphylococcal septicemia, and meningitis.

A disease may be acute or chronic, or it may be both, in which case it begins as an acute response that becomes a chronic illness. An *acute disease* has a rapid onset followed by a relatively rapid recovery, as with measles, mumps, and influenza. A *chronic disease* is one of slow onset and long duration such as tuberculosis, leprosy (Hansen's disease), and syphilis. Some diseases, such as gonorrhea, often have an acute inflammatory phase and then become chronic, causing slow deterioration of the infected tissues.

A disease may also reach a stage in which there are no symptoms (asymptomatic disease); it is then known as a *latent infection* or considered to be in a *latent stage;* tuberculosis, syphilis, gonorrhea, and herpes cold sores are examples. In tuberculosis, when the tubercles of *Mycobacterium tuberculosis* are successfully walled off in the lungs, the symptoms disappear temporarily; the infection then is said to be latent. Syphilis progresses through primary, secondary, latent, and tertiary stages (Fig. 7 – 2). The lesion, or chancre, of the primary stage appears at the site of entry of the spirochete *Treponema pallidum.* A few weeks after the spirochete has invaded the bloodstream, the chancre disappears, and the symptoms of the secondary stage arise, including rash, fever, and lesions of the mucous membranes. This is the most contagious stage of syphilis. These symptoms also disappear after a few weeks when the disease enters the latent stage, which may last from 1 to 50 years and may cause few or no symptoms. In tertiary syphilis, the organism causes destruction of the organ where it had been hiding—in the brain, heart, or bone tissue.

Gonorrhea, caused by *Neisseria gonorrhoeae,* is often an asymptomatic disease in the early phase when the organism lives in the mucous membranes of the urogenital tract, rectum, or throat. After many months, during which the organism causes damage, scarring, and destruction of the fallopian tubes or vas deferens, localized pain is experienced. Gonorrhea is very contagious by direct mucous membrane-to-mucous membrane transfer, usually sexual contact. It is particularly difficult to detect and control because it has been shown to be asymptomatic in most infected women and in many infected men.

Cold sores, or fever blisters, and genital herpes are caused by herpes viruses. After the initial local infection, the viruses remain dormant (latent) inside the ganglion cells of the nervous system and may not cause another local lesion until some type of stress acts as a trigger. These stresses may be fever caused by illness, heat and ultraviolet rays from the sun (sunburn), a bruise, extreme cold, or an emotional upset (stress).

An infectious disease that is the first or original illness is called the *primary disease,* such as a cold or influenza. A *secondary disease* or infection is caused by a pathogen that can invade only a weakened person with lowered resistance. Examples of secondary diseases are pneumonia and ear or sinus infections following a primary case of influenza.

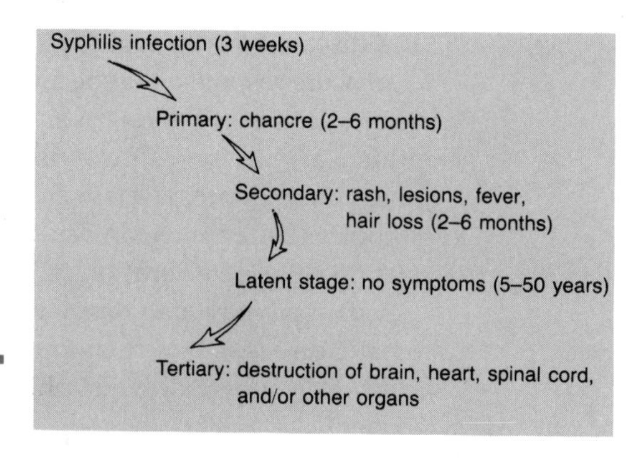

FIGURE 7 – 2
Stages of syphilis.

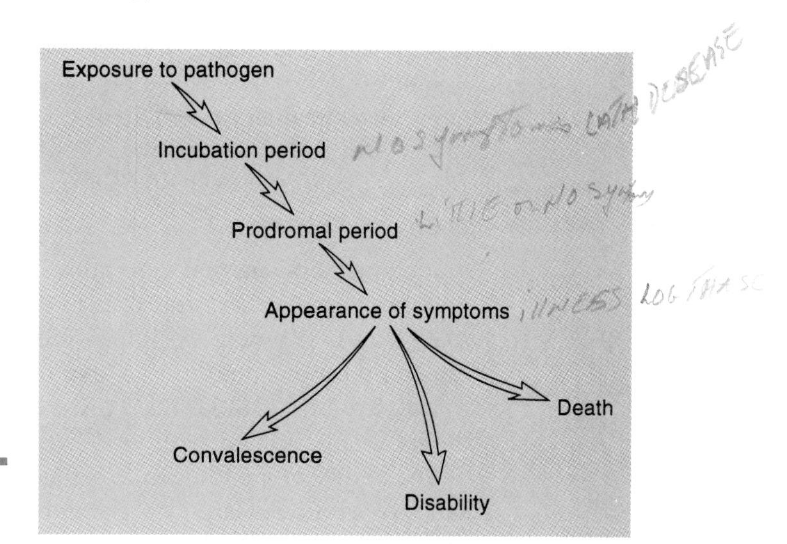

FIGURE 7 – 3
The course of an infectious disease.

Once exposure to the pathogen has occurred, the course of infectious disease has four periods or phases: (1) the incubation period when there are no symptoms, (2) the prodromal period, during which the patient feels "out of sorts," but actual symptoms have not as yet occurred, (3) the period of illness with progression of symptoms, and (4) the convalescent period, disability, or death (fortunately, the final stage is usually the convalescent period) (Fig. 7 – 3). Communicable diseases are most contagious during the third period. For instance, in measles, the incubation period lasts 10 to 12 days after the initial exposure to the virus. The early symptoms are much like those of a cold, with an eye infection, low-grade fever, and Koplik's spots (red with white centers) inside the cheeks. During this early stage, the virus is easily transmitted in secretions from the eyes, nose, and mouth. As the illness progresses, a skin rash appears; it disappears during the convalescent stage. Usually, measles is considered contagious only from the time the fever begins until the second day of the rash. Complications are frequent because body defenses are lowered by the initial virus infection; thus, the virus and possibly opportunistic bacteria can invade deeper to cause pneumonia, ear and sinus infections, or encephalitis.

Although the patient may recover from the illness itself, permanent damage may be caused by destruction of tissues in the affected area. Brain damage may follow encephalitis or meningitis; paralysis may follow poliomyelitis; deafness may follow ear infections.

Mechanisms of Disease Causation

VIRULENCE

The capacity of pathogens to cause disease (*pathogenicity*) is related to their abilities (1) to infect the host or protect themselves against the body's defenses,

(2) to invade and multiply in tissues, and (3) to cause damage or destruction to tissue. Virulence is a measure or degree of pathogenicity. Some pathogens are more virulent than others. Virulence can be defined as follows:

$$\text{virulence} = \text{infectivity} + \text{invasiveness} + \text{toxigenicity}$$

Each species of pathogenic microbe has specific characteristics, including its unique metabolism, that determine its pathogenicity and virulence. Properties or characteristics that contribute to the virulence of a pathogen are called *virulence factors.* It would be impossible to list every pathogen and its disease-causing mechanism(s) in this text; in fact, the mechanisms are frequently not completely understood. Keep these limitations in mind as we discuss some of the factors associated with virulence and pathogenicity.

The ability of a pathogen to settle on a susceptible tissue and to survive the shock of landing is largely a matter of chance. However, once it finds itself in a moist, warm environment, it must be able to attach to the site (so that it is not flushed away) and it must be able to resist the body's bacteriolytic enzymes, antibodies, and phagocytes if it is to multiply successfully. It should be noted that some pathogens are able to cause disease without attaching to or invading body tissues.

BACTERIAL MORPHOLOGICAL CHARACTERISTICS ASSOCIATED WITH INFECTION

Some structural features of pathogens enable them to attach to and/or invade tissues in certain areas of the host and multiply there, thus causing an infection. These structures include capsules, flagella, and pili.

Capsules Capsules are often regarded as a portion of the cell envelope (which includes the cell membrane, cell wall, and outer glycocalyx). Capsular constituents vary among the different species of procaryotes. Many bacteria have slimy capsule layers that consist of polysaccharides only; others have proteins within the polysaccharide capsule. Both types enable the bacteria to attach to tissues and to resist phagocytosis (Fig. 7 – 4). Most bacteria infecting the body have some type of capsule, but it is frequently not observable when they are grown on an artificial medium in the laboratory.

Some bacteria secrete polysaccharide fibers to increase their adherence to teeth and mucous membranes. This additional polysaccharide layer also protects the organisms from antibodies, phagocytes, and other antimicrobial agents.

The cell envelope of group A, β-hemolytic streptococci (*Streptococcus pyogenes*) contains a certain antigenic protein (protein M), which appears to serve an antiphagocytic function and aids the organism in adhering to pharyngeal cells.

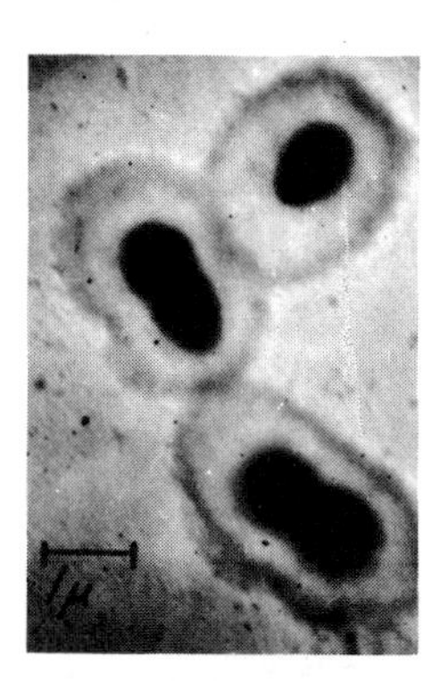

FIGURE 7 – 4
Electron micrograph of *Streptococcus pneumoniae,* type 1. The capsule has been treated with a specific antibody (Quellung reaction) to enhance its visibility. (Mudd S, et al: *J Exp Med* 78:327, 1943. By copyright permission, Rockefeller University Press)

Flagella Because flagella enable bacteria to be motile, they also aid bacteria to invade the aqueous areas of the body. Despite the flushing action and continuous peristaltic movement of the ureters, some flagellated pathogens are able to reach the kidney, where they cause serious infections and complications.

Pili (Fimbriae) Frequently, piliated pathogens are able to adhere to cells within mucous membranes to establish an infection when other pathogens cannot; for example, piliated *N. gonorrhoeae* cells attach to urethral cells and multiply, whereas nonpiliated pathogens are flushed away by urination. Enterotoxigenic *Escherichia coli,* with its many fimbriae (pili), is able to attach to cells in the intestine where it multiplies and secretes the exotoxins that cause gastroenteritis. (see Figure 2 – 9 for an example of piliated *E. coli*)

ENZYMES ASSOCIATED WITH INVASIVENESS

Some pathogens release enzymes (called exoenzymes) that increase their ability to invade body tissues. These substances include (1) *coagulase,* (2) *kinases,* (3) *hyaluronidase,* and (4) *collagenase.* Note that the "-ase" ending on these words indicates that these substances are enzymes, which you will recall are proteins that catalyze (speed up) particular chemical reactions.

Coagulase Pathogenic staphylococci are noted for their production of the exoenzyme, coagulase. Coagulase enables the organisms to clot plasma and thereby to form a sticky coat of fibrin around themselves for protection from phagocytes and other body defense mechanisms. In the laboratory, coagulase-positive staphylococci (*e.g., Staphylococcus aureus*) cause citrated blood *plasma* to clot. Plasma is the liquid portion of blood.

Kinases Sometimes referred to as *fibrinolysin,* kinase has the opposite effect of coagulase. *Streptokinase,* for example, lyses (dissolves) fibrin clots, thus, enabling streptococci to invade and spread throughout the body. A similar enzyme produced by staphylococci is called *staphylokinase.*

Hyaluronidase The spreading factor, as hyaluronidase is sometimes called, enables pathogens to spread through connective tissue by breaking down *hyaluronic acid,* the "cement" that holds tissue cells together. It is secreted by several pathogenic species of *Staphylococcus, Streptococcus,* and *Clostridium.*

Collagenase The enzyme collagenase breaks down *collagen,* the supportive protein found in tendons, cartilage, and bones. *Clostridium perfringens,* a major cause of gas gangrene, spreads deeply within the body by secreting both collagenase and hyaluronidase.

TOXINS AND ENZYMES ASSOCIATED WITH TOXIGENICITY

The ability to damage host tissues may depend on the production and release of *hemolysin, leukocidin,* or *exotoxins. Endotoxins* ("endo" = "within"), which are an integral part of the cell wall of gram-negative bacteria, may also be toxic to the host. Exotoxins ("exo" = "outside"), on the other hand, are poisons that are produced within, and then released from, the cell.

Hemolysin Hemolysins ("hemo" = "blood"; "lysis" = "breakdown or dissolution") are enzymes that cause damage to the host's red blood cells (*erythrocytes*). In the laboratory, hemolysis of the red blood cells in blood agar is useful for identifying α-hemolytic and β-hemolytic streptococci. The pathogenicity and invasiveness of streptococci may not depend on the amount of hemolysin they produce, although it is certainly related to the amount of damage the organisms can cause to the host. Lysis of erythrocytes also provides pathogens with a source of iron.

Leukocidin Leukocidin, an enzyme-like exotoxin secreted by some staphylococci and streptococci, causes destruction of white blood cells (*leukocytes*). The lytic effect of leukocidin can be observed on a laboratory slide of leukocytes in the presence of streptococci.

Lecithinase One of the toxins produced by *Clostridium perfringens* is an enzyme called *lecithinase,* which breaks down phospholipids collectively referred to as *lecithin.* It is destructive to cell membranes of red blood cells and other tissues.

Exotoxins Bacterial exotoxins are usually proteins secreted by living (vegetative) pathogens. Exotoxins are often described by the target organs they affect. The most toxic exotoxins are *neurotoxins;* they cause nerve destruction in botulism and diphtheria and spasmodic muscle contractions (tetany) in tetanus. *Enterotoxins,* those toxins that affect the vomiting centers of the brain and cause gastroenteritis, are secreted by *Staphylococcus aureus, Vibrio cholerae, Shigella dysenteriae,* and other enteric pathogens. Some strains of *E. coli* are known

INSIGHT
Microbes In The News: "Flesh-Eating" Bacteria

"Flesh-eating" bacteria, a term coined by the press, received worldwide attention in the news media, especially tabloid newspapers, following a small number of cases of infection with these organisms in an English town. However, medical scientists had been aware of such bacteria earlier.

The so-called "flesh-eating" bacteria are especially invasive strains of *Streptococcus pyogenes,* a gram-positive coccus also known as group A, beta-hemolytic streptococcus. This is the bacterium that causes "strep" throat and its various complications such as scarlet fever, rheumatic fever, and glomerulonephritis.

The "flesh-eating" strains of *S. pyogenes* produce proteases, enzymes capable of destroying proteins and enabling the bacteria to invade human epithelial cells. They also produce a toxin (pyogenic toxin) that causes a type of toxic shock syndrome. Such strains rely on genetic information obtained from a type of bacterial virus (bacteriophage) that infects the streptoccus.

Scientists have been aware of invasive group A streptococcal infections for many years, but the number of cases has increased significantly in the past few years. According to the Centers for Disease Control and Prevention (CDC), an estimated 10,000 to 15,000 cases of invasive strep infections occur in the United States per year, causing an estimated 2,000 to 3,000 deaths. Although this is a large number of infections and deaths, it does not constitute an epidemic. About 5–10% of the cases involve necrotizing fasciitis (the proper name for the "flesh-eating" cases). As one newspaper article stated, "the news coverage is more widespread than the bacteria."

Necrotizing fasciitis and toxic shock syndrome can occur following entry of the invasive strains into cuts, bruises, and even chicken pox lesions and often begins as a mild skin lesion. It can quickly spread, damaging nearby tissue and causing gangrene. Distant abscesses, pneumonia, multi-organ failure, shock, and death may occur. People with "strep" throat should be particularly careful not to contaminate any external lesions they might have with their own saliva. Laboratory workers and other health-care personnel should also be very careful when handling throat swabs, throat culture plates, and when working with *S. pyogenes* isolates. They should protect any open skin lesions on their hands by wearing gloves, avoid creation of aerosols when working with *S. pyogenes* cultures, and carefully sterilize all streptococcal-contaminated tissue, culture media, glassware, and any eating utensils used by patients with streptococcal infections.

---

to produce an enterotoxin causing a form of dysentery (see Insight Box in Chapter 10). Symptoms of toxic shock syndrome are caused by exotoxins C and F secreted by staphylococci. *Exfoliative toxins* of these organisms cause the epidermal layers of skin to slough away, and *erythrogenic toxins* cause rash and redness in the skin (as in scarlet fever). *Corynebacterium diphtheriae* (the cause of diphtheria) secretes toxins that affect heart muscles (cardiotoxins), nerves (neurotoxins), and kidneys (nephrotoxins).

Endotoxins The cell walls of gram-negative bacteria contain endotoxins that may cause disease when the bacteria are present in large numbers. With the destruction and death of the bacteria, the endotoxin molecules from the cell walls cause fever, vomiting, and diarrhea. Diseases associated with endotoxins include septic shock, dysentery, meningitis, typhoid fever, gonorrhea, and cholera.

Endotoxins may destroy phagocytes, upset the water and salt balance in the intestine, and weaken the infected host by dehydration (because of water loss in diarrhea); as a result, the person is a prime target for secondary infections.

Pathogenicity and Virulence

All of the terms presented thus far in this chapter concerning the pathogenicity and virulence of pathogens are useful in discussing general concepts of disease causation. Pathogenicity and virulence are sometimes used synonymously, but this is not entirely accurate; virulence actually refers to the degree of pathogenicity. Some pathogens are more virulent than others.

Pathogenicity refers to the ability of a species of microorganism to produce disease. However, some avirulent strains or species may exist within those groups. For instance, some avirulent strains of *N. gonorrhoeae, C. diphtheriae, Mycobacterium tuberculosis,* and *Streptococcus pneumoniae* have been identified.

Virulence depends on infectivity, invasiveness, and *toxigenicity* (the ability to produce toxins); therefore, one might observe that pathogens that are highly infective but low in invasiveness might produce a large number of carriers. This large group of pathogens would include the opportunists found among the indigenous microflora. Likewise, if the infectivity of a pathogen were low, but its invasiveness and toxigenicity were high, sporadic outbreaks of the disease would occur.

The production of toxic products by pathogens is the most influential factor in their pathogenicity. Toxin-producing (toxigenic) bacteria differ greatly in the amount of exotoxin they produce, depending on growth conditions. For example, virulent strains of some pathogens become avirulent when grown in the laboratory under artificial conditions; some virulent encapsulated pathogens lose their ability to produce capsules when grown on artificial media in the laboratory, as has been shown to occur in pneumococci and gonococci. Also, some toxin producers cease producing exotoxins when grown on artificial media.

These changes can be regarded as the pathogen's response to different environments. On a completely artificial medium in the incubator where a constant temperature is maintained, optimal conditions exist for the microorganisms. Plenty of nutrients are present; the temperature is ideal; no overcrowding occurs from other microbes; and no phagocytes or antibodies exist. Thus, pathogens have no need to arm themselves with capsules and toxins, as they do when attempting to grow in human tissues.

It should also be noted that virulence usually increases in pathogens as they are transmitted from animal to animal and human to human, which explains why virulence may increase during an epidemic. This factor may also account for the relatively high virulence of pathogens in a hospital environment, where there is a continuously changing population of susceptible persons. Rampaging hospital staphylococcal infections are caused by especially virulent staphylococci that continue to mutate and to become resistant to the antibiotics used in ever greater varieties and numbers to fight infections.

EPIDEMIOLOGY AND DISEASE TRANSMISSION

Epidemiology is the science that deals with the frequency and distribution of diseases and the factors that contribute to their spread. These contributing factors include the virulence of the pathogen; susceptibility of the population because of overcrowding, lack of immunization, or inadequate sanitation procedures; and the mode of transmission of the pathogen.

Several terms are used to describe the prevalence of disease in an area at a particular time: *endemic, epidemic, pandemic, sporadic,* and *nonendemic.* Each term is discussed in this section.

Final

Endemic Diseases

Endemic diseases are those that are constantly present in a population or community, usually involving relatively few people. The number of cases increases and decreases at various times, but the disease never dies out completely. Endemic infectious diseases that occur regularly in the United States include tuberculosis, staphylococcal and streptococcal infections, and viral diseases such as the common cold, influenza, chickenpox, and mumps. Tuberculosis remains endemic throughout the world, particularly among the poor, who live in crowded conditions. Although the number of tuberculosis deaths has been greatly reduced over the years by drug therapy, strains of multi–drug-resistant *M. tuberculosis* are now common in the United States and other countries. In some parts of the United States, plague (caused by *Yersinia pestis*) is endemic among rats, prairie dogs, and other rodents; as a result, plague in humans is occasionally observed. The actual incidence of an endemic disease at any particular time depends on a balance among several factors: environment, the genetic susceptibility of the population, behavioral factors, the number of people who are immune, the virulence of the pathogen, and the reservoir or source of infection (Fig. 7 – 5).

Epidemic Diseases

Infections that are usually endemic may on occasion become epidemic. An epidemic is defined as a greater than normal number of cases of a disease in an area within a particular period (Fig. 7 – 6).

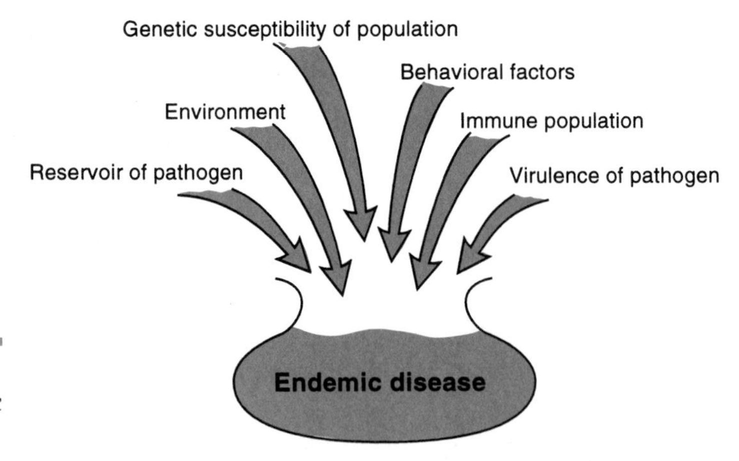

FIGURE 7 – 5
Factors influencing an endemic disease.

An example of an epidemic disease is the 1976 occurrence of a respiratory illness during an American Legion convention in Philadelphia, Pennsylvania (Legionnaires' disease or legionellosis). The organism was finally isolated, identified, and named *Legionella pneumophila* (Fig. 7 – 7). Other such epidemics have been identified through constant surveillance and accumulation of data by the U.S. Centers for Disease Control and Prevention (CDC). Epidemics usually follow a specific pattern in which the number of cases of a disease

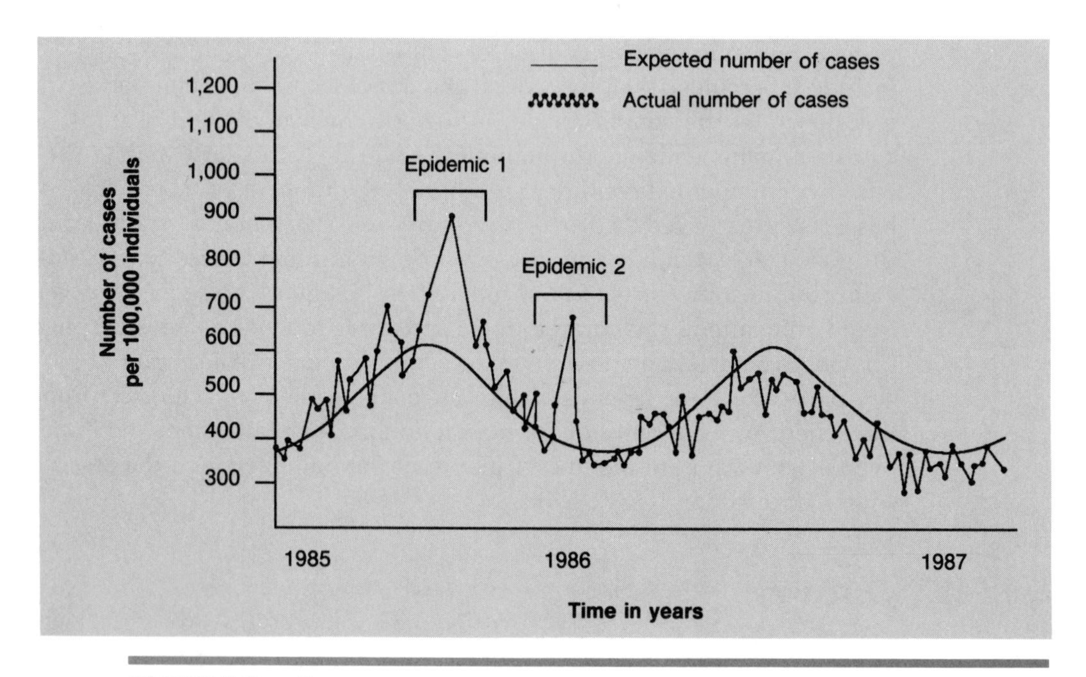

FIGURE 7 – 6
Graph illustrating two epidemics.

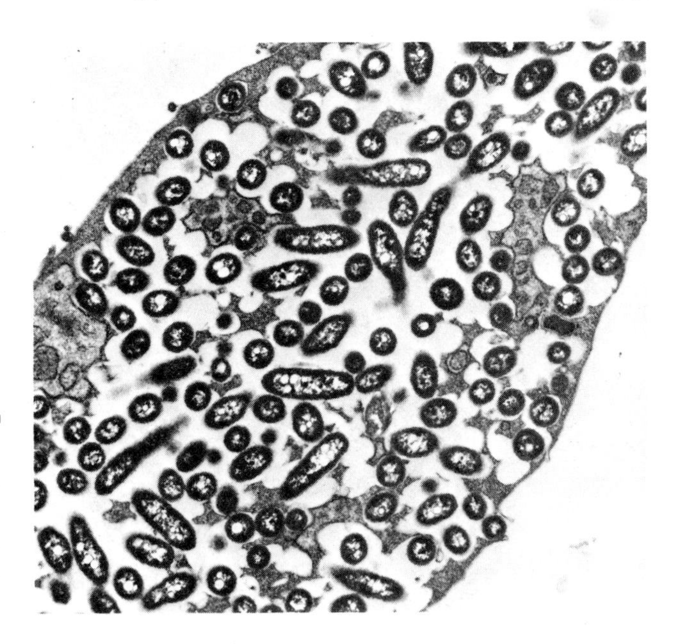

FIGURE 7 – 7
Legionella pneumophila cultured in human embryonic lung fibroblasts. (Courtesy of Mae C. Wong, Centers for Disease Control and Prevention, Atlanta, Georgia)

increases to a maximum and then decreases rapidly because the number of susceptible and exposed individuals is limited.

Epidemics may occur in communities that have not been previously exposed to a particular pathogen. People from populated areas who travel into isolated communities frequently introduce a new virulent pathogen to susceptible natives of that community; then the disease spreads like wildfire. There have been many such examples described in history. The syphilis epidemic in Europe in the early 1500s might have been caused by a highly virulent spirochete carried back from the West Indies by Columbus' men in 1492. Also, measles and tuberculosis introduced to Native Americans by early explorers and settlers almost destroyed many tribes. Recently, we have observed devastating outbreaks of measles and other contagious diseases in Australia, Africa, Greenland, and other relatively isolated areas.

In communities in which sanitation practices are relaxed, allowing fecal contamination of water supplies and food, epidemics of typhoid, cholera, giardiasis, and dysentery frequently occur. Visitors to these communities should be aware that they are more susceptible to these diseases because they never developed a natural immunity by being exposed to them during childhood.

In the 1970s, gonorrhea reached epidemic proportions in the United States and most of the world partly because no immunity remains after the disease is cured (Fig. 7 – 8). This disease runs rampant in a promiscuous, mobile society. The use of birth control pills makes women more susceptible by changing the environment of the vagina. Also, the careless use of penicillin allowed the gonococcus to mutate into penicillin-resistant strains (penicillinase-producing

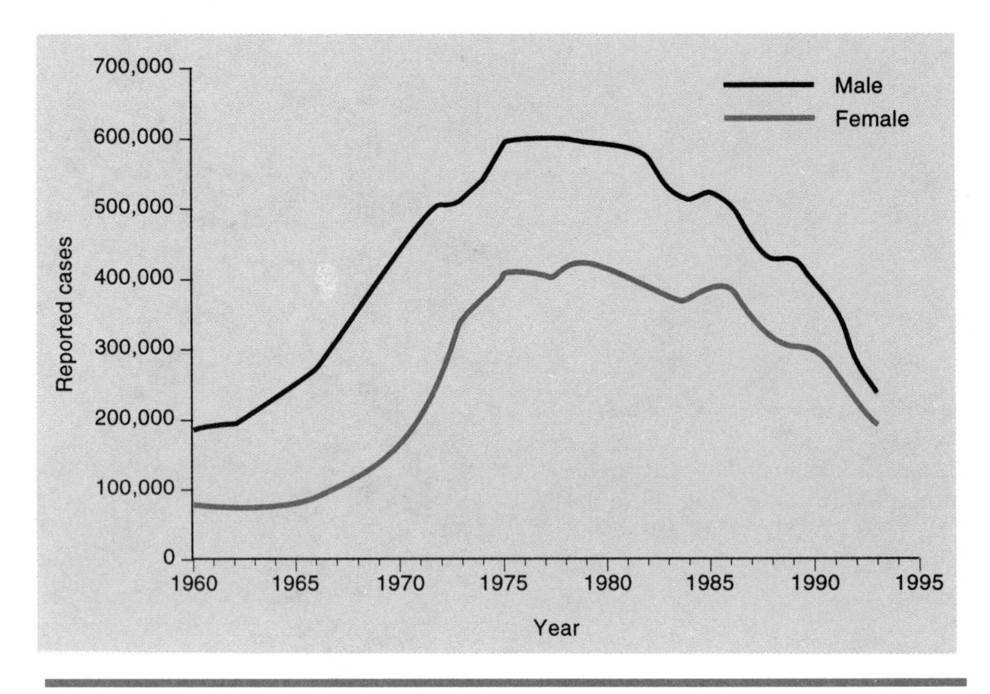

FIGURE 7 – 8
Gonorrhea in the United States. The graph shows the number of cases reported annually to the Centers for Disease Control and Prevention between 1960 and 1993. (Morbidity and Mortality Weekly Report, vol. 42, no. 53, 1994. U.S. Department of Health and Human Services, Public Health Service, Centers for Disease Control and Prevention, Atlanta, Georgia)

N. gonorrhoeae or PPNG), making control by penicillin more difficult. An effective vaccine might produce antibodies against pili to prevent the pathogen from adhering to mucous membranes. An additional complicating factor is that this disease is asymptomatic (without symptoms) in 80% of infected women and 20% of infected men; that is, the gonococcus lives and multiplies in the urogenital areas yet produces no symptoms.

Diseases such as the various types of influenza occur in many areas during certain times of the year and involve most of the population because the immunity produced against last year's strain(s) may not be effective against this year's strain(s). Thus, the disease recurs each year among those who are not revaccinated or naturally resistant to the infection.

Pandemic Diseases

A pandemic is a worldwide epidemic of a specific disease. Pandemics of influenza ("flu") have been recorded in 1898, 1917–1918, 1955, 1968, 1972, 1975, and 1978. Recent influenza pandemics were often named for the point of origin or

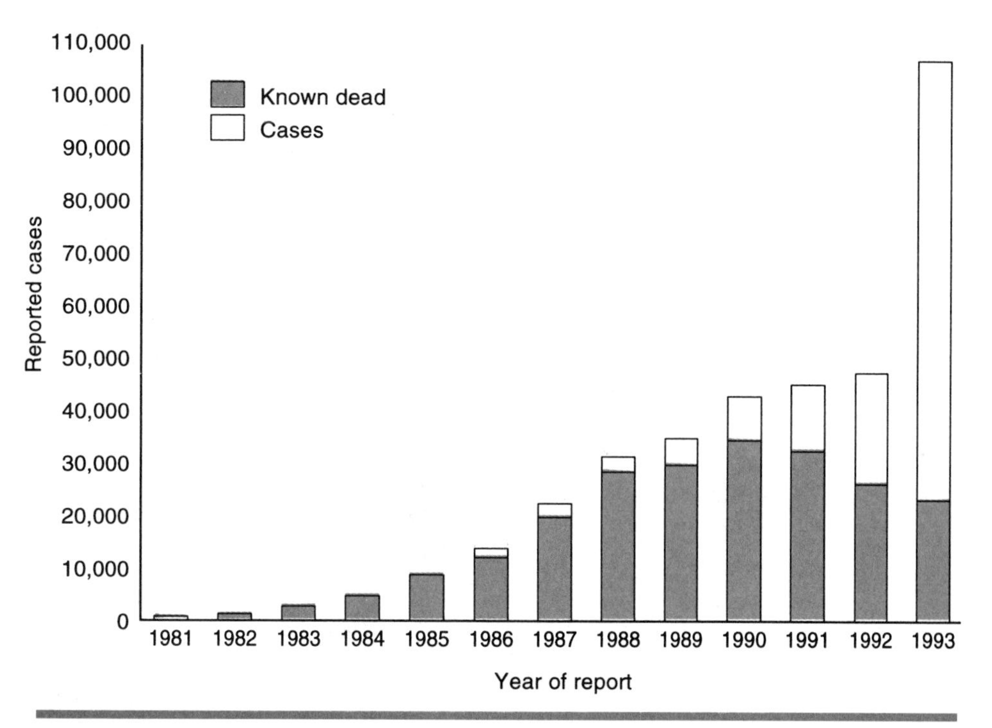

FIGURE 7 – 9
Acquired immunodeficiency syndrome (AIDS) in the United States. The graph shows the number of reported cases and known deaths between 1981 and 1993. (Morbidity and Mortality Weekly Report, vol. 42, no. 53, 1994. U.S. Department of Health and Human Services, Public Health Service, Centers for Disease Control and Prevention, Atlanta, Georgia)

first recognition, such as the Taiwan flu, Hong Kong flu, London flu, Port Chalmers flu, and Russian flu.

A current example is the expanding acquired immunodeficiency syndrome (AIDS) pandemic (Fig. 7 – 9). Various theories exist regarding the origin of the etiologic agent. One theory suggests that a mutant virus was somehow acquired from monkeys by Africans. It was then transmitted among Africans via mucous membranes, semen, and blood, usually by sexual contact or by blood transfusion. From Africa, the disease spread to Haiti, from where the virus was carried to New York, Miami, and many other parts of the United States. The virus then spread among homosexual males, within intravenous drug-using communities, and via blood banks to hemophiliacs. Finally, the occurrence of AIDS cases among the heterosexual population and children has increased throughout the world. The virus was quickly isolated and named human T-cell lymphotrophic virus, type 3 (HTLV–III) by one group of investigators and lymphadenopathy-associated virus (LAV) by another, and was later named human immunodeficiency virus (HIV) by international agreement.

The World Health Organization (WHO) estimated that as of mid-1993, more than 14 million persons worldwide had become infected with HIV and that more than 2 million of those persons had died. Approximately 104,000 new United States AIDS cases were reported to the CDC during 1993. The WHO estimates that as many as 40 million people worldwide will be infected with HIV by the year 2000. All persons who develop AIDS will probably die, unless new treatments are developed. This sexually transmitted or blood-borne virus has a long incubation period during which it destroys the $CD4^{+}$ T-helper lymphocytes, thus crippling the immune system so that most victims die of secondary infections (see Chapters 9 and 10). One significant outcome of the AIDS pandemic is the increased awareness of other sexually transmitted diseases (*STDs*) like gonorrhea, syphilis, and herpes.

Sporadic and Nonendemic Diseases

ONE IN A WHILE

Certain diseases follow neither the endemic nor the epidemic pattern but occur only occasionally. These sporadic diseases include tetanus, gas gangrene, and botulism. Diseases that are controlled as a result of immunization and sanitation procedures are termed nonendemic. These include smallpox, poliomyelitis, and diphtheria (in most parts of the world). Occasionally, outbreaks of these controlled diseases occur where vaccination programs have been neglected.

The communicability of pathogens relies entirely on the survival of pathogens during their transfer from one host to another. Thus, health professionals should be thoroughly familiar with pathways of transfer and sources of potential pathogens. For instance, a hospital staphylococcal epidemic begins when aseptic conditions are relaxed and a *Staphylococcus aureus* carrier introduces the organism to the many susceptible patients (babies, surgical patients, and debilitated persons). Such an epidemic may quickly spread from one person throughout the entire hospital population.

RESERVOIRS OF INFECTIOUS AGENTS

The sources of microorganisms that cause infectious diseases are many and varied. They are known as *reservoirs of infection* or simply *reservoirs.* A reservoir of infection is any site where the pathogen can multiply or merely survive until it is transferred to its host. Reservoirs of infection may be living hosts or inanimate objects or materials.

Living animal reservoirs include humans, horses, cattle, pigs, cats, dogs, wildlife, insects, ticks, mites, and many others. Humans may acquire pathogens from other humans and animals, which may or may not be diseased, through direct contact and bites; by eating meat or other products from diseased animals; or through the bites of mosquitoes, flies, fleas, mites, ticks, and other arthropod vectors. The most important reservoirs of human infections are other humans because many human pathogens are species-specific, meaning that they can cause disease in only one species of animal. Thus, human pathogens usually cause disease only in humans.

Some people who do not have the disease may harbor the pathogen and transmit it to susceptible people who then develop the disease symptoms. Those who carry the pathogen without manifesting symptoms are called carriers. Human carriers are infected with the pathogen, but they are asymptomatic. An *incubatory carrier* is a person who is capable of transmitting a pathogen during the incubation period of a particular infectious disease. *Convalescent carriers* harbor and can transmit the pathogen during their recovery from the disease, *i.e.,* during the convalescence period. *Active carriers* have recovered from the disease but continue to harbor the pathogen indefinitely. *Passive carriers* carry the pathogen without ever having had the disease. Usually, respiratory secretions and intestinal or urinary excretions are the vehicles by which the disease agent is transferred to food and water and directly to other persons. Human carriers are very important in the spread of staphylococcal and streptococcal infections, as well as in the spread of hepatitis, diphtheria, dysentery, meningitis, and STDs.

Inanimate reservoirs of infection include air, soil, food, milk, water, and *fomites.* Fomites include articles of clothing, bedding, eating and drinking utensils, and hospital equipment, such as bedpans, and urinals, that are easily contaminated by pathogens from the respiratory tract, intestinal tract, and the skin of patients. Air is contaminated by dust, smoke, and respiratory secretions of humans expelled into the air by breathing, blowing, sneezing, and coughing. The most highly contagious diseases include colds and influenza, in which the respiratory viruses can be transmitted through the air on droplets of water or by the hands to another person. Dust particles can carry spores of certain bacteria and dried bits of human and animal excretions that contain pathogens. Bacteria cannot multiply in the air but can be easily transported via airborne particles to a suitable nutrient site for growth. All personnel in hospitals and other facilities where housekeeping is conducted for large numbers of people must be especially aware of air currents that carry dust and pathogens throughout the facility. Great care must be taken by the hospital staff to prevent transmission of pathogens to patients.

MODES OF DISEASE TRANSMISSION

There are five principal modes of transfer of pathogens from an infected person to a susceptible person (Fig. 7 – 10 and Table 7 – 1).

- Direct person-to-person contact with the skin of a diseased person or carrier.
- Direct mucous membrane-to-mucous membrane contact by kissing or sexual intercourse.

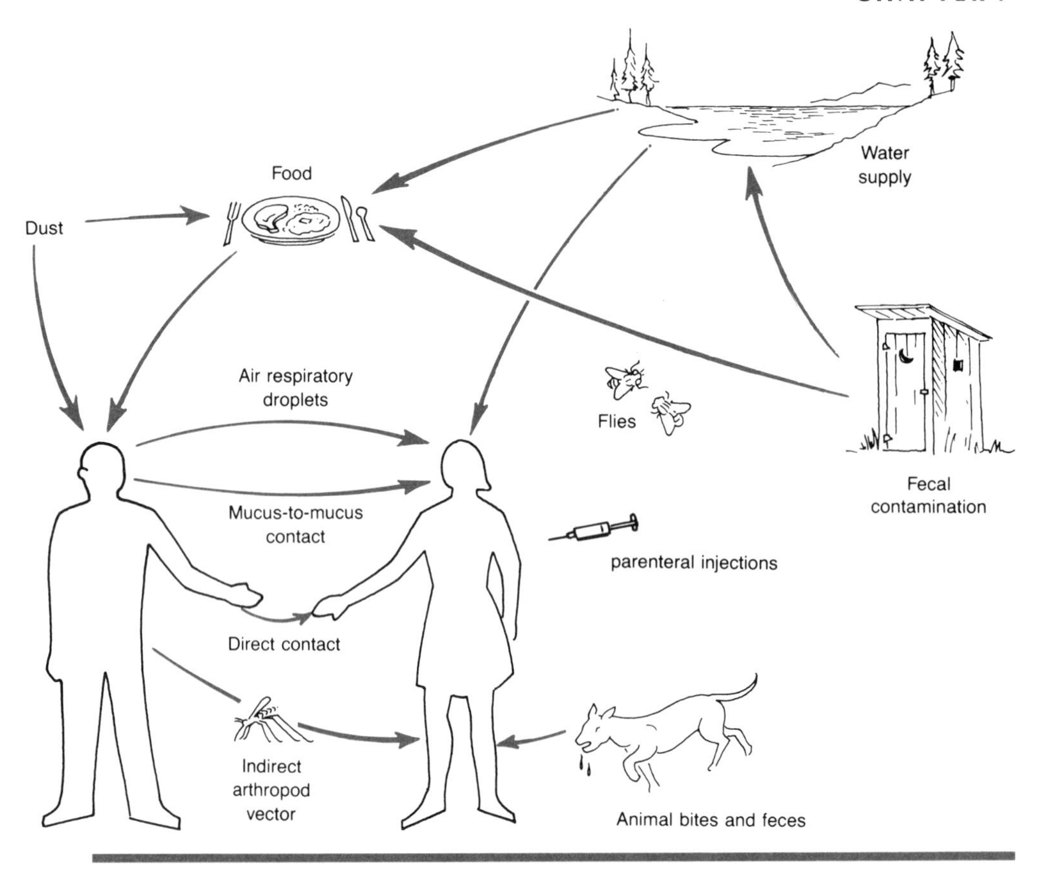

FIGURE 7 – 10
Modes of disease transmission.

- Indirectly through droplets of liquid or dust in the air.
- Indirect contamination of food and water by fecal material, dead or live animals, soil, and other sources.
- Blood contamination indirectly by arthropods and other animal vectors and directly by *parenteral* injection by nonsterile syringes and needles, as well as by intravenous transfusions and kidney dialysis.

Most diseases transmitted by direct contact are those in which the causative organism can be carried on the skin; usually the pathogen is transferred by the hands and face. Many viruses and opportunistic bacteria are thus transferred, causing colds, influenza, staphylococcal and streptococcal infections, pneumonia, polio, and diphtheria. In hospitals, this mode of transfer is particularly prevalent. Even the dysentery organisms, *Salmonella* and *Shigella,* can be transferred by fecal material on the hands of one person to the hands, mouth, and food of another.

A wide variety of diseases is transmitted via the mucous membrane-to-

TABLE 7 – 1.
Common Routes of Transmission

Route of Exit	Route of Transmission or Entry	Disease
Skin	Skin discharge → air → respiratory tract	Chickenpox, colds, influenza, measles, "staph" and "strep" infections
	Skin to skin	Impetigo, eczema, boils, warts, syphilis
Respiratory	Aerosol droplet inhalation Nose or mouth → hand or object → nose	Colds, influenza, pneumonia, mumps, measles, chickenpox, tuberculosis
Gastrointestinal	Feces → hand → mouth Stool → soil → food or water → mouth	Gastroenteritis, hepatitis, salmonellosis, shigellosis, typhoid fever, cholera, giardiasis, amebiasis
Salivary	Direct salivary transfer	Herpes cold sores, infectious mononucleosis, "strep" throat
Genital secretions	Urethral or cervical secretions	Gonorrhea, herpes, *Chlamydia* infection
	Semen	Cytomegalovirus infection, AIDS, syphilis, warts
Blood	Tranfusion or needle stick injury	Hepatitis B; cytomegalovirus infection; malaria, AIDS
	Insect bite	Malaria, relapsing fever
Zoonotic	Animal bite	Rabies
	Contact with carcasses	Tularemia, anthrax
	Arthropod	Rocky Mountain spotted fever; Lyme disease, typhus, viral encephalitis, yellow fever, malaria, plague

mucous membrane mode by kissing or sexual contact. Sexually transmitted diseases include syphilis, gonorrhea, chlamydia and herpes infections, and HIV.

Most contagious airborne diseases are due to respiratory pathogens carried in droplets of moisture to susceptible people. Some respiratory pathogens may dry as they settle in dust particles and be carried long distances through the air and into a building's ventilation or air conditioning system (as observed in the Legionnaires' disease epidemic). Improperly cleaned inhalation therapy equipment can easily transfer pathogens from one patient to another. Diseases that may be transmitted in this manner include colds, influenza, measles, mumps, chickenpox, smallpox, and pneumonia. Psittacosis or parrot fever is a respiratory infection that may be acquired from infected birds. Also, some fungal respiratory diseases (*e.g.*, cryptococcosis and histoplasmosis) are frequently transferred via dried bird feces.

Indirect transfer of organisms to a susceptible person frequently occurs through contamination of foods and water. Human and animal fecal matter from outhouses, cesspools, and feed lots often is carried into water supplies.

Improper disposal of sewage and inadequate treatment of drinking water contribute to the spread of fecal and soil pathogens. Food and milk may be contaminated by careless handling, which allows pathogens to enter from dust particles, dirty hands, hair, and respiratory secretions. If these vegetative pathogens and bacterial spores are not destroyed by proper processing and cooking, food poisoning can develop. Diseases frequently transmitted via foods and water are botulism, staphylococcal food poisoning, diarrhea caused by *Salmonella* and *Shigella* species, typhoid fever (caused by *Salmonella typhi*), infectious hepatitis (caused by hepatitis A virus), amebiasis (caused by the ameba, *Entamoeba histolytica*), giardiasis (diarrhea caused by the protozoan, *Giardia lamblia*), and trichinosis (caused by *Trichinella spiralis* worms in pork).

Blood is normally sterile; thus, the discovery of organisms in the blood often indicates the presence of an infection. A vector is often necessary to carry pathogens from the blood of one person to another. Common vectors are arthropods that bite an infected person or animal and then carry the pathogens to a healthy individual. Included among the arthropod vectors are insects such as mosquitoes, flies, fleas, and lice and arachnids such as mites and ticks. *Mechanical vectors* merely transport the pathogen from one host to another, but when the pathogen multiples or matures within the arthropod, it is known as a *biological vector*.

To better understand this mode of transmission, consider the tick. (If you own a dog, you probably already have an appropriate concern about ticks.) *Rickettsia rickettsii,* which causes Rocky Mountain spotted fever (see Chapter 2), is widespread throughout the animal population and their ticks. If an infected tick bites a person, the pathogen may be injected, causing an infection. Ticks may also carry the pathogens of Q fever, typhus, Lyme disease, babesiosis, tularemia, and the recently described diseases human granulocytic erlichiosis (HGE) and human monocytic erlichiosis (HME). In similar fashion, body lice and head lice are the vectors of epidemic typhus, trench fever, and relapsing fever. Blood-sucking fleas carry the pathogens of plague and endemic typhus. Deer flies transmit tularemia, tsetse flies carry African sleeping sickness, and certain species of mosquitoes transmit malaria, viral encephalitis, and yellow fever.

Many pets and other animals are important reservoirs of *zoonoses* (infections of animals transmissible to humans). Dogs, cats, bats, skunks, and other animals are known vectors of rabies, transmitting the rabies virus to a human via the saliva injected when they bite. Salmonellosis is frequently acquired from the feces of turtles and poultry. Cat and dog bites easily transfer *Pasteurella, Staphylococcus,* and *Streptococcus* species into tissues where severe infections may result. Toxoplasmosis, a protozoan disease carried by cats and other animals, may cause severe brain damage to the fetus when contracted by a woman during the first 3 months of pregnancy. This infective agent can be contracted by contact with cat feces in cat litter boxes as well as from ingestion of infected raw or undercooked meats.

The instruments and apparatus handled by medical personnel are common inanimate vectors of blood and internal infections. These most often include nonsterile syringes, needles, and solutions as well as blood-processing equipment such as kidney dialyzers and blood-transfusing apparatus (Fig. 7 – 11). One reason why disposable sterile tubes, syringes, and various other types of single-use hospital equipment have become very popular is that they are effective in preventing blood infections that result from the re-use of equipment. Any blood-borne disease can be transferred by improperly sterilized instruments and equipment. Hepatitis, syphilis, malaria, AIDS, and systemic staphylococcal infections are the diseases most often transmitted in this manner. Individuals using illegal intravenous drugs easily transmit these diseases to

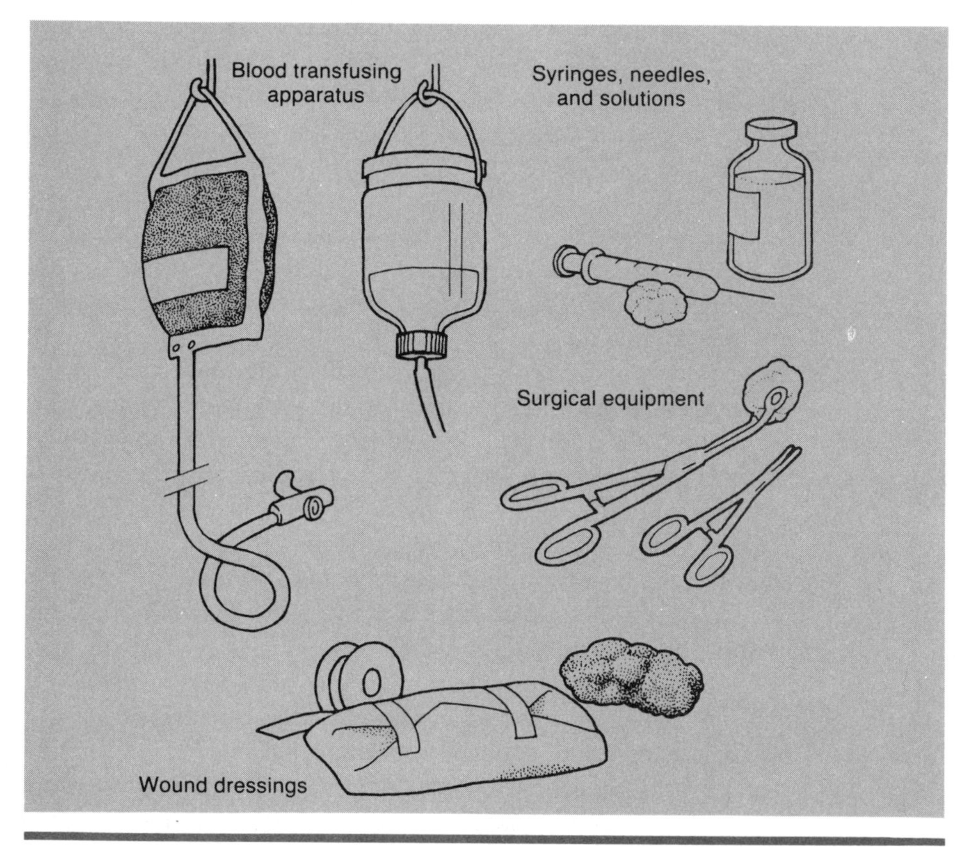

FIGURE 7 – 11
Various medical instruments and apparatus that may serve as inanimate vectors of infection (fomites).

each other by sharing needles and syringes, which easily become contaminated with the blood of an infected person.

CONTROL OF EPIDEMIC DISEASES

The WHO and the U.S. Public Health Service constantly strive to prevent epidemics and to identify and eliminate any that do occur. One way in which health personnel and community workers participate in this massive program is by reporting cases of communicable diseases to the proper agencies. They also can help by educating the public by describing how diseases are transmitted and explaining proper sanitation procedures, by identifying and attempting to eliminate reservoirs of infection, by carrying out measures to isolate diseased persons, by participating in immunization programs, and by helping to treat sick persons. Through measures like these, diphtheria, poliomyelitis, and smallpox have been totally or nearly eliminated in most parts of the world. Everyone in our society should contribute in whatever way possible to eliminate infectious diseases from the human environment.

SUMMARY

It is important to understand the factors that enable pathogens to invade tissues and cause disease. When individuals are exposed to pathogens, these microorganisms may or may not be able to cause disease. Disease causation depends on the natural resistance of the individual and the virulence, infectivity, invasiveness, and toxicity of the pathogens. Natural defenses of the human body include the skin, mucous membranes, bacteriolytic secretions, indigenous microflora, phagocytes, antibodies, and the process of inflammation. Pathogenic microorganisms are better able to resist phagocytosis and the inflammatory response than nonpathogens. Also, many secrete enzymes that enable them to invade and cause damage to tissues. Thus, they may overwhelm and win the battle against the body defenses and cause disease in localized areas or throughout the body.

The frequency and distribution of cases of infectious diseases and the factors that contribute to the spread of these diseases constitute the field of epidemiology. The factors that contribute to the occurrence of epidemics include virulence of the pathogen, susceptibility of the population, sanitation practices, and ways by which pathogens are transmitted. To eradicate certain diseases and to prevent epidemics, all of these factors must be considered.

PROBLEMS AND QUESTIONS

1. Define infectious, communicable, and contagious diseases. Why are some diseases more contagious than others?
2. List several ways that the human body resists infection by foreign invaders.
3. Define and illustrate the following terms used to describe infectious diseases: acute, chronic, latent, primary, secondary, local, and systemic.
4. Name four phases in the course of infectious diseases.
5. On what general properties do virulence and pathogenicity depend?
6. Compare exotoxins with endotoxins. List some diseases produced by each.
7. Name some general types of exotoxins.
8. Define and give examples of epidemic, endemic, pandemic, sporadic, and nonendemic diseases.
9. In what way do carriers influence epidemics?
10. Discuss how reservoirs of infection could be reduced or destroyed.
11. How could each of the five modes of transmission be interrupted to prevent diseases transmitted in that manner?
12. If you were called to a distant island to stop an epidemic, what measures would you take?

Self Test

After you have read Chapter 7, examined the objectives, reviewed the chapter outline, studied the new terms, and answered the problems and questions above, complete the following self test.

Matching Exercises

Complete each statement from the list of words provided with each section.

INFECTIOUS DISEASES

communicable
contagious
infectious
inflammation
virulence

1. The ability of a pathogen to produce disease may depend on the pathogen's ______________.
2. The body tissues respond to damage or infections by a process called ______________.
3. A disease that results from dietary deficiencies, such as rickets, is not a/an ______________ disease.
4. A disease caused by the growth of microorganisms is a/an ______________.

5. Diseases that are very easily transmitted, such as colds, are called ________________.
6. Diseases that result from close contact with the infected person are termed ________________ diseases.
7. The symptoms fever, redness, swelling, and pain are indications of a/an ________________.

TYPES OF DISEASE STATES

local	chronic	asymptomatic
generalized	primary	carrier
latent	secondary	systemic
acute		

1. When an infection is confined to a single area it is ________________.
2. During the course of a disease, the pathogen may be walled off or hidden in certain organs; then the infection is ________________.
3. If the pathogen is present and detectable but there are no symptoms of the disease, it is ________________.
4. A disease that has a rapid onset and a rapid recovery period is said to be ________________.
5. If the disease is characterized by slow onset and long duration, it is said to be ________________.
6. A person who has recovered from a disease but who still harbors the pathogens and may transmit them to others is known as a ________________.
7. An infection that has invaded the bloodstream has become ________________ and thus is a ________________ infection.
8. A ________________ infection, such as influenza, may leave the individual in a weakened condition so that a ________________ infection may occur because the body defenses are at a low level.

DISEASE CAUSATION

coagulase	hyaluronidase	leukocidin
streptokinase	collagenase	endotoxin
fibrinolysin	exotoxins	hemolysins

1. Very toxic proteins that are secreted by living pathogens that cause diseases such as botulism, tetanus, and diphtheria are ________________.
2. The protective coating outside the cell wall of an organism, which may protect it from phagocytosis, is a ________________.
3. An enzyme secreted by a pathogen that enables it to spread through connective tissue by breaking down hyaluronic acid is ________________.
4. Some gram-negative bacteria cause disease because their cell walls are very toxic to humans. This toxic material, which is part of the bacterial cell, is an ________________.

5. An enzyme secreted by some bacteria that clots plasma around the bacterial cell is ________________.
6. Streptococci can break down a fibrin blood clot by secreting ______________, which could also be called ___________.
7. Red blood cells (erythrocytes) are lysed by ________________ when α- and β-hemolytic streptococci are grown on blood agar.
8. Some streptococci and staphylococci are able to destroy white blood cells (leukocytes) by secreting ____________.
9. An enzyme that destroys collagen in tendons and cartilage is ____________.

EPIDEMIOLOGY

epidemic
endemic
pandemic
sporadic
nonendemic

1. An epidemic that spreads throughout the world is a/an ________________.
2. When a greater than usual number of cases of a disease occur in a specific area during a certain time, the disease has reached ________________ proportions.
3. Those diseases, such as smallpox, that do not occur in most countries are ________________.
4. Certain diseases, such as botulism, that occur only occasionally are ________________.
5. Some diseases are usually present in an area, but the numbers of cases vary with the season and month of the year. These diseases are ________________.
6. In most areas, colds and influenza are ________________.
7. In the United States diphtheria has been controlled as a result of immunization programs; it is a/an __________ disease.
8. In recent years gonorrhea has become ________________ in the United States and ________________ in the world.

RESERVOIRS OF INFECTION

carriers
STDs
vectors
arthropod
fomites
respiratory
fecal material
pathogens

1. Arthropods such as mosquitoes, ticks, fleas, mites, and lice that transfer pathogens by their bites are _________.
2. Pathogens may be transferred to a susceptible patient via bed clothes, urinals, towels, or sheets of another sick patient. These items are referred to as ________________.
3. Food, milk, and water supplies are easily contaminated with human ________________ containing enteric pathogens.
4. Contaminated syringes, needles, and other hospital equipment frequently serve as inanimate ________________ of disease agents.
5. The most contagious diseases are usually transferred by droplets of moisture in the air and by dust; these particles often carry ________________.

6. Persons who transmit pathogens to other susceptible people but who have no symptoms of the disease themselves are called ______________.
7. Pathogens that must be transmitted by mucous membrane-to-mucous membrane contact are the ones that cause ______________, such as syphilis and gonorrhea.
8. Rocky Mountain spotted fever and tularemia are transmitted from wildlife to humans by the bite of infected ticks, which are a type of ______________.
9. Hepatitis and syphilis are often transmitted by blood transfers in which the blood as well as the needles are considered to be ______________.

True or False (T or F)

___ 1. Infectious diseases are those caused by the growth of pathogenic microorganisms.
___ 2. The virulence of a pathogen depends on its ability to infect, invade, and damage the host.
___ 3. A communicable disease is one that can be transmitted from one person to another by direct contact.
___ 4. When a pathogen lands on the skin, it immediately starts to grow and cause a sore.
___ 5. A carrier is a person who harbors a pathogen and may transmit it to others but does not have symptoms of the disease caused by that pathogen.
___ 6. An acute disease, such as gonorrhea, may become a chronic one.
___ 7. In the latent stage, herpes cold sores are not apparent.
___ 8. Gonorrhea is never asymptomatic.
___ 9. Pneumonia and ear infections are usually primary diseases.
___ 10. Streptokinase helps staphylococci break down blood clots.
___ 11. The actual number of cases of an endemic disease depends on many factors, including the environment.
___ 12. Hospital staphylococcal epidemics are usually caused by the hospital personnel.
___ 13. Gonorrhea has never been epidemic in the United States.
___ 14. Living hosts or inanimate objects may be reservoirs of infection.
___ 15. Humans are the most important reservoir of infection for human diseases.
___ 16. Arthropod vectors are all insects that carry pathogens.
___ 17. Body lice may be vectors of epidemic typhus, trachoma, and impetigo.
___ 18. Improperly sterilized syringes, needles, and blood transfusion and dialysis equipment are frequently the sources of blood infections such as systemic staphylococcus and hepatitis.
___ 19. An endemic disease can become an epidemic.
___ 20. *Legionella pneumophila* was the causative agent of an endemic disease.
___ 21. Toxoplasmosis is an example of a zoonosis.

Multiple Choice

1. Measles is a disease considered to be
   a. avirulent
   b. eliminated
   c. contagious
   d. secondary
2. Syphilis is caused by
   a. *Staphylococcus aureus*
   b. *Yersinia pestis*
   c. *Treponema pallidum*
   d. *Streptococcus syphilitis*
3. Leprosy would be an example of a disease that is
   a. acute
   b. local
   c. chronic
   d. viral
4. *Clostridium botulinum* is pathogenic because of its
   a. exotoxin
   b. endospore
   c. endotoxin
   d. capsule
5. An example of a systemic disease would be
   a. herpes cold sores
   b. "staph" septicemia
   c. ear infections
   d. sinusitis
6. Diseases are usually most contagious during the
   a. convalescent period
   b. latent period
   c. period of illness
   d. incubation period
7. Enterotoxins that cause gastroenteritis are secreted by
   a. *Haemophilus ducreyi*
   b. *Vibrio cholerae*
   c. *Pseudomonas aeruginosa*
   d. *Bordetella pertussis*
8. Virulence indicates
   a. the ability to resist disease
   b. the production of toxins
   c. the ability to produce disease
   d. high rate of reproduction
9. Enterotoxins tend to affect the cells of the
   a. skin
   b. heart
   c. vascular system
   d. intestinal tract
10. An example of a sporadic disease is
    a. tuberculosis
    b. tetanus
    c. influenza
    d. poliomyelitis
11. A passive carrier
    a. is recovering from the disease
    b. is an inanimate vector
    c. has never had the disease but carries the pathogen
    d. has recovered from the disease but still carries the pathogen
12. Tularemia, typhus, and Rocky Mountain spotted fever are transmitted by
    a. inanimate vectors
    b. direct skin contact
    c. poor hygiene
    d. arthropod vectors
13. Smallpox is nonendemic primarily because of
    a. the avirulence of the organism
    b. the mutagenic properties of the organism
    c. immunization programs
    d. the increased number of passive carriers
14. Reservoirs of infection are
    a. only inanimate objects
    b. any site where pathogens survive
    c. only living hosts
    d. only fomites

CHAPTER 8

Preventing the Spread of Communicable Diseases

OBJECTIVES

After studying this chapter, you should be able to

- List six factors that have contributed to an increase in hospital-acquired (nosocomial) infections
- List areas in the hospital where nosocomial infections are most probable
- List several types of patients who are extremely vulnerable to infectious diseases
- Write a brief description of reverse isolation and source isolation
- Briefly describe important procedures to follow in universal precautions
- Discuss the role of health-care workers in the collection of clinical specimens
- List types of specimens that usually must be collected from patients
- Discuss general precautions that must be observed during collection and handling of specimens
- Describe proper procedures for obtaining specimens

- Discuss the importance of quality control in a microbiology laboratory
- List sources of water contamination
- Describe how water and sewage are usually treated
- Discuss how epidemics are controlled and prevented

CHAPTER OUTLINE

Prevention of Hospital-Acquired Infections
How Hospital-Acquired Infections Develop
General Control Measures
Prevention of Airborne Contamination
Handling Food and Eating Utensils
Handling of Fomites
Hand-Washing
Infection Control Procedures
Medical and Surgical Asepsis
Isolation of Patients
Hospital Infection Control
Medical Waste Disposal
Specimen Collecting, Processing, and Testing
Role of Health Care Personnel
Proper Collection of Specimens
Types of Specimens Usually Required
Blood
Urine
Cerebrospinal Fluid
Sputum
Mucous Membrane Swabs
Feces
Shipping Specimens
Identification and Antimicrobial Susceptibility Testing of Pathogens
Types of Tests Used for Identification
Antimicrobial Susceptibility Testing
Quality Control in the Laboratory
Environmental Disease Control Measures
Public Health
Water Supplies and Sewage Disposal
Sources of Water Contamination
Water and Sewage Treatment

NEW TERMS

Bacteremia
Bacteriuria
Community-acquired infection
Hepatitis B virus (HBV)
Hospital-acquired infection
Human immunodeficiency virus (HIV)
Iatrogenic infection
Medical asepsis
Medical aseptic technique
Nosocomial infection
Reverse isolation
Septicemia
Source isolation
Surgical asepsis
Surgical aseptic technique
Universal precautions

PREVENTION OF HOSPITAL-ACQUIRED INFECTIONS

How Hospital-Acquired Infections Develop

The importance of microbiology to those who work in health-related occupations can never be overemphasized. Whether working in a hospital, nursing home, medical or dental clinic, or caring for sick persons in their homes, all health-care workers must follow the same procedures to prevent the spread of communicable diseases.

Thoughtless or careless actions when providing patient care can cause preventable infections. Infections associated with hospitalization are of two types: *community-acquired infections* and *hospital-acquired infections*; the latter are also called *nosocomial infections.* According to the Centers for Disease Control and Prevention (CDC), community-acquired infections are those present or incubating at the time of hospital admission. All other hospital infections are considered nosocomial, including those that erupt within 14 days of hospital discharge. *Iatrogenic infections* are caused by surgeons or other physicians.

A nosocomial infection often adds several weeks to the patient's hospital stay and may cause serious complications and even death. Cross-infections transmitted by hospital personnel, including physicians, are all too common; this is particularly true when hospitals and clinics are overcrowded and the staff is overworked. However, these infections *can* be avoided through proper care and the disciplined use of aseptic techniques and precautions.

Nurses and physicians play major roles in preventing nosocomial infections; nevertheless, respiratory therapists, physical therapists, laboratory technicians, phlebotomists, occupational therapists, radiologic technologists, dental hygienists, dentists, and all others who deal with patients must be equally knowledgeable about methods of preventing the spread of pathogens. Members of the hospital housekeeping staff and central supply department; those who prepare, dispense, and dispose of food; administrative personnel; and those who dispose of medical wastes can help prevent cross-contamination among patients or, through their carelessness, contribute to the spread of diseases. Health workers must protect themselves from infectious agents in all situations. Of course, sick and debilitated (weakened) hospitalized patients are much more susceptible to even minor opportunistic pathogens than are the healthy people who are caring for them.

The number of nosocomial infections has increased during the past 25 years to more than 2 million cases per year despite the availability of new disinfectants and antibiotics. There are many reasons for this situation.

1. Indiscriminate use of broad-spectrum antibiotics, which may allow opportunistic pathogens to mutate and become resistant to antibiotics.

2. A false sense of security about the effectiveness of antibiotics, with a corresponding neglect of aseptic techniques and precautions.
3. Lengthy, more complicated types of surgery.
4. Overcrowding of hospitals and shortages of staff.
5. Increased numbers and types of hospital workers who are often not aware of the importance of routine infection control and aseptic and sterile techniques.
6. Increased use of anti-inflammatory and immunosuppressant agents such as radiation, steroids, anticancer chemotherapy, and antilymphocyte serum.
7. Use of indwelling medical devices.

The presence of even one of these factors can enable opportunistic pathogens to cause problems; taken together, they are the cause of approximately 60% of hospital-acquired infections. The most common causes of nosocomial infections are *Staphylococcus aureus, Escherichia coli, Enterococcus* species, and *Pseudomonas* species, which collectively cause most urinary tract, wound, burn, respiratory, and surgical infections.

Medical devices that support or monitor basic body functions contribute greatly to the success of modern treatment. However, by bypassing normal defensive barriers, these devices provide microorganisms access to normally sterile body fluids and tissues. The risk of bacterial or fungal infection is related to the degree of debilitation of the patient and the design and management of the device. The most common nosocomial infections are urinary tract infections associated with the use of urinary catheters. Thus, it is advisable to discontinue the use of urinary catheters, vascular catheters, respirators, and hemodialysis as soon as medically feasible.

The emergency room, operating room, delivery room, nursery, and central supply area are the most critical areas for disease transmission. In the emergency room, many patients who are carrying unknown pathogens are rushed in and treated quickly to halt life-threatening situations; in haste, the emergency room attendants may neglect to protect themselves and others by failing to use recommended precautions. In operating and delivery rooms, particular care must be taken because portals of entry into the body are readily accessible to pathogens. The nursery is critical because newborns have very little resistance to disease. The central supply department must take great care not to distribute contaminated materials or supplies, because these might expose patients to a myriad of pathogens.

The most vulnerable patients in a hospital are final

- premature infants and newborns;
- women in labor and delivery;
- surgical and burn patients;

INSIGHT
Nosocomial Infections

A nosocomial infection is an infection that a person acquires while they are hospitalized. The person was not infected at the time they entered the hospital, but became infected during his or her hospital stay. Of the approximately 40 million patients who are hospitalized in the United States per year, an estimated 2 million (5%) acquire nosocomial infections. An estimated 60,000 to 70,000 deaths per year are related to nosocomial infections, and nosocomial infections add an estimated minimum of $4.5 billion to the cost of health care in the United States each year.

The hospital setting harbors many pathogens and potential pathogens. They live on and in health-care workers, other hospital employees, and patients themselves. Some live in dust, whereas others live in wet or moist areas like sink drains, shower heads, whirlpool baths, mop buckets, flowerpots, and even food.

The most common pathogens associated with nosocomial infections are *Escherichia coli, Staphylococcus aureus, Pseudomonas aeruginosa,* and *Enterococcus* species. Although some of these pathogens come from the external environment, most come from the patients themselves—their own indigenous microflora that enters a surgical incision or otherwise gains entrance to the body. Urinary catheters, for example, provide a "superhighway" for organisms to gain access to the urinary bladder. In fact, urinary tract infections are the most common type of nosocomial infections, followed by surgical wound infections, lower respiratory tract infections, and bacteremia.

About 50–60% of nosocomial infections involve drug-resistant bacteria, which are common in the hospital environment as a result of all the antimicrobial agents that are used. The drugs have put selective pressure on the microbes, ensuring that only those that are resistant to the drugs will survive. *Pseudomonas* infections are especially hard to treat, as are infections caused by vancomycin-resistant *Enterococcus* species and methicillin-resistant strains of *Staphylococcus aureus* and *Staphylococcus epidermidis.*

Health-care workers and other hospital employees must be aware of the problem of nosocomial infections and must take appropriate measures to minimize the number of such infections that occur within the hospital. Handwashing remains the most effective means of control; hands must be washed before and after working with every patient. Other means of reducing the incidence of nosocomial infections include sterilization techniques, air filtration, use of ultraviolet lights, isolating especially infectious patients, and wearing gloves, masks, and gowns whenever appropriate.

- diabetic and cancer patients;
- those receiving treatment with steroids, anticancer drugs, antilymphocyte serum, and radiation;
- those with a deficient immune response, *e.g.*, patients with acquired immunodeficiency syndrome (*AIDS*); and
- patients who are paralyzed or are undergoing renal dialysis or catheterization; these patients' normal defense mechanisms are not working properly.

The greatest risk to health-care workers who handle blood and body fluids is the transmission of *hepatitis B virus* (HBV) and/or *human immunodeficiency virus* (HIV). Control of these viruses requires meticulous attention to procedures, such as the use of gloves and gowns, that prevent direct contact with blood and other body substances.

Nosocomial viral infections are frequently ignored because most hospitals lack adequate viral diagnostic laboratories. Influenza virus, respiratory syncytial virus (RSV), and other respiratory viruses have been shown to spread in hospitals via direct contact or droplet inhalation. Other highly infectious viruses that cause measles and chickenpox may cause outbreaks among susceptible patients and hospital staff.

General Control Measures

A general state of cleanliness must be maintained throughout the hospital or health-care institution using general principles of sanitation, disinfection, and sterilization. Each area, from the trash disposal area to the operating rooms, must be thoroughly cleaned and disinfected to prevent the growth and spread of microorganisms.

Hospitals vary greatly in their specific requirements for the maintenance of medical asepsis, but there is general agreement about sanitary methods for handling food and eating utensils, proper cooking and storing of food, proper disposal of waste products and contaminated materials, proper hand-washing and personal hygiene of hospital personnel, proper use of gowns and masks in isolation rooms, proper washing and sterilizing of hospital equipment, proper use of disposable equipment, and proper disinfection of a room after a patient has been discharged.

Prevention of Airborne Contamination Respiratory infections are most often transmitted through the air. The following measures should be taken to decrease the number of pathogens transmitted by this means:

- Cover the mouth and nose when coughing or sneezing.
- Limit the number of persons in a room.
- Remove the dirt and dust from the floor and furniture by dusting with a damp cloth.
- Open the room to fresh air and sunlight whenever possible.
- Roll linens together carefully to prevent dispersal of microbes in the air.
- Remove bacteria from the air with a filtered air-conditioning system.

Handling Food and Eating Utensils Contaminated food provides an excellent environment for the growth of pathogens. Most often, human carelessness—such as failing to take precautions when handling fecal material, flies, insects, dust, dirt, and domestic animals and pets, as well as neglecting the practice of hand-washing—is responsible for this contamination. The pathogens most frequently carried in food are *Staphylococcus* species (from skin and dust); *Clostridium botulinum* (from dust and dirt); *Clostridium perfringens* (from dust, dirt, and hands); *Salmonella, Shigella,* and *Proteus* species (from feces, hands, flies, and pets); and *Pseudomonas* species (from dirt, hands, and contaminated equipment).

Regulations for safe handling of food and eating utensils are not difficult to follow. They include

- using high-quality fresh food;
- properly refrigerating and storing food;
- properly washing, preparing, and cooking food;
- properly disposing of uneaten food;
- thoroughly washing hands and fingernails before handling food and after using a restroom;
- properly disposing of nasal and oral secretions in tissues;
- covering hair and wearing clean clothes and aprons;
- providing periodic health examinations for kitchen workers;
- prohibiting anyone with an infection or intestinal upset from handling food or eating utensils;
- keeping all kitchen equipment and all other equipment scrupulously clean; and
- rinsing and then washing eating utensils in a dishwasher in which the temperature is above 80° C (176° F).

Handling of Fomites As previously described, fomites are any articles or substances other than food that may harbor and transmit microbes. Examples of fomites are eating utensils, bedpans, urinals, thermometers, washbasins, bed linen, and clothing and other personal items. Transmission of pathogens by these items may be prevented by observing the following rules:

- Use disposable equipment and supplies wherever possible.
- Disinfect or sterilize equipment as soon as possible after use.
- Use individual equipment for each patient.
- Use an individual thermometer for each patient and store each thermometer in a disinfectant solution.
- Empty bedpans and urinals, wash them in hot water, and store them in a clean cabinet between uses.
- Place bed linen and soiled clothing in bags to be sent to the laundry.

Hand-Washing Hand-washing is the most important aseptic precaution. Hands should be kept clean at all times and should be washed before and after contact with every patient to preclude carrying pathogens from one patient to another. A disinfectant soap containing hexachlorophene or some other effective disinfectant is often used to control staphylococci. Some common-sense rules include the following:

- Use gloves or tongs to handle contaminated materials.
- Wash hands with disinfectant soap before and after contact with every patient.

- Rinse hands under running water.
- Dry hands and apply antiseptic lotion to prevent chapping.

Infection Control Procedures

From the discoveries and observations of Semmelweis and Lister in the 19th century, we know that wound contamination is not inevitable and that we must prevent microorganisms from reaching susceptible areas, a concept referred to as asepsis. Asepsis may be medical or surgical in nature. The techniques used to achieve asepsis depend on the site, circumstances, and environment.

MEDICAL AND SURGICAL ASEPSIS

Once basic cleanliness is achieved, it is not difficult to maintain asepsis. The goal of *medical asepsis* is to exclude all pathogenic microorganisms from the immediate environment. *Medical aseptic techniques* include all precautionary measures necessary to prevent direct transfer of pathogens from person to person and indirect transfer of pathogens through the air or on instruments, bedding, equipment, and other inanimate objects (fomites).

In the hospital, medical asepsis is practiced using sterile equipment, dressings, medications, and any items that could transfer microorganisms to susceptible sites. Invasive procedures—such as drawing blood, giving injections, inserting urinary and cardiac catheters, and performing lumbar punctures—must be performed while taking strict aseptic precautions. The most controlled and strict aseptic procedures must be applied in the operating room and other surgical areas. The goal of *surgical asepsis* is to exclude *all* microorganisms from the immediate environment. *Surgical aseptic techniques* include those practices that make and keep all objects and the area itself sterile. These practices are necessary during all surgical procedures, and any other procedure that involves exposure of the deep body tissues, to prevent the entry of any microorganisms into those tissues.

The surgical area of the patient's skin must be shaved and thoroughly cleansed and scrubbed with soap and antiseptic. If the surgery is to be extensive, the surrounding area is covered with a sterile plastic film or sterile cloth drapes so that a sterile surgical field is established. The surgeon and all surgical assistants must scrub for 10 minutes with a disinfectant soap and cover their clothes, mouth, and hair because these might shed microorganisms onto the operative site. These coverings include sterile gloves, gowns, caps, and masks (Fig. 8 – 1). All instruments, sutures, and dressings must be sterilized; as soon as they are contaminated, they must be discarded and replaced with sterile ones. Any instruments or equipment that cannot be autoclaved must be properly disposed of. All needles, syringes, and sharps must be placed in appropriate containers. Remember that surgical asepsis is a sterile technique, and medical asepsis is a clean technique.

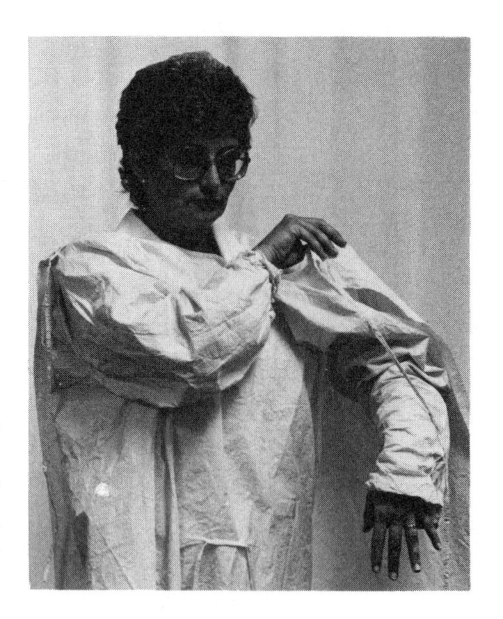

FIGURE 8 – 1
Nurse donning a sterile gown. (Timby BK, Lewis LW: Fundamental Skills and Concepts in Patient Care, 5th ed. Philadelphia, JB Lippincott, 1992)

Floors, walls, and all equipment in the operating room must be cleaned and disinfected before each use. Proper ventilation must be maintained to ensure that fresh, filtered air is circulated throughout the room at all times.

Universal Body Substance Precautions Hepatitis B virus and HIV are occasionally transmitted from patients to medical staff, with transmission of HBV being more common than transmission of HIV. Blood and blood products are the primary modes of transmission of these viruses in a medical setting; the overall most common means of transmission are by sexual contact, contaminated needles, organ transplants, transfusions, and transmission of the virus from infected mothers to their infants.

In a medical setting, *all* patients are considered potentially infectious for HIV, HBV, and other bloodborne pathogens. For this reason, *universal precautions* or *universal body substance precautions* have been recommended by the CDC to prevent transmission of these bloodborne pathogens in health-care facilities. This means that the *same* safety measures are taken whenever potential contact with body substances from *any* patient is possible. Contact with the patient's blood or mucosal surfaces or penetration of the skin by a contaminated "sharp," *e.g.*, needle-stick injuries, is the primary concern. Universal precautions demand the use of gloves, gowns, masks, and sometimes goggles during high-risk procedures such as suctioning, bronchoscopy, surgery, and many emergency or dental procedures.

Universal precautions include, but are not limited to, the following procedures:

- Hands should always be washed before and after contact with patients. Hands should be washed even before and after gloves are used. If hands come into contact with blood, body fluids, or human tissue, they should be immediately washed with soap and water.

- Gloves should be worn when contact with blood, body fluid, body tissues, or contaminated surfaces is anticipated.
- Gowns are worn if blood splattering is likely.
- Masks and protective goggles should be worn if aerosolization or splattering is likely to occur, such as in certain surgical procedures, wound irrigations, and bronchoscopy.
- The need for emergency mouth-to-mouth resuscitation and mouthpieces should be minimized by strategically locating resuscitation bags or other ventilation devices in areas where they may be needed.
- Sharp objects (sharps), such as needles and broken glass, should be handled cautiously to prevent accidental cuts or punctures. Used needles should never be bent, broken, reinserted into their original sheath, or unnecessarily handled. They should be discarded intact immediately after use into an easily accessible, impervious needle disposal container. All needle-stick accidents, mucosal splashes, or contamination of open wounds with blood or body fluids should be reported immediately.
- Blood spills should be cleaned up promptly with a disinfectant solution, such as a 1:10 dilution of chlorine bleach.
- All patients' blood specimens should be considered biohazardous.

Although universal precautions apply to blood and other body fluids containing visible blood, they should also be applied to other tissues and fluids, such as semen and vaginal secretions, cerebrospinal fluid, synovial fluid, pleural fluid, peritoneal fluid, pericardial fluid, amniotic fluid, and breast milk. The risk of transmission from these fluids is unknown, and results of epidemiologic studies are currently inadequate to assess the potential risk to health-care workers from occupational exposure to them.

Universal precautions do not apply to feces, nasal secretions, sputum, sweat, tears, urine, and vomitus, unless they contain visible blood. The risk of transmission of HIV and HBV from these fluids and materials is extremely low; however, some of these specimens are potential sources for nosocomial and community infections. For this reason, care should be exercised when handling *any* type of clinical specimen (Fig. 8 – 2).

Although universal precautions do not apply to saliva, general infection control practices should be observed, including wearing gloves for examination of mucous membranes and for endotracheal suctioning as well as handwashing after exposure to saliva to further reduce the minute risk of salivary transmission of HIV and HBV. However, special precautions, including the use of gloves and masks, must be observed for dental procedures.

The implementation of universal precautions does not eliminate the need for other category-specific or disease-specific isolation precautions, such as enteric precautions for infectious diarrhea or isolation for pulmonary tuberculosis.

FIGURE 8 – 2
Body substance isolation sign (Taylor C, et al.: Fundamentals of Nursing, 2nd ed. Philadelphia, JB Lippincott, 1993)

Universal precautions are intended to supplement rather than to replace other recommendations for routine infection control, such as hand-washing and using gloves to prevent gross microbial contamination of hands. These guidelines are published by the CDC, Atlanta, Georgia, in the *Morbidity and Mortality Weekly Report* (published in August 1987 and June 1988, in *MMWR*; available through the National AIDS Information Clearinghouse, P.O. Box 6003, Rockville, MD 20850; see also *Federal Register,* Vol. 59, no. 214, pp. 55552–55570, Draft Guideline for Isolation Precautions in Hospitals).

Isolation of Patients

Patients are placed in isolation for one of two reasons: (1) to prevent the spread of pathogens to other susceptible people or (2) to protect a very susceptible patient from exposure to pathogens.

Reverse Isolation Certain patients are especially vulnerable to infection; among them are patients with severe burns, those who have leukemia, patients who have received a transplant, immunodeficient persons, and those receiving radiation treatments. Premature babies are also highly susceptible. All such patients are protected through isolation procedures. This type of isolation is referred to as *reverse isolation* (also referred to as protective or neutropenic isolation). The room must be thoroughly cleaned and disinfected before the patient is admitted. Those entering the room must wear sterile gowns and masks to prevent depositing microorganisms into the room from their clothes or respiratory tracts. Proper hand-washing procedures must also be followed before entering the room.

Source Isolation Whenever possible, patients with contagious diseases should be isolated in private air-conditioned rooms with a private bath (Table 8 – 1, Fig. 8 – 3). In this manner, infectious agents are isolated within a definite area and the spread of pathogens to other patients is prevented; this is referred to as *source isolation*. Isolation procedures should be determined by

- mode of transmission of the disease from one person to another;
- location of the pathogen causing the infection, its portal of exit (intestinal, respiratory, or wound drainage), and its portal of entry; and
- virulence of the pathogen and its susceptibility to antibiotics.

Isolation Techniques Isolation techniques are designed to immediately destroy pathogenic organisms in the infectious discharges of a patient. Nondisposable items are disinfected, with a minimum of handling, then cleaned thoroughly and resterilized. If the patient's infection involves spore-forming

TABLE 8 – 1.
Category-Specific Isolation System

Type	Purpose	Specifications						Diseases Requiring Isolation
		Private Room	Hand-washing	Gowns	Masks	Gloves	Articles	
Strict	Prevents transmission of highly contagious or virulent infections spread by air and contact	X	X	X	X	X	Discard or bag and label and send for decontamination and reprocessing	Diphtheria (pharyngeal) Lassa fever Smallpox Varicella
Contact	Prevents transmission of highly transmissible infections that do not require strict isolation	X	X	Wear if soiling is likely	Wear in close contact with patient	Wear if touching infective material	Discard or bag and label and send for decontamination and reprocessing	Acute respiratory tract infection in infants and young children Herpes simplex Impetigo Multi-resistant bacterial infections
Respiratory	Prevents transmission of infectious diseases primarily over short distances by air droplets	X	X	–	Wear in close contact with patient	–	Discard or bag and label and send for decontamination and reprocessing	Measles Meningitis Pneumonia, *Haemophilus influenzae* in children Mumps
Tuberculosis	For a patient with pulmonary tuberculosis who has positive sputa or chest x-ray that indicates active disease	X	X	Wear if soiling is likely	Wear if patient is coughing and does not consistently cover mouth	–	Rarely involved in transmission of tuberculosis. Should still be thoroughly cleansed and disinfected	Tuberculosis

Enteric precautions	To prevent infections that are transmitted by direct or indirect contact with feces	Indicated if patient's hygiene is poor and there is risk of contamination with infective materials	X	Wear if soiling is likely	–	Wear if touching infective material	Discard or bag and label and send for decontamination and reprocessing	Hepatitis, viral (type A) Gastroenteritis caused by highly infectious organism Cholera Diarrhea, acute with infectious cause
Drainage-secretion precautions	To prevent infections that are transmitted by direct or indirect contact with purulent material or drainage from infected site	–	X	Wear if soiling is likely	–	Wear if touching infective material	Discard or bag and label and send for decontamination and reprocessing	Abscess Burn infection Conjunctivitis Pressure ulcer Skin or wound infection
Blood–body fluid precautions	To prevent infections that are transmitted by direct or indirect contact with blood or body fluid	Only if patient's hygiene is poor	X	Wear if soiling with blood or body fluids is likely	–	Wear if touching blood or body fluids	Discard or bag and label and send for decontamination and reprocessing	AIDS Hepatitis, viral (type B) Malaria Syphilis, primary and secondary

X = necessary; – = not necessary. (Category-specific isolation system. From Taylor, C., Lillis, C., and LeMone, P.: Fundamentals of Nursing, 2nd ed, Philadelphia, J. B. Lippincott Co., 1993.)

Strict Isolation

Visitors—Report to Nurses' Station
Before Entering Room

1. Private Room—*necessary;* door must be kept closed.
2. Gowns—must be worn by all persons entering room.
3. Masks—must be worn by all persons entering room.
4. Hands—must be washed on entering and leaving room.
5. Gloves—must be worn by all persons entering room.
6. Articles—must be discarded, or wrapped before being sent to Central Supply for disinfection or sterilization.

Enteric Precautions

Visitors—Report to Nurses' Station
Before Entering Room

1. Private Room—necessary for children only.
2. Gowns—must be worn by all persons having direct contact with patient.
3. Masks—not necessary.
4. Hands—must be washed on entering and leaving room.
5. Gloves—must be worn by all persons having direct contact with patient or with articles contaminated with fecal material.
6. Articles—special precautions necessary for articles contaminated with urine and feces. Articles must be disinfected or discarded.

Protective Isolation

Visitors—Report to Nurses' Station
Before Entering Room

1. Private Room—*necessary;* door must be kept closed.
2. Gowns—must be worn by all persons entering room.
3. Masks—must be worn by all persons entering room.
4. Hands—must be washed on entering and leaving room.
5. Gloves—must be worn by all persons having direct contact with patient.

Wound & Skin Precautions

Visitors—Report to Nurses' Station
Before Entering Room

1. Private Room—desirable.
2. Gowns—must be worn by all persons having direct contact with patient.
3. Masks—not necessary except during dressing changes.
4. Hands—must be washed on entering and leaving room.
5. Gloves—must be worn by all persons having direct contact with infected area.
6. Articles—special precautions necessary for instruments, dressings, and linen.

Respiratory Isolation

Visitors—Report to Nurses' Station
Before Entering Room

1. Private Room—*necessary;* door must be kept closed.
2. Gowns—not necessary.
3. Masks—must be worn by all persons entering room if susceptible to disease.
4. Hands—must be washed on entering and leaving room.
5. Gloves—not necessary.
6. Articles—those contaminated with secretions must be disinfected.
7. Caution—all persons susceptible to the specific disease should be excluded from patient area; if contact is necessary, susceptibles must wear masks.

FIGURE 8 – 3
Hospital precaution signs used for the protection of patients, staff, and visitors. (Rosdahl CB: Textbook of Nursing, 5th ed. Philadelphia, JB Lippincott, 1991)

bacteria such as those causing gas gangrene and tetanus, then contaminated equipment must be autoclaved to destroy the spores before the items can be reused. Disposable equipment must be incinerated or thoroughly autoclaved before disposal.

Hand-washing Doctors, nurses, and other hospital personnel entering the room must be particularly careful to wash their hands before and after caring for a patient. Hand contact with door knobs, telephones, elevator buttons, and furniture easily transmits pathogens to and from the patient. Fingernails should be kept short and clean, and jewelry should not be worn in an isolation room.

Gowns In every isolation room, gowns should be worn. A clean fresh gown should be available for each person who enters the isolation room on each occasion and discarded after each use. To be protective, gowns should be large enough to be fastened securely. They must be kept dry—remember that a moist environment is required for growth of pathogens, so if an area of the gown is moist, pathogens may gain access to clothing worn under the gown.

Masks Masks generally are not worn outside the operating room. However, if a patient is isolated because of a respiratory infection, the patient must be masked whenever it is necessary for him or her to go for treatment in other parts of the hospital. Personnel working closely with an isolated respiratory patient must wear masks for self-protection. Remember that wet masks are not effective. Used masks should be promptly discarded or laundered and autoclaved. Hands should be washed again after touching a used mask.

Gloves The type of latex or vinyl gloves selected should be appropriate for the task being performed.

- Use sterile gloves for procedures involving contact with normally sterile areas of the body.
- Use examination gloves for procedures involving contact with mucous membranes and for other patient care or diagnostic procedures not requiring sterile gloves, unless otherwise indicated.
- Change gloves between patient contacts.
- Do not wash or disinfect surgical or examination gloves for reuse. Disinfectants may cause deterioration, and washing with surfactants may allow liquids to penetrate the barrier through undetected holes in the glove.
- Use general-purpose utility gloves for housekeeping chores involving potential blood contact and for instrument cleaning and decontamination procedures. Utility gloves may be decontaminated and reused but should be discarded if there is evidence of deterioration.

Hospital Infection Control

All hospitals are required by the regulatory agencies to have a formal infection control program. This program may vary slightly from hospital to hospital, but all must have an infection control committee and an epidemiology service. The infection control committee consists of representatives from the medical and surgical services, as well as the laboratory, pathology, nursing, hospital administration, housekeeping, food services, central supply, and other hospital departments. The chairperson is usually an infectious disease specialist, pathologist, microbiologist, or one knowledgeable about infection control. The committee periodically reviews the hospital's infection control procedures and incidence of nosocomial infections. It is a policy-making and review body that may take drastic action when epidemiologic circumstances warrant.

The epidemiology service is the part of the infection control committee that is charged with patient surveillance, environmental surveillance, investigation of outbreaks and epidemics, and education of the hospital staff regarding infection control. In many hospitals, these activities are actually performed by the infection control nurse or epidemiologist.

Although every department of the hospital or institution endeavors to maintain aseptic conditions, the total environment is constantly bombarded with microbes from the outside. These must be controlled for the protection of patients.

Hospital personnel (members of the infection control committee or nurses) entrusted with this aspect of health care diligently and constantly work to maintain the proper environment. The microbiology laboratory cooperates to keep constant watch over asepsis within the hospital and consults, as necessary, with various personnel to ensure the maintenance of aseptic conditions throughout the hospital.

Laboratory tests can demonstrate the presence and location of contaminating microbes as well as the effectiveness of disinfection processes. The laboratory should also be able to trace the sources of infections and make recommendations to prevent recurrence. In case of an epidemic, the infection control committee notifies city, county, and state health authorities so they can concentrate their efforts toward ending the epidemic.

Medical Waste Disposal

General Regulations According to the Occupational Safety and Health Administration (OSHA) Standards, medical wastes must be disposed of properly. These standards include the following:

- Any receptacle used for putrescible (decomposable) solid or liquid waste or refuse must be constructed so that it does not leak and must be maintained in a sanitary condition. This receptacle must be equipped with a solid, tight-fitting cover, unless it can be maintained in a sanitary condition without a cover.

- All sweepings, solid or liquid wastes, refuse, and garbage shall be removed to avoid creating a menace to health and shall be removed as often as necessary to maintain the place of employment in a sanitary condition.
- The infection control program must address the handling and disposal of potentially contaminated items.

Sharp Instruments and Disposables These items must be disposed of in the following manner:

- Needles shall not be recapped, purposely bent or broken by hand, removed from disposable syringes, or otherwise manipulated by hand.
- After use, disposable syringes, scalpel blades, and other sharp items must be placed in puncture-resistant containers for disposal of "sharps".
- These containers must be easily accessible to all personnel needing them and must be located in all areas where needles are commonly used, as in areas where blood is drawn, including patient rooms, emergency rooms, intensive care units, and surgical suites.
- The containers must be constructed so that the contents will not spill if knocked over and will not cause injuries.

Laboratory Specimens All specimens of body fluids must be placed in a well-constructed container with a secure lid to prevent leaking during transport and must be disposed of in an approved manner. Contaminated materials used in laboratory tests should be decontaminated before processing or be placed in bags and disposed of in accordance with institutional policies for disposal of infectious waste.

SPECIMEN COLLECTING, PROCESSING, AND TESTING

It would not be feasible in a book of this size to provide a complete discussion of collection, handling, and processing of specimens. Only a few important concepts are discussed here.[1]

[1]For more detailed information, refer to the *Manual of Clinical Microbiology,* 6th ed, 1995, and *Cumitech 9: Collection and Processing of Bacteriological Specimens,* 1979, both of which are published by The American Society for Microbiology, Washington, D.C.

Role of Health-Care Personnel

Extreme care must be taken by those involved in collecting, handling, and processing specimens that are to be examined for the presence of microorganisms. The microbiological analysis can be only as reliable as the quality of the specimen allows. When specimens are improperly collected and handled, (1) the causative microorganisms may not be found or may be destroyed, (2) overgrowth by indigenous microflora may mask the pathogen, and (3) contaminants may interfere with the identification of the pathogens.

A close working relationship among the members of the health-care team is essential for the proper identification of pathogens. When the attending physician recognizes the clinical symptoms of a possible infectious disease, certain specimens and clinical tests may be requested. The clinical microbiologist who performs the laboratory microbial analysis must provide adequate collection materials and instructions for their proper use. The doctor, nurse, medical technologist, or other qualified health-care worker must perform the collection procedure properly, and then the specimen must be transported quickly to the laboratory where it is cultured, stained, analyzed, and identified. Laboratory findings must then be conveyed to the attending physician as quickly as possible to facilitate the prompt diagnosis and treatment of the infectious disease.

Proper Collection of Specimens

When collecting specimens, these general procedures should be followed:

- All specimens should be collected in a sterile manner. This requires that they be placed into a sterile container to prevent contamination by indigenous microflora and airborne microorganisms.
- The material should be collected from the site where the suspected organism is most likely to be found and where the least contamination is likely to occur.
- Specimens should be obtained before antimicrobial therapy has begun. If this is not possible, the laboratory should be informed about which antimicrobial agent(s) is/are being used.
- The acute stage of the disease is the appropriate time for the collection of most specimens. Some viruses, however, are more easily isolated during the onset stage of disease.
- Specimen collection should be performed with care and tact to avoid harming the patient, causing discomfort, or causing undue embarrassment. If a specimen, such as sputum or urine, is to be collected by the patient, clear and detailed instructions should be given.
- A sufficient quantity of the specimen should be obtained to provide

enough material for all diagnostic tests that need to be performed. The amount should be indicated by the physician or laboratory microbiologist.

- Specimens should be protected from heat and cold and promptly delivered to the laboratory so the results of the analysis will be a valid representation of the organisms present at the time of collection. If delivery is delayed, some delicate pathogens will die. Obligate anaerobes die when exposed to air. Also, indigenous microflora may overgrow, inhibit, or kill pathogens. Delay of delivery considerably decreases the chances of isolating pathogens.
- Dangerous specimens must be handled with even greater care to avoid contamination of the ward messenger, patients, nurses, and other hospital personnel. Such dangerous specimens are usually placed in a sealed plastic bag for immediate transport to the laboratory.
- All specimen containers must be cleaned, sterilized, and properly stored to avoid contamination of the specimen by microbes and harmful chemicals from the container.
- After the specimen is collected, the container must be properly labeled and accompanied by appropriate laboratory instructions written on the requisition. The label must identify the patient and the source of the specimen (*e.g.*, throat, wound). The requisition must give the date, time of collection, doctor's name and address, and laboratory tests requested. The laboratory also should be given any additional clinical information that will aid in performing the appropriate analyses.
- Specimens should be collected and delivered to the laboratory as early in the day as possible to allow the technicians time to process the material, especially if the hospital or clinic does not have 24-hour laboratory service. Many hospitals require laboratory technicians to collect specimens for immediate evaluation.

Types of Specimens Usually Required

Special techniques in collecting and handling are required for specific types of specimens.

Blood Blood is usually sterile. Organisms found in blood usually indicate a disease condition, although temporary, transient bacteremias may occur following oral surgery, tooth extraction, etc. To prevent contamination with indigenous skin flora, extreme care must be taken to use sterile techniques when collecting blood for culture. After locating a suitable vein, disinfect the skin with 70% ethyl alcohol and then with 2% tincture of iodine or similar antiseptic and allow it to dry. Apply a tourniquet and withdraw 10 ml to 20 ml of blood with a 21-gauge needle into a sterile blood culture bottle, containing an anticoagulant. The blood culture bottle(s) should be transported promptly to the laboratory for incubation at 37°C.

In many infectious diseases, *bacteremia* (bacteria in the blood) may be found during certain stages. These diseases include bacterial meningitis, typhoid fever and other salmonella infections, pneumococcal pneumonia, urinary infections, endocarditis, brucellosis, tularemia, plague, anthrax, syphilis, and wound infections caused by β-hemolytic streptococci, staphylococci, and other invasive bacteria. A severe *septicemia* caused by gram-negative rods is frequently found in patients after gastrointestinal or urological surgery.

Urine In the urinary bladder, urine, like blood, is ordinarily sterile; however, it is usually contaminated by indigenous microflora of the urethra during urination. Contamination can be reduced by collecting a "midstream, clean-catch" urine specimen. The area around the external opening of the urethra is cleansed by washing with soap and rinsing with water. With the glans penis or labial folds separated, the urethra is flushed with the first portion of urine; then the middle portion of the urine is collected in a sterile container. In some circumstances, the physician may prefer to collect a catheterized specimen or to use the suprapubic needle aspiration technique to obtain a sterile sample of urine.

A urinary tract infection is indicated if the number of bacteria in a midstream clean-catch urine specimen exceeds 100,000 (1×10^5) organisms per ml (expressed as colony-forming units [CFU] per ml); healthy urine (contaminated during collection) may contain less than 10,000 CFU/ml.

To prevent continued bacterial growth, urine specimens must be processed within 1 hour or refrigerated at 4° C (39.2° F) until they can be analyzed (within 5 hours). The presence of two or more bacteria per ×1,000 microscopic field of a Gram-stained smear indicates *bacteriuria* (bacteria in the urine) with 100,000 or more CFU/ml.

Cerebrospinal Fluid Meningitis, encephalitis, and meningoencephalitis are rapidly fatal diseases that can be caused by a variety of microbes, including bacteria, fungi, protozoa, and viruses. To diagnose these diseases, spinal fluid specimens must be collected into a sterile tube by a lumbar puncture (spinal tap) under surgically aseptic conditions (Fig. 8 – 4). This difficult procedure is performed by a physician. Because *Neisseria meningitidis* (meningococci) are susceptible to cold temperatures, the specimen must be cultured immediately and not refrigerated. Specimens to be further examined for viruses may be kept frozen at −20° C (−4° F).

Sputum Sputum (pus within the lungs) may be collected by allowing the patient to spit the coughed-up specimen into a sterile wide-mouthed bottle

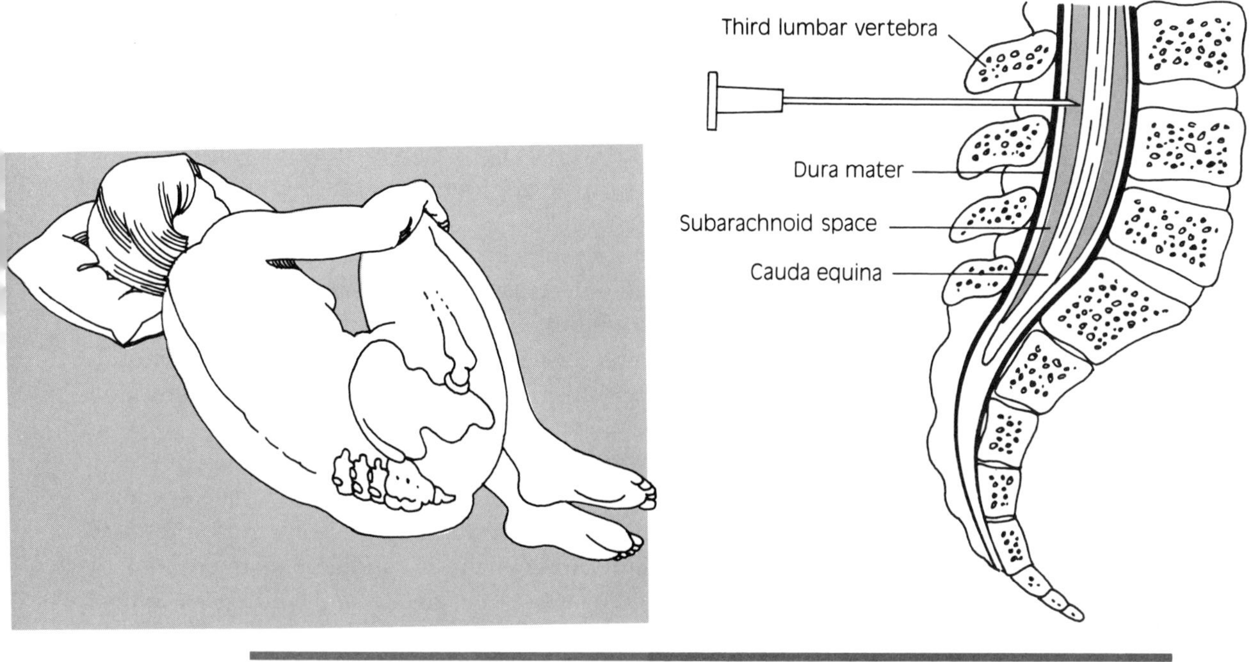

FIGURE 8 – 4
Technique of lumbar puncture. (Taylor C, et al.: Fundamentals of Nursing, 2nd ed. Philadelphia, JB Lippincott, 1993)

with a lid, after warning the patient not to contaminate the sputum with saliva. If proper mouth hygiene is maintained, the sputum may not be severely contaminated with oral microorganisms. If tuberculosis is suspected, extreme care in handling the specimen should be exercised because one could easily be infected with the pathogens. Sputum specimens usually may be refrigerated for several hours without loss of the pathogens.

However, the physician may wish to obtain a better specimen by bronchial aspiration through a bronchoscope or by inserting a needle into the trachea below the glottis (transtracheal aspiration). Needle biopsy of the lungs may be necessary for diagnosis of *Pneumocystis carinii* pneumonia (as in patients with AIDS) and for mycobacterial, fungal, and viral pathogens.

Mucous Membrane Swabs Sterile polyester swabs are used to collect specimens of exudates and secretions of the throat, nose, ear, eye, urogenital openings, rectum, wounds, operative sites, and ulcerations. Cotton swabs are no longer used because fatty acids in the cotton inhibit the growth of some microorganisms. Handy, sterile, disposable collection units can be obtained from many medical supply houses. Each unit contains a sterile polyester swab and transport medium in a sterile tube. With this set-up, the organisms are kept alive and protected during transportation to the laboratory.

Genital swabs should be inoculated immediately onto Thayer-Martin chocolate agar plates and incubated in a carbon dioxide (CO_2) environment, or they should be inoculated into a tube or bottle (*e.g.*, Transgrow®) that contains CO_2, while the bottle is held in an upright position to prevent loss of the CO_2. These cultures should be incubated at 35° C (95° F) overnight and then shipped to a public health diagnostic facility for positive identification of gonococci.

Feces Ideally, fecal specimens should be collected at the laboratory and processed immediately to prevent a decrease in temperature, which allows the pH to drop, causing the death of many *Shigella* and *Salmonella* species; or the specimen may be placed in a container with a buffer that maintains a pH of 7. In some cases, the specimens are transported anaerobically if the presence of *Clostridium* species is suspected.

The majority of fecal bacteria are obligate and facultative anaerobes, but fecal specimens are cultured anaerobically only when *Clostridium difficile*-associated disease is suspected or for diagnosing clostridial food poisoning. In intestinal infections, the pathogens frequently overwhelm the indigenous microflora so that they are the predominant organisms seen in smears and cultures. A combination of culture, direct microscopic examination, and immunological tests may be performed to identify gram-negative and gram-positive bacteria (*e.g.*, enterotoxigenic and enteroinvasive *E. coli*, *Salmonella* spp., *Shigella* spp., *C. perfringens*, *C. difficile*, *Vibrio cholerae*, *Campylobacter* spp., and *Staphylococcus* spp.), fungi (*Candida*), intestinal protozoa (*Giardia*, *Entamoeba*), and intestinal helminths.

Shipping Specimens

Occasionally, a specimen must be sent to a laboratory in another city for identification. Such specimens should be shipped by the fastest means available. If they are sent by mail, viral cultures must be quick-frozen and shipped in dry ice. Pathogenic bacteria must be so labeled on the outside of the package. They should be cultured on solid agar in a tube that is capped, with tape around the cap. The glass tube should be wrapped in cotton or bubble wrap and placed in a metal can and then in a cardboard mailing box. These specimens should be sent by first-class mail.

Identification and Antimicrobial Susceptibility Testing of Pathogens

After a pathogen from a specimen has been isolated, the main remaining tasks are to identify the organism and determine its susceptibility (sensitivity) to antimicrobial agents.

TYPES OF TESTS USED FOR IDENTIFICATION

By performing a few simple tests on a pure culture of a pathogen, the genus or species of the organism can be determined. The Gram stain enables the laboratory worker to describe the Gram reaction (positive or negative), morphology (cocci, bacilli, curved, or spiral-shaped), grouping (pairs, chains, tetrads, or clusters), and whether it produces spores. By observing an isolate on various types of culture media, one can describe the shape, consistency, size, and color of the colonies, as well as whether the organism is an anaerobe or an aerobe and the type of hemolysis it produces. The use of certain selective and differential media—such as MacConkey agar—and a few simple biochemical tests—such as those for catalase or oxidase—substantially helps determine the genus and species of the organisms (see Color Figures 14, 15, 16). The final determination may be made by serotyping, if necessary.

Many laboratories make use of miniaturized multiple-test systems (minisystems) available from Analytab Products, Inc., Roche Diagnostics, BBL-Bioquest, and many other companies (see Color Figures 17, 18). Each system has certain advantages and disadvantages, but in general, they are faster, more comprehensive, and easier to use than conventional test-tube methods. Care should be taken to evaluate each before one is adopted as the accepted technique. Many of the test results are read by instruments and evaluated by computers, greatly increasing the speed and accuracy of the tests.

ANTIMICROBIAL SUSCEPTIBILITY TESTING

One of the most helpful tests for the physician is the disk-diffusion antimicrobial susceptibility testing method originally described by Drs. Kirby and Bauer. In the Kirby–Bauer test, a pure culture of the organism in Mueller–Hinton broth is streaked evenly for complete coverage on a Mueller–Hinton agar plate. Small, filter paper disks that have been permeated with various antimicrobial agents are then placed on the agar plate, and the plate is incubated at 37° C (98.6° F) under proper conditions (Figs. 8 – 5 and 8 – 6).

For gram-positive bacteria, disks of amoxicillin / clavulanic acid, ampicillin, ampicillin / sulbactam, cephalosporins, chloramphenicol, ciprofloxacin, clindamycin, erythromycin, gentamicin, imipenem, nitrofurantoin, oxacillin, penicillin, tetracycline, trimethoprim / sulfamethoxazole, and vancomycin might be used. For gram-negative bacteria, disks of amikacin, amoxicillin / clavulanic acid, ampicillin, ampicillin / sulbactam, azlocillin, aztreonam, carbenicillin, cephalosporins, chloramphenicol, ciprofloxacin, gentamicin, imipenem, mezlocillin, nitrofurantoin, piperacillin, tetracycline, ticarcillin, ticarcillin / clavulanic acid, trimethoprim / sulfamethoxazole, and tobramycin, as well as other newer antimicrobial agents, might be used. After 18 hours of incubation, the diameter of the zones of inhibition around each disk are measured in millimeters. These measurements must then be compared with zone sizes listed on published charts to determine whether the organism is susceptible or resistant to the various drugs tested. The test procedure, which must be strictly followed, and

FIGURE 8 – 5
An antimicrobial disk dispenser. (Photograph courtesy Difco Laboratories, Detroit, Michigan)

charts for interpretation of zone sizes are published by the National Committee for Clinical Laboratory Standards (NCCLS).

Quality Control in the Laboratory

A quality control program is necessary in any laboratory to monitor the reliability and quality of the work performed. In the microbiology laboratory, all media, reagents, and staining solutions should be evaluated frequently for effectiveness. Also, all equipment should be properly maintained and monitored for performance. Refrigerators, freezers, incubators, and water baths should be checked for accuracy of temperature control. By constant surveillance and frequent checking, the efficiency and reliability of the laboratory work can be maintained so that other members of the hospital team have confidence in the information supplied to them by the laboratory.

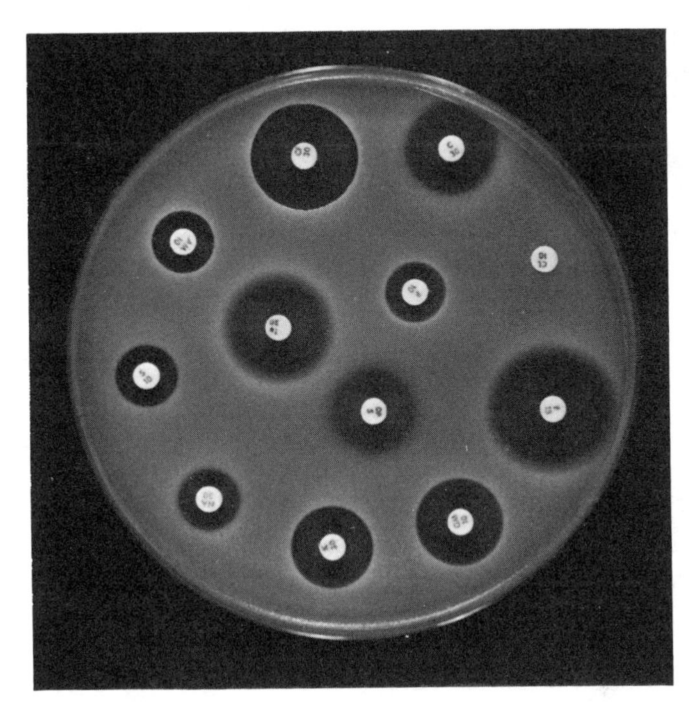

FIGURE 8 – 6
Paper disks impregnated with various antimicrobial agents are placed on the surface of an inoculated plate of Mueller-Hinton agar. The agents diffuse into the medium, inhibiting growth of organisms sensitive to them. The sizes of the zones of inhibition of growth determine if the organism is susceptible or resistant to the various agents. (Davis BD, et al.: Microbiology, 4th ed. Philadelphia, JB Lippincott, 1990)

ENVIRONMENTAL DISEASE CONTROL MEASURES

Public Health

Massive networks involving the World Health Organization (WHO), public health agencies at all levels, and community groups work together to coordinate preventive health programs and to maintain constant surveillance of sources and causes of epidemics. The WHO also develops international regulations for disease control and standardization of drugs and plays an active part in the distribution of technical information. When an epidemic strikes, teams of epidemiologists are sent to the scene to investigate the situation and assist in bringing the outbreak under control. Because of this assistance, many countries have been successful in their fight to control smallpox, diphtheria, trachoma, and numerous other diseases. In fact, it appears that smallpox has been eradicated; therefore, smallpox vaccination is no longer required. The WHO is currently attempting to eradicate polio by the year 2000.

In the United States, the Department of Health and Human Services administers the U.S. Public Health Service and the CDC, which assist state and local health departments in the application of all aspects of epidemiology. Through the efforts of these agencies working with local physicians, nurses, other health-care workers, educators, and community leaders, many diseases are no longer endemic in the United States. A list of diseases that no longer pose a serious threat to U.S. communities includes diphtheria, smallpox, typhoid fever, and cholera.

The prevention and control of epidemics is a never-ending community goal. To be effective, it must include measures to

- increase host resistance through development and administration of vaccines that induce active immunity and maintain it in susceptible persons;
- ensure that persons who have been exposed to a pathogen are protected against the disease, *e.g.*, injections of gamma globulin or antisera are effective against outbreaks of diphtheria;
- segregate, isolate, and treat those who have contracted a contagious infection to prevent the spread of pathogens to others; and
- identify and control potential reservoirs and vectors of infectious diseases; this control may be accomplished by prohibiting healthy carriers from working in restaurants, hospitals, and other institutions where they may transfer pathogens to susceptible people, and by instituting effective sanitation measures to control diseases transmitted through water supplies, sewage, and food (including milk).

WATER SUPPLIES AND SEWAGE DISPOSAL

Water is the most essential resource necessary for the survival of humanity. The main sources of community water supplies are surface water from rivers, natural lakes, and reservoirs, as well as groundwater from wells. However, two types of water pollution are present in our society that are making it increasingly difficult to provide safe water supplies; these come from chemical and biological sources.

Chemical pollution of water occurs when industrial installations dump waste products into local waters without proper pretreatment, when pesticides are used indiscriminately, and when chemicals are expelled in the air and carried back to the earth by rain ("acid rain"). The main source of biological pollution is waste products of humans—fecal material and garbage—that swarm with pathogens. The pathogens of cholera, typhoid fever, bacterial and amebic dysentery, giardiasis, cryptosporidiosis, infectious hepatitis, and poliomyelitis can all be spread through contaminated water.

Waterborne epidemics today are the result of failure to make use of the available existing knowledge and technology. In those countries that have established safe sanitary procedures for water purification and sewage disposal, outbreaks of typhoid fever, cholera, and dysentery occur only rarely.

Sources of Water Contamination Rain water falling over a large area collects in lakes and rivers and, thus, is subject to contamination by soil microbes and raw fecal material. For example, an animal feed lot located near a community water supply source harbors innumerable pathogens, which are washed into lakes and rivers. A city that draws its water from a local river, processes it, and uses it, but then dumps inadequately treated sewage into the river at the other side of town may be responsible for a serious health problem in another city downstream on the same river. The city downstream must then find some way to rid its water supply of the pathogens. In many communities, untreated raw sewage and industrial wastes are dumped directly into local waters; also, a storm or a flood may result in contamination of the local drinking water with sewage (Fig. 8 – 7).

Ground water from wells can also become contaminated. To prevent such contamination, the well must be dug deep enough to ensure that the surface water is filtered before it reaches the level of the well. Outhouses, septic tanks, and cesspools must be situated in such a way that surface water passing through these areas does not carry fecal microbes directly into the well water. With the growing popularity of trailer homes, a new problem has arisen because of trailer sewage disposal tanks that are located too near the water supply.

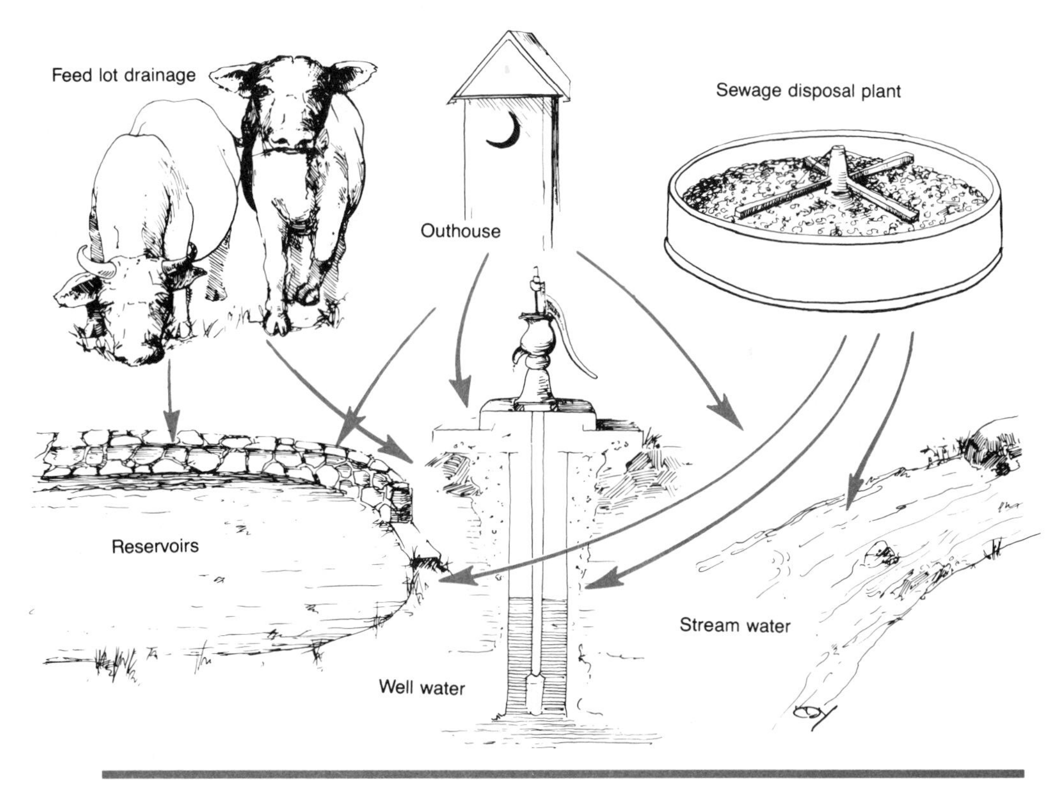

FIGURE 8 – 7
Sources of water contamination

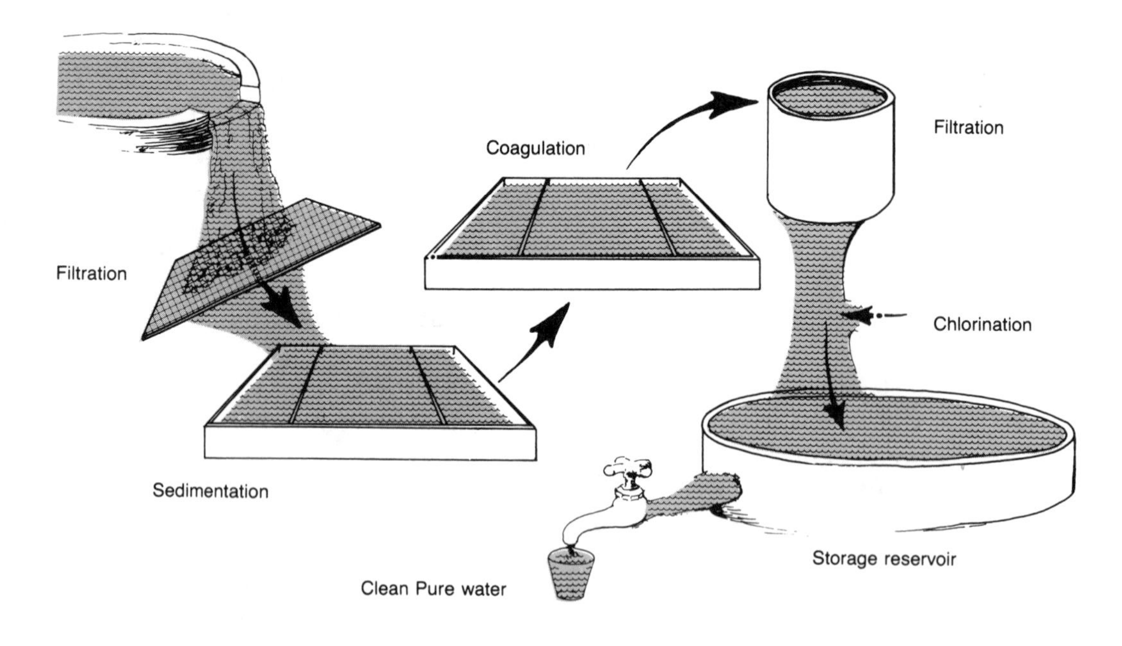

FIGURE 8 – 8
Steps in water treatment

In some very old cities where water pipes are cracked and sewage pipes leak, sewage has easy access to water pipes wherever there is a break in the pipe at a point just before it enters a dwelling.

Water and Sewage Treatment Water must be properly treated to make it safe for human consumption. It is interesting to trace the many steps involved in such treatment (Fig. 8 – 8). The water is first filtered to remove large pieces of debris such as twigs and leaves. Next, it is held in sedimentation ponds where the addition of alum coagulates bacteria and organic materials, which then settle more rapidly. The water is then passed through sand filters to remove the remaining bacteria and other small particles. Finally, chlorine gas or sodium hypochlorite is added to a final concentration of 0.2 to 1.0 part per million; this kills most remaining bacteria.

In the laboratory, water can be tested for fecal contamination by checking for the presence of coliform bacteria (*E. coli*). These bacteria normally live in the human intestine; thus, their presence in drinking water is a sure sign that the water was fecally contaminated.

If one is unsure about the purity of drinking water, boiling it for 20 minutes destroys any pathogens present. It can then be cooled and used.

When sewage is adequately treated in a disposal plant, the water it contains can be returned to lakes and rivers to be recycled. Raw sewage consists mainly of water, fecal material including intestinal pathogens, and garbage and bacteria from the drains of human habitations. In the sewage disposal plant, large debris is first filtered out, and then the bacteria break down some of the organic material. Next, the activated sludge, which includes solid matter and bacteria, is settled out in a settling tank. In some communities, the sludge is heated to kill bacteria, then dried and used as fertilizer. The remaining liquid is filtered and chlorinated so that the effluent water can be returned to rivers or oceans. In some desert cities where water is at a premium, the effluent water from the sewage disposal plant is distilled so that it can be returned directly to the drinking water system. In some other cities, effluent water is used to irrigate lawns; however, it is expensive to install a separate water system for this purpose.

SUMMARY

Persons, such as health-care personnel, who are in contact with many infectious diseases must be aware of the necessary preventive measures to inhibit the spread of communicable diseases. They must prevent cross-infections from themselves and patients who are hospitalized with contagious diseases to susceptible patients such as newborns and women in delivery, as well as surgical, diabetic, cancer, and paralyzed patients. Health-care workers must also use precautions to protect themselves against bloodborne pathogens from patients.

Collecting laboratory specimens properly is of utmost importance in identifying causative agents. Special techniques are necessary for the handling of blood, urine, spinal fluid, sputum, and mucous swabs to prevent contamination and assure survival of the pathogens. The microbiology laboratory plays an important role in the identification of infectious diseases, selection of antimicrobial agents, and maintenance of quality control in the hospital. National, state, and local health authorities help to prevent and control epidemic diseases. Spread of enteric and waterborne pathogens may be prevented by appropriate water and sewage treatments.

PROBLEMS AND QUESTIONS

1. What groups of people are responsible for the development of nosocomial infections?
2. How can hospital-acquired infections be eliminated or reduced?
3. Which patients are most susceptible to nosocomial infections?
4. Which pathogens are most frequently found in nosocomial infections?
5. Why are universal body substance precautions necessary and how have they changed nursing procedures?
6. What are the differences between surgical and medical asepsis?
7. Who must be notified to help control an epidemic?
8. How can you be hurt by collecting and handling laboratory specimens?
9. In what ways can a specimen be contaminated?
10. Why should a nurse not use palpation to locate the vein after the skin has been disinfected when he or she is taking a blood sample?
11. Which pathogens might be found in a blood sample?
12. How would you prepare a pathogenic specimen for shipping to a distant laboratory?
13. How would you determine which antimicrobial agent would be most effective against a certain pathogen?
14. How could you determine which pathogen you had isolated from a throat culture?
15. How is water treated to make it safe for human consumption?
16. If a laboratory test showed that *E. coli* was present in a sample of drinking water, what would this indicate?

Self Test

After you have read Chapter 8, examined the objectives, reviewed the chapter outline, studied the new terms, and answered the problems and questions above, complete the following self test.

Matching Exercises

Complete each statement from the list of words provided

aseptic	sterile	medical
surgical	sanitation	disinfection
sterilization	reverse isolation	

1. The technique used to avoid *all* microorganisms is the ________________ technique and is accomplished by ________________.
2. The ________________ technique is used to exclude *all* microorganisms from surgical areas to maintain ________________ asepsis.
3. To maintain ________________ asepsis, ________________ technique is employed to exclude *all* pathogens from the area.
4. ________________ of dressings and ________________ of the skin is used to maintain ________________ asepsis while dressing a wound.
5. The practical application of sanitary measures and cleanliness is termed ________________.
6. Patients with severe burns and organ transplants are protected from hospital infections by ________________.

True or False (T or F)

__ 1. Hospital infections can be avoided by proper awareness and the use of aseptic techniques.

__ 2. Sick and debilitated patients are much more susceptible to opportunistic pathogens than are healthy individuals.

__ 3. Because of the new disinfectants and antibiotics, the incidence of hospital-acquired infections has decreased.

__ 4. Many health-care workers are not adequately aware of the importance of aseptic and sterile techniques and universal precautions.

__ 5. The operating and delivery rooms are always clean, so no special precautions are necessary to protect the patient.

__ 6. Patients receiving steroids, anticancer drugs, antilymphocyte sera, and radiation treatments are usually resistant to hospital-acquired infections.

__ 7. Medical asepsis includes all precautionary measures necessary to prevent the transfer of pathogens from person to person, including indirect transfer of pathogens through the air or on inanimate objects.

__ 8. Contaminated foods provide an excellent growth medium for pathogens.

__ 9. *S. aureus* is one of the main pathogens spread by the hands and nasal secretions of hospital workers.

__ 10. Isolation techniques are used to protect everyone in the hospital from contagious diseases.

__ 11. Damp or wet masks are just as effective as dry ones.

__ 12. The microbiology laboratory should be able to monitor aseptic conditions in the hospital.

__ 13. Smallpox is thought to be totally eradicated due to the efforts of public health authorities.

___ **14.** All pathogens can be identified in the clinical laboratory, even in the presence of many contaminants.

___ **15.** If a specimen is improperly collected, unnecessary contamination may mask the disease-causing agent.

___ **16.** Laboratory findings must be conveyed to the attending physician as soon as possible to aid in the diagnosis and treatment of infectious diseases.

___ **17.** The specimens most likely to yield pathogens are collected before antimicrobial therapy begins.

___ **18.** Aerobic microbes die when exposed to the air.

___ **19.** Dangerous specimens should be placed in a sealed container for immediate transport to the laboratory.

___ **20.** The label on the specimen need only identify the patient.

___ **21.** Staphylococci are a part of the indigenous microflora of the blood and spinal fluid.

___ **22.** A septicemic condition indicates that there are pathogens in the blood.

___ **23.** Sputum specimens may be refrigerated for several hours without loss of pathogens.

___ **24.** Genital swabs for gonorrhea must be inoculated immediately onto Thayer–Martin media and incubated in the presence of carbon dioxide.

___ **25.** Only the people who work in hospitals are responsible for the prevention and control of epidemics.

Multiple Choice

1. Nosocomial refers to
- **a.** the patients' condition on hospital admission
- **b.** the physician's health status
- **c.** infection that a patient develops during hospitalization
- **d.** infection that develops in staff
- **e.** none of the above

2. "Fomite" refers to
- **a.** inanimate materials
- **b.** sputum contamination of a fork
- **c.** bandages from an infected surgical site
- **d.** contaminated bedpan
- **e.** all of the above

3. Reverse isolation would be appropriate for
- **a.** a patient with tuberculosis
- **b.** a patient who has had minor surgery
- **c.** a patient with glaucoma
- **d.** a patient with leukemia

4. The microorganism most apt to cause nosocomial infections is
- **a.** *Yersinia pestis*
- **b.** *Bacillus anthracis*
- **c.** *S. aureus*
- **d.** *E. coli*

5. Housekeeping and central supply personnel contribute to hospital asepsis by
- **a.** restricting contact with patients
- **b.** protecting themselves from infectious organisms
- **c.** using techniques to prevent cross-contamination
- **d.** all of the above

6. During strict isolation
   a. a ward room is desirable
   b. gloves must be worn by everyone having contact with the infected area
   c. masks are necessary only during dressing changes
   d. masks must be worn by all persons entering the room
7. Using a clean-catch midstream urine specimen, a urinary tract infection is indicated by
   a. less than 100 org/ml
   b. more than 1000 org/ml
   c. any bacteria present
   d. more than 100,000 org/ml
8. For an antibiotic sensitivity test, the test plate should be
   a. streaked for isolated colonies
   b. incubated at 35° C
   c. read after 48 hours
   d. stored in an anaerobic chamber
9. Spinal fluid specimens
   a. are easily obtained
   b. must be inoculated onto Thayer–Martin agar plates
   c. are seldom submitted to the laboratory
   d. must be cultured immediately for evidence of meningococci
10. All specimens should be
    a. refrigerated
    b. promptly delivered to the laboratory
    c. given time to grow
    d. obtained after onset of drug therapy

CHAPTER 9

Human Defenses Against Infectious Diseases

OBJECTIVES

After studying this chapter, you should be able to

- List and describe the nonspecific defenses of the human body
- Define the first and second lines of defense
- Define phagocytosis
- List the various types of phagocytic cells
- Describe the process of inflammation
- Describe the immune response, or third line of defense
- Define antigen, antibody, and immunoglobulin
- Differentiate between active and passive immunity
- Compare natural and artificial immunity
- List three ways in which vaccines are prepared
- Draw a graph representing primary and secondary antibody responses to antigens
- Differentiate between immediate and delayed hypersensitivity
- Define autoimmunity and give examples
- List five serological tests used to determine the presence of a specific antibody in human serum

CHAPTER OUTLINE

Nonspecific Mechanisms of Defense
First Line of Defense
Second Line of Defense
- *Fever Production*
- *Iron Balance*
- *Cellular Secretions*
- *Blood Proteins*
- *Phagocytosis*
- *Inflammation*

Immune Response to Disease: Third Line of Defense
Immunity
Acquired Immunity
- *Active Acquired Immunity*
- *Passive Acquired Immunity*

Immunology
The Immune System
The Immune Response
- *Humoral Immunity*
- *Antibody Structure and Function*
- *Hypersensitivity*

Immunodiagnostic Procedures
- *Agglutination tests*
- *Precipitin tests*
- *Lysis by Complement*
- *Fluorescent Antibody Technique*
- *Radioimmunoassay*
- *Enzyme-Linked Immunosorbent Assay*
- *Opsonization*
- *Capsular Swelling*
- *Immobilization*
- *Toxin Neutralization*

NEW TERMS

Acquired immunity
Active acquired immunity
Agammaglobulinemia
Agglutination
Agglutination tests
Allergen
Anaphylactic shock
Anaphylaxis
Antibody
Antigen
Antigenic determinant
Antiserum
Antitoxin
Atopic person
Attenuation
Autogenous vaccine
Autoimmune disease
Bacteriocins
Basophil
B-cell
Beta-lysin (β-lysin)
Blocking antibodies
Chemotaxis
Colicin
Complement
Cytokines
Cytotoxic
Edema
Eosinophil
Epitope
Exudate
Fibronectin
Granulocyte
Hapten
Histamine
Histiocyte or histocyte
Hybridoma
Hypersensitivity
Hypogammaglobulinemia
Immunodiagnostic procedures
Immunogen
Immunogenic
Immunoglobulin
Immunology
Inflammation
Interferon
Interleukins
Lymphokines
Macrophage
Mast cell
Microbial antagonism
Monoclonal antibodies
Monocyte
Natural killer cell
Neutrophil
Opsonin
Opsonization
Passive acquired immunity
Phagolysosome
Phagosome
Plasma cell
Precipitate
Precipitin tests
Properdin
Prostaglandin
Pus
Pyrogenic
Reticuloendothelial system (RES)
Serology
Serum (pl. sera)
T-cell
Toxoid
Vasodilation
Wandering macrophages

NONSPECIFIC MECHANISMS OF DEFENSE

Humans and lower animals have survived on earth for millions of years because they have many built-in mechanisms of defense against harmful microorganisms and the infectious diseases they cause. The ability of any animal to resist these invaders and recover from disease is due to many complex interacting functions within the body.

Humans have three lines of defense against bacteria, viruses, fungi, and other pathogenic organisms. The first two lines of defense are nonspecific; these are ways in which the body attempts to destroy *all* types of substances that are foreign to it. The third line of defense, the immune response, is very specific. Special proteins called *antibodies* are formed in response to the presence of particular foreign substances. These foreign substances are called *antigens* because they cause the production of specific antibodies; they are *anti*body *gen*erating substances. The immune response is discussed in more detail later in this chapter.

Nonspecific defense mechanisms are general in nature and serve to protect the body against many harmful substances. One of the nonspecific defenses is the innate, or inborn, resistance observed among some species of animals, some races of humans, and some persons who have a natural resistance to some diseases. Innate or inherited characteristics make these persons and animals more resistant to some diseases than to others. Resistance to cholera is an example of species resistance. Human beings can contract human cholera but not chicken cholera. The exact factors that produce this innate resistance are not well understood but are probably due to chemical, physiological, and temperature differences between the species as well as the general state of physical and emotional health of the person and environmental factors that affect certain races and not others.

Although we are usually unaware of it, our bodies are, more or less, constantly in the process of defending against microbial invaders. Nonspecific defense mechanisms include such things as mechanical and physical barriers to invasion, chemical factors, microbial antagonism by our indigenous microflora, phagocytic host cells, fever, and the inflammatory response (*inflammation*) (Fig. 9 – 1).

First Line of Defense

One of the ways in which the body is protected from invasion by foreign microorganism and other substances is the presence of mucous membranes at the openings to the respiratory, digestive, urinary, and reproductive systems. The mucous membranes entrap the invaders.

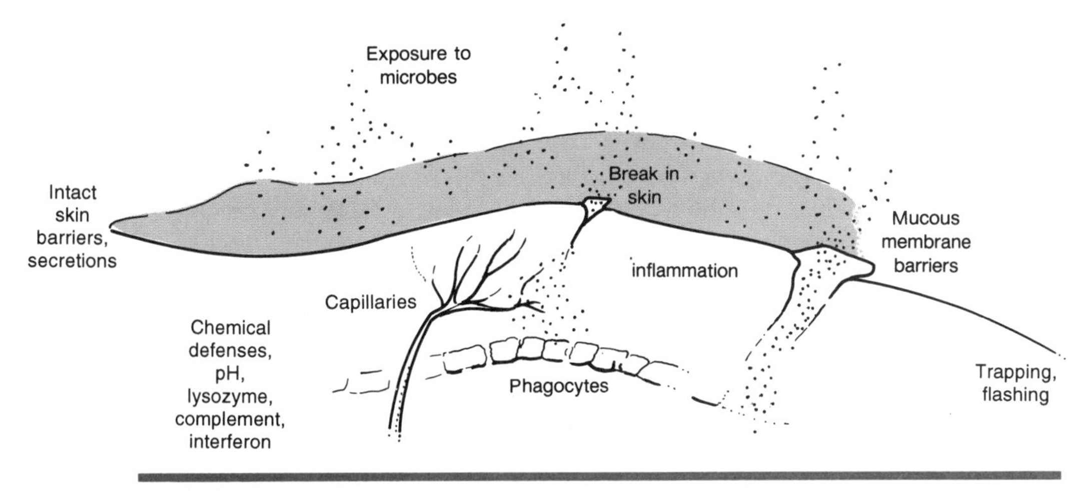

FIGURE 9 – 1
Nonspecific defenses against microbial invasion.

Another way in which the body is protected is by intact skin providing a complete external covering for all other parts of the body. The unbroken skin acts as a physical or mechanical barrier to pathogenic microorganisms; only when it is cut, abraded (scratched), or burned can they gain entrance. There are several factors that account for the skin's ability to resist pathogens. One is the skin's normal secretions, which destroy bacteria or inhibit their growth on its surface. These bactericidal secretions include acidic perspiration from sweat glands and fatty acids from oil glands. In addition, pathogens may not be able to establish themselves and multiply because of the many microorganisms already living in the pores and in moist parts of the head, underarms, hands, feet, and perianal and urethral regions.

The respiratory system would be particularly accessible to invaders that could ride on dust or other particles inhaled with each breath, were it not for the hair, mucous membranes, and irregular chambers of the nose that serve to trap much of the inhaled debris. The cilia (mucociliary covering) of the epithelial (surface) cells of the posterior nasal membranes, nasal sinuses, bronchi, and trachea sweep the trapped dust and microbes toward the throat, where they are swallowed or expelled by sneezing and coughing. Phagocytic white blood cells (phagocytes) in the mucous membranes may also be involved in this mucociliary clearance mechanism. Lysozyme and other enzymes that lyse or destroy bacteria are present in nasal secretions, saliva, and tears. The microflora usually residing in the nose and mouth may also serve a protective function.

To a large degree, the digestive system is protected by the process of digestion, digestive enzymes, acidity of the stomach, and alkalinity of the intestine. Bile, which is secreted from the liver into the intestine, lowers the surface tension and causes chemical changes in bacterial cell walls and membranes that make bacteria more digestible. Many invading microorganisms are trapped in the mucous lining of the digestive tract where they may be destroyed by bactericidal enzymes and phagocytes. The indigenous microflora that use available nutrients and occupy space in the intestines include *Escherichia coli, Enterobacter aerogenes, Enterococcus faecalis,* and many other anaerobic or facultatively anaerobic enteric (intestinal) bacteria. Peristalsis and the expulsion of feces serve to remove bacteria from the intestine.

The urinary tract is usually sterile in healthy persons; therefore, there should be no microorganisms present in urine drawn from the urinary bladder in an aseptic manner, *i.e.,* via needle and syringe. Also, the reproductive system of the male and most of the reproductive organs in the female lack an indigenous microflora. However, the indigenous microflora and pathogens from the anus and perianal skin may enter the vagina and invade the urethra. Microorganisms (which may be opportunistic pathogens) are continually flushed from these areas by frequent urination and expulsion of mucous secretions. Many bladder infections occur simply as a result of infrequent urination, especially failure to urinate after intercourse. Normally, the acidic urine and vaginal secretions also inhibit microbial growth. Many women who are taking certain oral contraceptives are particularly susceptible to some infections because those chemicals reduce the acidity of the vagina.

The prevention of colonization of potential microbial pathogens by the indigenous microflora of a given anatomical site is called *microbial antagonism;* this is another example of a nonspecific defense mechanism. The inhibitory capability of these microflora has been attributed to a competition for nutrients and the production of certain inhibitory substances. Examples of such inhibitory substances are the *colicins* produced by certain strains of *E. coli.* Similar substances are produced by some strains of *Pseudomonas* and *Bacillus* species and by other bacteria. These antibacterial substances (proteins), which are produced by other bacteria, are collectively known as *bacteriocins.* The effectiveness of microbial antagonism is frequently decreased following prolonged administration of broad-spectrum antibiotics. The antibiotics reduce or eliminate certain members of the microflora (*e.g.,* the vaginal and gastrointestinal flora) and permit overgrowth by bacteria and / or fungi that are resistant to the antibiotic being administered. This overgrowth of organisms, by *Candida albicans* or *Clostridium difficile* for example, is called a superinfection.

Second Line of Defense

The nonspecific cellular and chemical responses to microbial invasion are considered to be the second line of defense. Virulent pathogens that penetrate the first line of defense usually are destroyed by a series of defense mechanisms, including the inflammatory response. A complex sequence of events develops involving fever production, iron balance, cellular secretions (*interferon, fibronectin, β-lysin, interleukins, prostaglandins, histamine*), activation of blood proteins (*complement, properdin*), *chemotaxis*, phagocytosis, neutralization of toxins, and clean-up and repair of damaged areas. Each of these responses is discussed in this section.

FEVER PRODUCTION

A fever may be triggered by various *pyrogenic* (fever-producing) secretions of many infecting pathogens. The increased body temperature augments the host's defenses by (1) stimulating white blood cells (leukocytes) to deploy and destroy invaders, (2) reducing available free plasma iron, which limits the growth of pathogens that require iron for replication and synthesis of toxins; and (3) inducing the production of interleukin-1 (IL-1), which causes the proliferation, maturation, and activation of lymphocytes in the immunological response.

IRON BALANCE

The virulence of many bacteria is enhanced in the presence of free iron, which is used for the synthesis of exotoxins. In response to pathogenic invasion, some of the host's leukocytes produce IL-1. Among other functions, this substance induces the release of lactoferrin, which stimulates iron storage in the liver and thus reduces the amount of free iron available for the pathogen. Interleukin-1 is also known as endogenous pyrogen, a fever-producing substance produced within the body.

CELLULAR SECRETIONS

Interferons The α-, β-, and γ- (gamma) interferons are small proteins produced by certain body cells (leukocytes, fibroblasts, and T-lymphocytes) when they are infected or stimulated by certain viruses, chlamydias, rickettsias, or protozoa and by the presence of some tumors and cancers. Interferons enter surrounding cells, where they inhibit synthesis of certain essential proteins that are necessary for multiplication of viruses and other pathogens within those cells. Thus, the spread of the infection is inhibited, allowing other body defenses to fight the disease more effectively. In this way, many viral diseases (*e.g.*, colds, influenza, and measles) are self-limiting in duration. Similarly, the acute phase of herpes simplex cold sores is of limited duration. Then the herpes virus enters a latent phase and hides in nerve ganglion cells where it is protected until the person's defenses are down; thus, the cycle of cold sores is repeated.

Interferons are not pathogen-specific, meaning that they are effective against a variety of pathogens, not just the pathogen that stimulated their production. Interferons are species-specific, however, meaning that they are effective only in the species of animal that produce them. Thus, human interferon is only effective for humans. Human interferons are industrially produced by bacteria containing recombinant DNA (with interferon genes inserted) and are used experimentally to treat some viral infections, cancers, tumors, and immunodeficiency diseases.

Fibronectin Fibronectin is an epithelial tissue secretion that enables cells to bind with collagen and other components of the extracellular matrix. It can also interact with certain bacteria (*e.g.*, staphylococci, streptococci) to block their attachment to epithelial cells. Thus, it aids in clearing and flushing these pathogens from the body.

β-lysin β-lysin is a polypeptide that is released from blood platelets during an infection. It destroys gram-positive bacteria by disrupting their plasma membranes, causing lysis. It is also found inside phagocytes where it aids in the digestion of microbes.

Interleukins Interleukins (IL-1, IL-2, IL-3) are polypeptides that are secreted by antigen-stimulated *macrophages* (large, phagocytic leukocytes). They enhance T-lymphocyte activation, proliferation, and activity during the immune response.

BLOOD PROTEINS

Complement Complement is actually a group of 25 to 30 proteins (including C1 through C9) found in normal blood plasma that constitute the "complement system." It is so named because it is complementary to the action of antibodies in immune and allergic reactions. Complement is a nonspecific defense mechanism because it binds to many different antigen–antibody complexes (*immune complexes*). Once bound, it becomes activated (1) to enhance the inflammatory response, (2) to aid in the destruction (lysis) of cells and microorganisms, (3) to attract phagocytes into the region (chemotaxis), and (4) to aid in neutralizing the toxins of certain microorganisms.

Properdin The blood protein properdin works in tandem with complement components C3 and C5 in the absence of an antigen–antibody complex. It also enhances phagocytosis, inflammation, and the destruction of bacteria and certain viruses.

Prostaglandins Prostaglandins are membrane-associated lipids that act much like local hormones. They are biologically active in controlling platelet aggregation, the immune response, inflammation, increased capillary permeability, pain production, diarrhea, autoimmune responses, and many other conditions in health and disease.

PHAGOCYTOSIS

The process by which phagocytes surround and engulf (ingest) foreign material is called phagocytosis. Foreign materials not needed by the body include dead cells, unused secretions (such as milk), dust, debris, and microorganisms. Phagocytes serve as the clean-up crew to rid the body of these unwanted and often harmful substances.

Leukocytes that act as phagocytes include phagocytic *granulocytes* and macrophages (Fig. 9 – 2). Phagocytic granulocytes include *neutrophils* and *eosinophils.* Neutrophils are more efficient at phagocytosis than eosinophils and are the body's most efficient and abundant phagocytes. Eosinophils become more plentiful and play a phagocytic role in allergic responses. A third type of granulocyte, *basophils,* (discussed later in the chapter) are also involved in allergic and inflammatory reactions, although they are not phagocytes. Macrophages develop from a type of leukocyte called *monocytes* during the inflammatory response to infections. Those that leave the bloodstream and migrate to infected areas are called *wandering macrophages. Fixed macrophages,* or *histiocytes,* remain in tissues and organs and serve to trap foreign debris. Macrophages are extremely efficient phagocytes. They are found in tissues of the *reticuloen-*

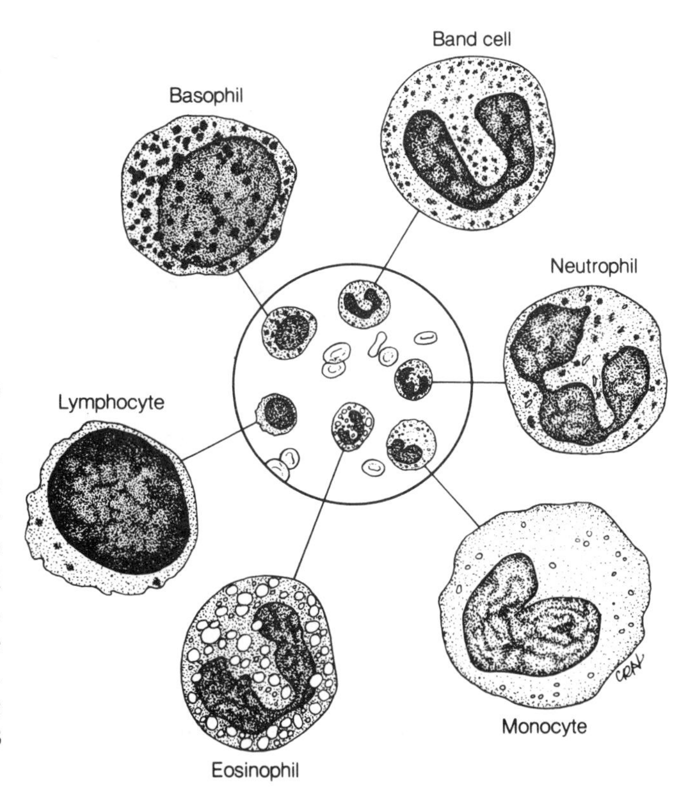

FIGURE 9 – 2
Types of blood cells. When blood smears are stained with Wright's stain, eosinophils have reddish-orange granules, basophils have purple granules, and neutrophil granules are colorless. In this way, the different granulocytes can be distinguished from one another. (Rosdahl CB: Textbook of Basic Nursing, 6th ed. Philadelphia, JB Lippincott, 1995)

dothelial system (RES); this nonspecific defensive system includes cells in the liver (Kupffer cells), spleen, lymph nodes, bone marrow, lungs (dust cells), blood vessels, intestines, and brain (microglia). Thus, it appears that macrophages are more efficient scavenger cells than neutrophils, although both cell types are able to destroy most of the microorganisms they ingest.

Cellular Elements of the Blood

- Erythrocytes (red blood cells)
- Thrombocytes (platelets)
- Leukocytes (white blood cells)
  - Granulocytes
    - Basophils
    - Eosinophils
    - Neutrophils
  - Monocytes/Macrophages
  - Lymphocytes
    - B-cells
    - T-cells
      - Helper T-cells
      - Suppressor T-cells
      - Cytotoxic T-cells
      - Delayed Hypersensitivity T-cells

The principal function of the entire RES is the engulfment and removal of foreign and useless particles, living or dead, such as excess cellular secretions, dead and dying leukocytes, erythrocytes, and tissue cells as well as foreign debris and microorganisms that gain entrance to the body.

Phagocytosis begins when a phagocyte moves to within 100 μm of a foreign object such as a bacterium. The phenomenon that causes the phagocyte to be attracted to the bacterium is called chemotaxis. This chemical attraction is not well understood but is the result of *lymphokines* produced by T-lymphocytes and the activation of complement in many circumstances (discussed later).

When the phagocyte touches a pathogen, the phagocyte's membrane indents (invaginates), and the phagocyte moves around the microbe until the microbe is completely surrounded (Fig. 9 – 3). The membrane-bound vesicle that surrounds the pathogen, which is now inside the phagocyte, is called the *phagosome* or phagocytic vacuole. The phagosome fuses with nearby lysosomes to form a digestive vacuole (*phagolysosome*), and killing and digesting of the microbe begins. Within 10 to 30 minutes after ingestion, the microorganism is killed by lactic acid and hydrogen peroxide from the lysosome. Then lysosomal digestive enzymes, including lysozyme and β-lysin, digest and degrade the microbe's carbohydrates, lipids, proteins, and nucleic acids. The products of this digestion, which may be used by cells as nutrients, are then absorbed into

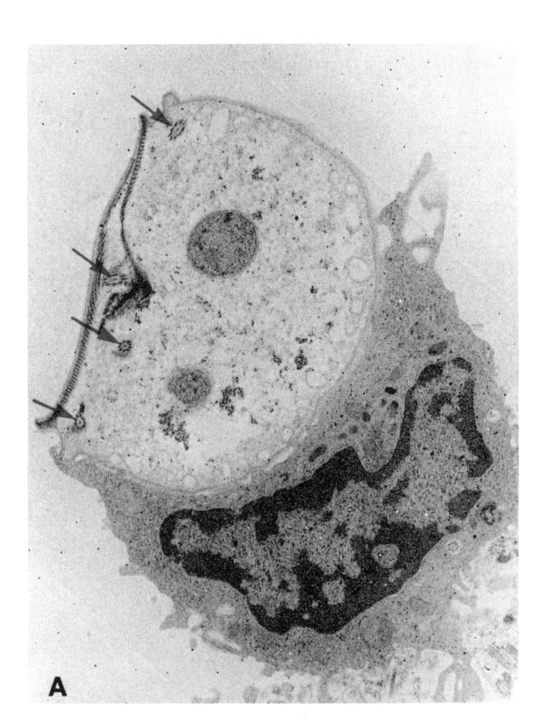

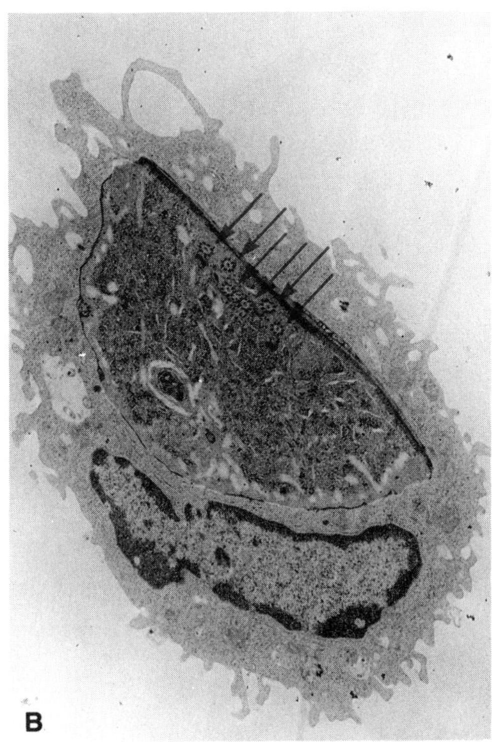

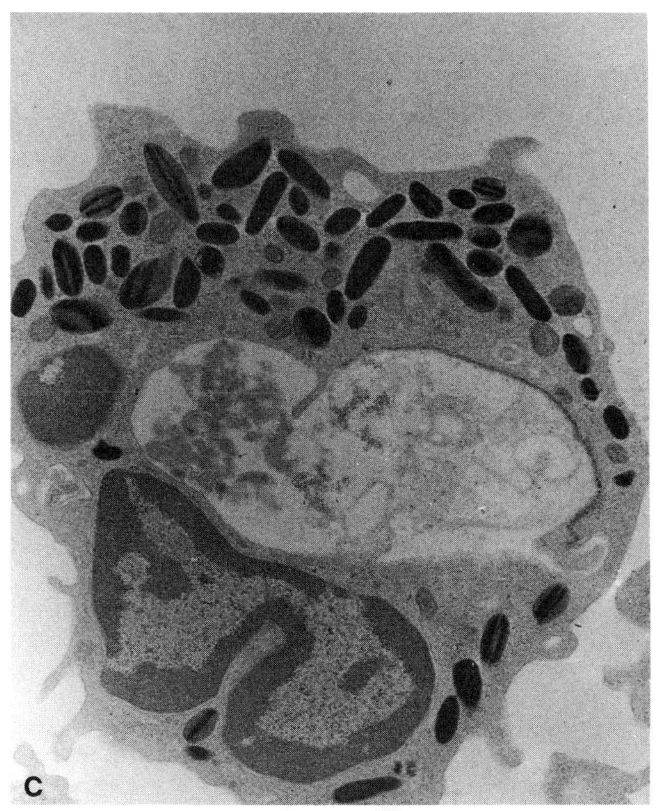

FIGURE 9 – 3
Phagocytosis of *Giardia* trophozoites. (*A*) Attachment. (*B*) Ingestion. (C) Digestion. Note the cross sections of flagella (arrows) in *A* and *B*, and the lysosomes in *C*. (Photographs from S. Koester and P. Engelkirk)

the cytoplasm. Undigested materials are retained within the membrane; thus, the digestive vacuole becomes a residual body until the undigested wastes are expelled from the phagocyte.

During the initial phases of infection, capsules may serve to protect encapsulated bacteria from being phagocytized. It should also be noted that not all bacteria engulfed by phagocytes are destroyed by the phagocyte's digestive enzymes. Some may be transported within phagocytes to other parts of the body before the phagocytes expel them. Some pathogens, such as certain *Mycobacterium* species, are even able to multiply within the phagosome and destroy the phagocyte. The causative agents of brucellosis and tularemia may remain dormant within phagocytes for months or years before they escape to cause disease. These types of virulent pathogens usually win the battle with phagocytes. Unless antibodies are present to activate complement to aid in the destruction of these pathogens, the infection may progress unchecked (Fig.

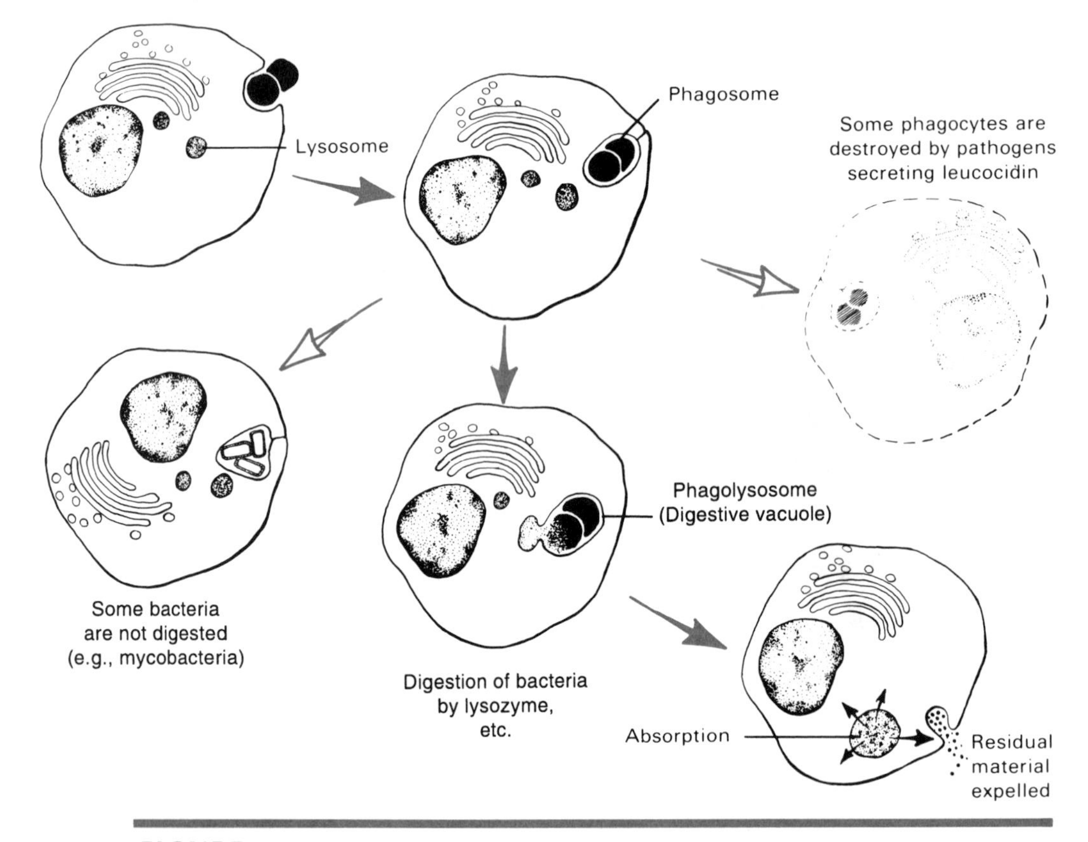

FIGURE 9 – 4
Phagocytosis. Not all pathogens are destroyed by phagocytes. Encapsulated bacteria can escape phagocytosis and bacteria that secrete leukocidin can destroy phagocytic cells.

9 – 4). Also, some pathogens secrete an enzyme—leukocidin (discussed in Chapter 7)—that destroys leukocytes, including the phagocytes that ingest them.

INFLAMMATION

The body normally responds to any local injury, irritation, microbial invasion, or bacterial toxins by a complex series of events called inflammation. The purposes of the inflammatory response are to localize an infection, to prevent the spread of microbial invaders, to neutralize toxins, and to aid in the repair of damaged tissue (Fig. 9 – 5). In this process, all of the nonspecific defenses come into play. These interrelated physiological reactions result in characteristic signs and symptoms of inflammation: *edema* (swelling of the area), redness, heat, pain, and often pus formation as well as occasional loss of function of the damaged area.

A complex series of physiological events occurs immediately after the initial damage to the tissue. Some of the injured cells (*mast cells*, basophils, and platelets) release certain chemicals (histamine, bradykinin, other kinins, and heparin) that increase the permeability of capillaries and venules and prevent immediate blood clotting. *Vasodilation* (an increase in the diameter of blood vessels) and increased capillary permeability allow more blood (plasma, clotting agents, and cells) to enter the area, including more leukocytes for phagocytosis and antibody production. Macrophages are also attracted to the region by chemotaxis due to the secretions from damaged cells.

The surrounding tissue becomes engorged with fluids, and edema results. Red blood cells collecting in the irritated area cause redness. Metabolic heat is generated, and fever is produced by increased cellular activity in the destruction and detoxification of foreign materials, microbes, dead cells (tissue and phagocytic), and toxic chemicals, including waste products released by the invading microorganisms.

Pain or tenderness usually accompanies inflammation. It may result from

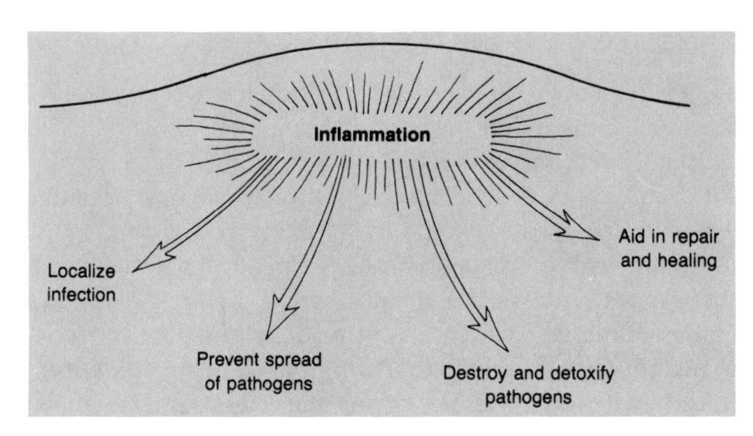

FIGURE 9 – 5
The purposes of inflammation.

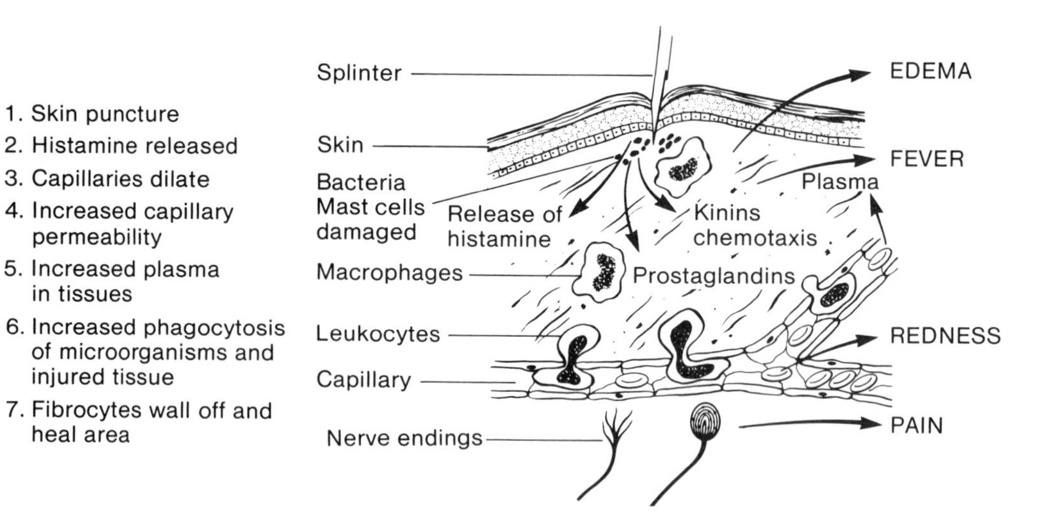

FIGURE 9 – 6
The inflammatory response.

actual damage of the nerve fibers because of the injury, from irritation by microbial toxins or other cellular secretions (such as prostaglandins), or from the edema causing increased pressure on the nerve endings (Fig. 9 – 6).

The accumulation of fluid and cells within the inflamed site is referred to as the inflammatory *exudate*. If the exudate is thick and greenish-yellow with many dead leukocytes, it is known as a purulent exudate or *pus*. However, it should be noted that in many inflammatory responses, such as arthritis or pancreatitis, there is no exudate and no invading microorganisms. When *pyogenic* (pus-producing) microorganisms, such as some staphylococci and streptococci, are present, even more pus is produced as a result of the *cytotoxic* (cell-killing) effect of the bacterial toxins on phagocytes and tissue cells. Most pus is greenish-yellow, but in infections caused by *Pseudomonas aeruginosa*, the exudate may be greenish-blue due to a certain pigment (pyocyanin) produced by the bacteria.

When the inflammatory response is over and the body has won the battle, the phagocytes continue cleaning up the area and helping to restore order. The cells and tissues can then repair the damage and begin to function normally again in a homeostatic (equilibrated) state, although some permanent damage and scarring may have occurred.

The lymphatic system—including the lymph, lymphatic vessels, lymph nodes, and lymphatic organs (tonsils, spleen, and thymus gland)—also plays an important role in defending the body against invaders. The primary functions of this system include draining and circulating intercellular fluids from the tissues and transporting digested fats from the digestive system to the blood. Also, macrophages and B-lymphocytes, and T-lymphocytes in the lymph nodes serve to filter the lymph by removing foreign matter and microbes and by producing antibodies and other factors to aid in the destruction and detoxification of any invading microorganisms.

The body continually wages war against damage, injury, malfunction, and microbial invasion. The outcome of each battle depends on the person's age, hormonal balance, genetic resistance, and overall state of physical and mental health as well as the virulence of the pathogens involved.

IMMUNE RESPONSE TO DISEASE: THIRD LINE OF DEFENSE

The immune response is the third line of defense against pathogens. Usually, in this protective type of immunity, antibodies are produced by lymphocytes to recognize, bind with, inactivate, and destroy specific microorganisms. These humoral (circulating) antibodies are found in blood plasma, lymph, and other body secretions where they readily protect the body against those specific pathogens that stimulated their formation. Thus, a person has an immunity to a particular disease because of the presence of specific protective antibodies that are effective against the causative agent of that disease. In addition, there are protective cell-mediated immune responses that do not involve the presence of antibodies.

The study of the immune system is called *immunology*. Although the immune system protects against disease and aids in fighting cancer, it may also cause damage to its host, as you will learn from the following discussions on hypersensitivity and autoimmunity.

Immunity

Immunity is resistance to disease that is usually due to the presence of antibodies to the causative pathogens (etiologic agents) of that disease. The innate, or native, resistance to disease found in certain individuals, races, and species of animals is not a type of immunity conferred by antibodies; rather, it is a resistance resulting from natural nonspecific factors. A person who is susceptible to a disease usually has inadequate levels of protective antibodies or insufficient nonspecific defenses, which may simply reflect a very poor state of health or the presence of an immunodeficiency disease.

ACQUIRED IMMUNITY

Immunity that results either from the active production or transfer of antibodies is termed *acquired immunity*. If the antibodies are actually produced by the lymphocytes of the protected person, the immunity is called *active acquired*

TABLE 9 – 1.
Types of Acquired Immunity

Active		Passive	
Natural	Artificial	Natural	Artificial
Clinical or subclinical disease	Vaccines: Inactivated (killed) pathogens Attenuated (weakened) pathogens Extracts (parts of pathogens) Toxoids	Congenital (across placenta) Colostrum	Antiserum Antitoxin Gamma globulin

immunity. In *passive acquired immunity,* antibodies are transferred from one person (or from an animal) to another person to temporarily protect the latter against infection (Table 9 – 1).

Active Acquired Immunity People who have had a specific infection usually have some resistance to reinfection by that causative pathogen because of the presence of antibodies and stimulated lymphocytes; this is called natural active acquired immunity. Symptoms of the disease may or may not be present when these antibodies are formed. Such resistance to reinfection may be permanent (as with mumps, measles, smallpox, diphtheria, whooping cough, poliomyelitis, plague, and typhoid fever) or it may be temporary (as with pneumonia, influenza, gonorrhea, and streptococcal and staphylococcal infections). There is no immunity to reinfection following recovery from syphilis and tuberculosis.

Artificial active acquired immunity is the second type of actively acquired immunity. It occurs when a person receives a vaccination, which is the administration of a vaccine that causes specific antibodies to be produced. Sufficient antigens of a pathogen are contained in the vaccine to enable the person to form antibodies against that pathogen. A good vaccine is one that (1) contains enough antigens to protect against infection by the pathogen, (2) contains antigens from all of the strains of the pathogen that cause that disease (*e.g.*, three strains of the virus that causes polio), (3) is not too toxic, and (4) does not cause disease in the vaccinated person.

A successful vaccine for colds has not been developed because so many different viruses cause colds. Maintaining a successful vaccine for influenza is also difficult because the viruses continually change by mutation.

Vaccines are made from living or dead (inactivated) pathogens or from certain toxins they excrete (Fig. 9 – 7). In general, vaccines made from living organisms are most effective, but they must be prepared from harmless organ-

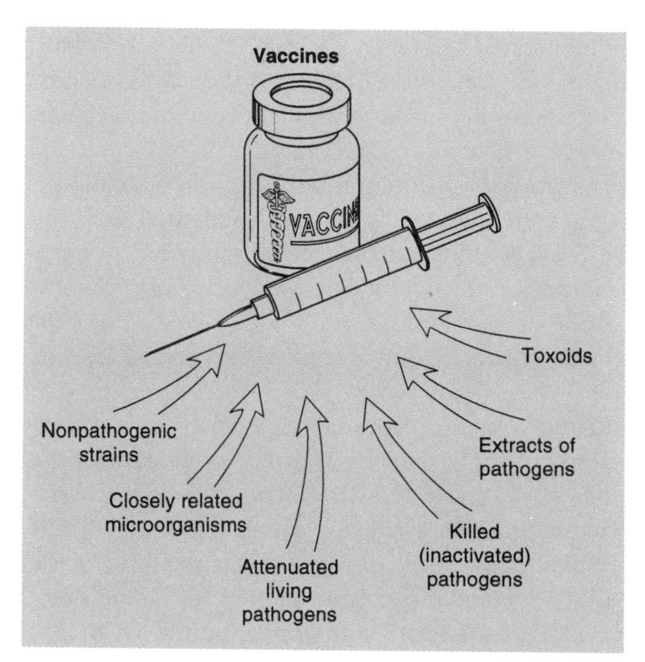

FIGURE 9 – 7
Sources of vaccines.

isms that are antigenically closely related to the pathogens or from weakened pathogens that have been genetically changed so they are no longer pathogenic. The process of weakening pathogens is called *attenuation.* The smallpox (variola) vaccine is derived from the cowpox virus (vaccinia), which causes a mild pox infection in cattle and humans. Most other live vaccines are avirulent (nonpathogenic) mutant strains of pathogens that have been derived from the virulent organisms; this is accomplished by growing them for many generations under various conditions or by exposing them to mutagenic chemicals or radiation. Pasteur developed the attenuated vaccine for rabies, and Sabin developed the attenuated oral vaccine for poliomyelitis. Other attenuated living vaccines include those for measles (rubeola), mumps, German measles (rubella), yellow fever, and typhus.

Vaccines made from dead pathogens, which have been killed by heat or chemicals, can be produced faster and more easily, but they are less effective than live vaccines. This is because the antigens on the dead cells are usually less effective and produce a shorter period of immunity. Inactivated vaccines are safer to use during the experimental phase of vaccine production because of the remote possibility that an avirulent strain of a living organism may revert to the virulent strain before the strain's stability has been established. For example, the first poliomyelitis virus vaccine, which was developed by Jonas Salk, contained the killed organisms of three strains of polioviruses, which had to be injected to produce adequate immunity against the disease. A few years later, it was found that the live virus vaccine, developed by Sabin and others, was safe when taken by mouth. Killed pathogens or their extracts are used to vaccinate against whooping cough (pertussis), typhoid fever, paratyphoid fever, cholera, plague, Rocky Mountain spotted fever, and many respiratory diseases including influenza. The use of such vaccines illustrates a very important and practical application of the principles of microbiology and immunology.

Another type of vaccine used to prevent tetanus and diphtheria, called a *toxoid* vaccine, is prepared from exotoxins that have been inactivated or made nontoxic by heat or chemicals. These toxoids can be injected safely to cause the formation of antibodies that neutralize the exotoxins of the pathogens, such as those that cause tetanus and diphtheria. A serum containing specific antibodies is called an *antiserum,* and an antiserum containing antibodies against toxoids is called an *antitoxin.*

As microbiologists made further studies of the characteristics of vaccines, they found that it was practical to vaccinate against several diseases at one time by combining specific vaccines in a single injection. Thus, the diphtheria-tetanus-pertussis (DTP) vaccine contains toxoids for diphtheria and tetanus and killed organisms for whooping cough (pertussis). Another example is the measles-mumps-rubella (MMR) vaccine that is recommended for all children. According to the Centers for Diseases Control and Prevention (CDC), U.S. children should receive the following vaccines between birth and entry into school:

- diphtheria-tetanus-pertussis (DTP) vaccine;
- live oral polio vaccine;
- measles-mumps-rubella (MMR) vaccine;
- *Haemophilus influenzae* type b (Hib) vaccine; and
- hepatitis B (HepB) vaccine.

(*Source:* MMWR Recommendations and Reports, vol. 44, RR-5, June 16, 1995.)

An *autogenous vaccine* is one prepared using bacteria isolated from a localized infection, such as a staphylococcal boil. The pathogens are killed and then injected into the same person to induce production of more antibodies.

Passive Acquired Immunity Passive immunity differs from active immunity in that antibodies formed in one person are transferred to another to protect the latter from infection. Because the person receiving the antibodies did not produce them, the immunity is temporary, lasting only about 3 to 6 weeks. The antibodies of passive immunity may be transferred naturally or artificially.

In natural passive acquired immunity, small antibodies (like Ig G, which is described later in this chapter) present in the mother's blood cross the placenta to reach the fetus while it is in the uterus (*in utero*). Also, the milk or colostrum (secreted for a few days after delivery) contains maternal antibodies to protect the infant during the first months of life.

Artificial passive acquired immunity is accomplished by transferring antibodies from an immune person to a susceptible person. After a patient has been exposed to a disease, the length of the incubation period usually does not allow sufficient time for vaccination to be an effective preventive measure. This is because a span of at least 2 weeks is needed before sufficient antibodies are formed to protect the exposed person. To provide temporary protection in these situations, the patient is given human gamma globulin or "pooled" immune serum globulin (ISG), that is, antibodies from the serum of many immune people. Thus, the patient receives some antibodies to all of the infectious diseases to which the donors are immune. The ISG may be given to provide temporary protection against measles, mumps, polio, diphtheria, and hepatitis in persons, especially infants, who are not immune and have been exposed to these diseases. Serum is the liquid portion of blood that has been allowed to clot; the liquid remaining after the clotting factors have been removed from plasma.

Hyperimmune serum globulin (or specific immune globulin) has been prepared from the serum of persons with high antibody levels (titers) against certain diseases. For example, hepatitis B immune globulin (HBIG) is given to protect those who have been, or are apt to be, exposed to hepatitis; tetanus immune globulin (TIG) is used for nonimmunized patients with deep, dirty wounds; and rabies immune globulin (RIG) may be given following a bite by a rabid animal. Other examples include chickenpox immune globulin, measles immune globulin, pertussis immune globulin, poliomyelitis immune globulin, and zoster immune globulin. In botulism, potentially lethal food-poisoning cases, antitoxin antibodies are used to neutralize the toxic effects of the botulinal toxin. Remember that passive acquired immunity is always temporary because the antibodies are not produced by the lymphocytes of the protected person.

Immunology

Immunology is the scientific study of immune responses. It is a huge and complex field of study, and only the basic fundamentals of immunology can be presented in this book. The topics briefly discussed in this chapter include active and passive immunity to infectious agents, processes involved in antibody production, cell-mediated immune responses, allergies and other types of hypersensitivities, autoimmunity, and serologic testing for antibodies and antigens.

An *antigen* can be any foreign organic substance that is large enough to stimulate the production of antibodies; in other words, it is *immunogenic.* Antigens (or *immunogens*) may be proteins of more than 10,000 daltons molecular weight, polysaccharides larger than 60,000 daltons, large molecules of DNA or RNA, or any combination of biochemical molecules (*e.g.,* glycoproteins, lipoproteins, and nucleoproteins) that are cellular components of either microorganisms or macroorganisms. Foreign proteins are the best antigens. An antigen must have one or more antigenic sites, known as *antigenic determinants* (or *epitopes*), to which antibodies or lymphocytes can bind. The important point is that antigens must be *foreign* materials that the human body does not recognize as *self* antigens. Certainly, all microorganisms fall into this category. Some small molecules called *haptens* may act as antigens if they are coupled with a large carrier molecule such as a protein. Then the antibodies formed against the antigenic determinant(s) of the hapten may combine with the hapten molecules when they are not coupled with the carrier protein. As an example, penicillin and other low-molecular-weight chemical molecules may act as haptens, causing some people to become allergic (or hypersensitive) to them.

Antibodies are glycoproteins produced by lymphocytes in response to the presence of an antigen. Antibodies bind specifically with the antigen that stimulated their production. (Actually, the antibody-producing cells are a type of lymphocyte called B-lymphocytes [or *B-cells*], which often work in coordination with T-lymphocytes [*T-cells*] and macrophages, as described later.) A bacterial cell has numerous antigenic determinant sites on its cell membrane, cell wall, capsule, and flagella that stimulate the production of many antibodies. Usually, those antibodies are considered to be specific in that they will recognize and bind to only those antigenic determinants that stimulated their production. Occasionally, similar antigenic sites (heterophile antigens) are found on other microorganisms or even on tissue cells, and those antibodies that can bind to these similar antigenic sites are referred to as cross-reacting antibodies.

All antibodies are in a class of proteins called *immunoglobulins* (Ig); they are globular glycoproteins in the blood that participate in immune reactions. We usually use the term antibodies to refer to immunoglobulins with particular specificity for an antigen. In addition to being found in the blood, immunoglobulins are also found in lymph, tears, saliva, and colostrum (Fig. 9 – 8). Colostrum is the clear fluid secreted by the mother's mammary glands after giving birth. This fluid contains a large number of antibodies and some lymphocytes from the mother that serve to protect the newborn during the first few months of life (natural passive acquired immunity).

The amount and type of antibodies produced by a given antigenic stimulation depends on the nature of the antigen, the site of antigenic stimulus, the amount of antigen, and the number of times the person is exposed to the antigen. Figure 9 – 9 shows that after the first exposure to an antigen (such as a vaccine), there is a delayed primary response in the production of antibodies. During this lag phase, the antigen is processed by macrophages, T-cells, and B-cells or by B-cells only. Some antigens, called T-dependent antigens, require the involvement of all three cell types whereas others, called T-independent antigens, require only B-cells. Ultimately, small B-cells develop into large B-cells (or *plasma cells*) that are capable of producing antibodies by protein

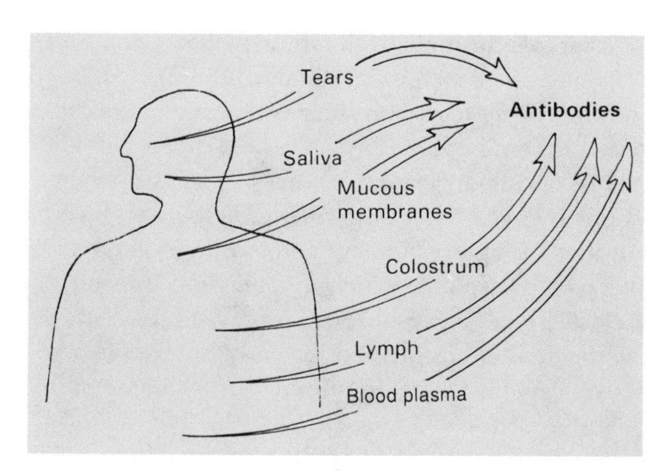

FIGURE 9 – 8
Body fluids and sites where antibodies can be found.

synthesis. This initial immune response to a particular antigen is called the primary response. When the antigen is used up, the number of antibodies in the blood declines as the plasma cells die off. Other antigen-stimulated B-cells become "memory cells", which are small lymphocytes that can be stimulated to produce antibodies quickly when later exposed to the same antigens. This increased production of antibodies following the second exposure to the antigen (*e.g.*, a booster shot) is called the secondary response, anamnestic response, or memory response. A second booster shot of antigen many months later returns the antibody concentration to the level of the secondary response. This is the reason why booster shots are given during the first year of life, before children

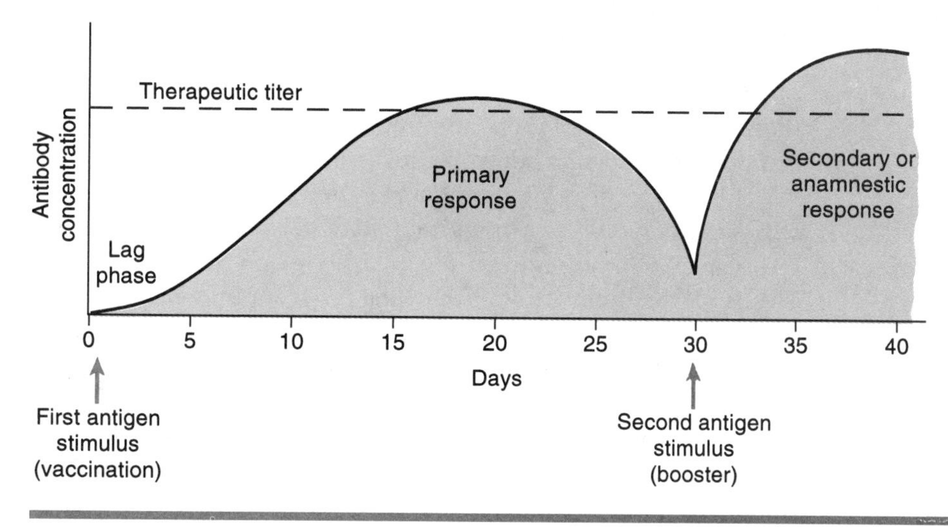

FIGURE 9 – 9
Antibody production following exposure to antigen.

attend school, and at specific intervals throughout life if the person is continually exposed to certain pathogens as in the case of tetanus.

Some persons are born without the ability to produce protective antibodies. Because they are unable to produce antibodies, they have no gamma globulin in their blood. This abnormality is called *agammaglobulinemia.* These persons are very susceptible to infections by even the least virulent microorganisms in their environment. One treatment for agammaglobulinemia that is often successful consists of a bone marrow transplant, which involves the transfer of precursor leukocytes from a closely related person. Some of these cells become lymphocytes. These lymphocytes may be implanted in the lymph nodes and become immunocompetent, *i.e.,* capable of being stimulated by antigens to produce antibodies.

Persons who produce an insufficient amount of antibodies are said to have *hypogammaglobulinemia.* Their resistance to infection is lower than normal, so they usually do not recover from infectious diseases as readily as most other persons.

Some patients are immunosuppressed (unable to make antibodies) after the administration of immunosuppressive drugs or agents, such as the antilymphocyte serum given before organ transplant surgery. Others are infected with a virus (such as the human immunodeficiency virus [HIV], which causes acquired immunodeficiency syndrome [AIDS]), which destroys the helper T-cells (T_H cells) that are required in the processing of T-dependent antigens and are also involved in cell-mediated immune responses. These patients usually die of secondary infections to which they have little resistance.

The Immune System

The immune system encompasses the whole body, but the lymphatic system is the site and source of most immune activity. The cells involved in the immune responses originate in bone marrow (from which most blood cells develop; Fig. 9 – 10). Three lines of lymphocytes—*B-cells, T-cells,* and *natural killer (NK) cells*—are derived from lymphoid stem cells of bone marrow. About half of these stem cells migrate to the thymus gland where they differentiate into T-cells (T for thymus) of the helper (T_H), suppressor (T_S), cytotoxic (T_C), or delayed hypersensitivity (T_D) type. Thymus processing of T-cells begins shortly before birth. T-cells are small lymphocytes found in the blood, lymph, and lymphoid tissues. They do not produce humoral antibodies but do aid in the control of antibody production and are involved in cell-mediated immune responses (*e.g.,* tissue transplant rejection; cellular immunity to mycobacteria, fungi, and viruses; and cytotoxicity of viral-infected cells and tumor cells).

Other lymphocytic stem cells differentiate in the liver and intestinal lymphoid areas into B-cells (named for the bursa lymphoid area in birds). B-cells

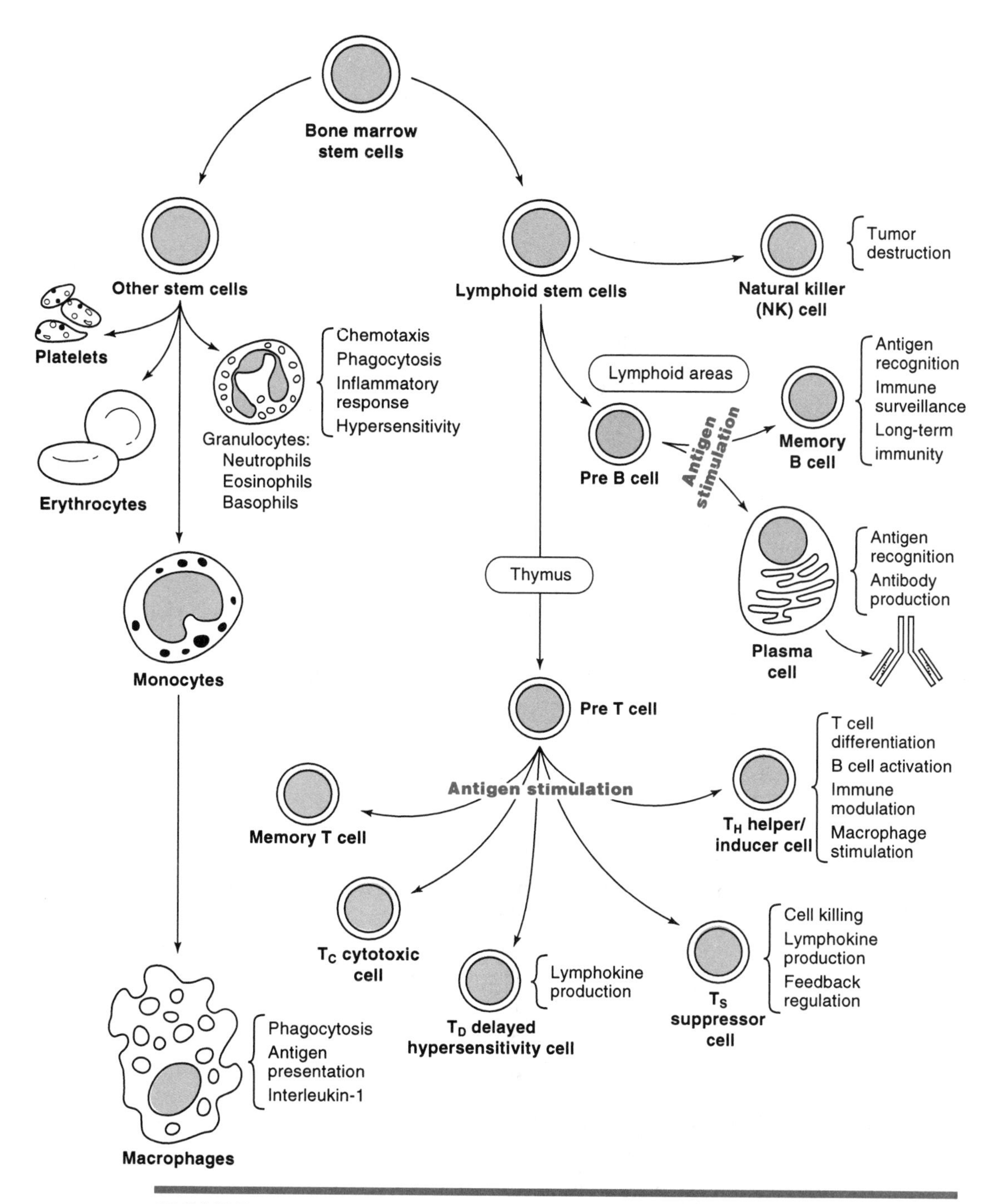

FIGURE 9 – 10
Differentiation of blood and lymph cells from bone marrow cells; development of cell-mediated immunity and antibody production.

migrate to lymphoid tissues where they produce antibodies that circulate through lymph and blood to protect the individual (humoral immunity). Thousands of B-cells exist in the body, although these cells live only about 1 to 2 weeks. When stimulated by an antigen, each B-cell is capable of producing thousands of specific antibodies per minute. Cells of the immune system are shown in Figure 9 – 10.

The Immune Response

Many chemicals, cells, and reactions are involved in immune responses. Some chemicals and complexes are nonspecific and yet depend on antigen–antibody complexes for activation (the complement system, for example). The specific humoral response always depends on the presence of specific antibodies for each antigen or antigenic determinant. Although some types of cell-mediated responses occur in the presence of antibodies, others do not involve antigen–antibody complexes.

HUMORAL IMMUNITY

For antibodies to be produced, a complex series of events must occur, some of which are not completely understood. It is known that macrophages, T-cells, and B-cells often are involved in a cooperative effort.

Most antigens are T-dependent antigens; they are processed as follows. When bacteria invade, they are first ingested and processed by macrophages. It appears that macrophages prepare and present bacterial antigens to T_H cells. These antigens appear on the surface of the macrophage and, at this point, the macrophage is referred to as an antigen-presenting cell. T_H cells are sensitized or primed by the antigen on the macrophage and serve to stimulate the production of antibody by B-cells; the T_H cells do not manufacture antibodies themselves. Helper T-cells induce B-cells to produce antibodies, whereas suppressor T-cells (T_S cells) inhibit antibody production. In this way, the level of antibodies is neither excessive nor insufficient if the control mechanisms are working properly. T_H and T_S cells are referred to as regulatory T-cells. After the antigen is presented to the T-cells by the macrophages, some B-cells are activated and stimulated to enlarge, differentiate, and divide into clones of antibody-producing plasma cells (large B-cells). Antibodies are expelled in gunfire fashion for several days until the plasma cells die. Each plasma cell clone makes only one type of antibody—the one that will bind with the antigen that was responsible for activating the B-cell. This specificity may be determined by the IgD-type antibodies (discussed later) on the surface of the B-cell. Those activated B-cells that did not become plasma cells, and some primed T-cells, remain as memory cells to respond more quickly when the antigen appears again.

When an antibody combines with an antigen, an antigen–antibody complex (or immune complex) is formed. When an antigen–antibody complex is formed, complement may be activated to destroy invading cells, increase phagocytosis, and/or neutralize bacterial toxins. Thus, acute, extracellular bacterial infections are controlled almost entirely by antibody-mediated immunity (AMI).

Cell-Mediated Immunity Antibodies are unable to enter cells, including cells containing intracellular pathogens. Fortunately, the macrophage/T-cell cell-mediated immunity (CMI) serves the body by controlling chronic infections of intracellular parasites (bacteria, protozoa, fungi, viruses). For example, in cytolytic viral infections (*e.g.*, herpes), the virus moves in the body fluids from a lysed cell to an intact cell, during which time the virus can be neutralized and destroyed by the antibody–complement complex; thus, viral infections can be prevented in this manner. However, when the virus is established within body cells, the cell-mediated immune response (macrophages, T_C cells) can destroy many virus-infected cells and damaged tissues. If the virus is not completely destroyed, it may become latent in nerve ganglion cells, as is the case in herpes infections.

Cell-mediated immunity results from several types of antigen-stimulated T-cells. Helper T-cells cooperate with B-cells to initiate the antibody-mediated response; T_S cells reduce the intensity of antibody-mediated response and moderate the activities of T_C cells, usually ensuring that the immune response is effective but not destructive.

Cytotoxic T-cells and NK cells kill infected host cells when pathogens are established inside the cells. Thus, infected liver cells are destroyed in hepatitis infections during the body's battle against the disease. The AIDS virus (HIV) that targets T_H cells is particularly destructive because it destroys the very cells that would have helped fight the infection. The lack of T_H cells impairs both humoral and cell-mediated immunity; thus, patients with AIDS are very susceptible to many opportunistic infections and malignancies.

Another type of T-cell (T_D cell) is involved in delayed hypersensitivity reactions. Its action is delayed because it is formed after antigen sensitization and reacts to produce *lymphokines* after a subsequent antigen stimulation. This response is discussed later in this chapter.

Nonspecific NK Cells Some specialized lymphocytes (neither T- nor B-cells) become nonspecific NK cells. These cells leave the lymphoid area and migrate to the inflammatory site of microbial invasion or an area of tumor growth. There they attach to and destroy abnormal cells, including those infected with intracellular agents such as chlamydias and rickettsias, transplanted cells, and tumor or malignant cells, whereas cytotoxic T_C cells target and destroy antigen-bearing, virus-infected cells. Natural killer cells are not dependent on macrophages, antigens, or antibodies; they strive to rid the body of abnormal or foreign cells. Both NK and T_C cells function by secreting lethal cytotoxic proteins (certain *cytokines*).

ANTIBODY STRUCTURE AND FUNCTION

Antibodies are Y-shaped glycoprotein immunoglobulin molecules produced by plasma cells in response to stimulation of B-cells by foreign antigens. Antibodies found in the blood are called humoral or circulating antibodies. All antibodies are immunoglobulins, but not all immunoglobulins are antibodies.

Studies of the gamma globulin component of human blood have revealed that five classes (or isotypes) of antibodies exist. These antibodies, or immunoglobulins, have been designated IgA, IgD, IgE, IgG, and IgM. Each may consist of several subclasses. The functions of each of these classes are listed in Table 9 – 2. The typical Y shape of an immunoglobulin is referred to as a monomer.

Immunoglobulin A (IgA) is designated the secretory antibody because these immunoglobulins are found in saliva, tears, colostrum, and other body secretions, as well as in the bloodstream and intestine. IgA molecules usually consist of two monomers held together by a J-chain (J for joining) and a secretory piece; in this configuration, they are referred to as dimers. They serve to protect the external openings and mucous membranes from invasion by pathogens. Not all IgA molecules are dimers; monomeric IgA molecules are found in internal body cavity secretions.

The IgA in colostrum and breast milk helps protect nursing newborns. In the

TABLE 9 – 2.
Immunoglobulin Classes and Functions

Ig Class	Molecular Weight	% in Serum (Approx.)	Functions
IgA	405,000 or 170,000	10–15	Protects the mucous membranes and internal cavities against infection; found as secretory antibodies in tears, saliva, colostrum, and other secretions
IgD	175,000	0.2	Fetal antigen receptor; controls antigen stimulation of B-cells; found in blood and on lymphocytes
IgE	190,000	0.002	Causes allergies, drug sensitivity, anaphylaxis, and immediate hypersensitivity; combats parasitic diseases
IgG	150,000	80	Protects against disease; attaches to phagocytes and tissues; fixes complement; crosses placental barrier; causes certain immunological diseases; found in blood and lymph
IgM	970,000	5–10	Protects against early infection; bactericidal to gram-negative bacteria; fixes complement; found in blood and lymph

intestine, IgA attaches to viruses, bacteria, and protozoal parasites, such as *Entamoeba histolytica,* and prevents the pathogens from adhering to mucosal surfaces, thus preventing invasion.

Immunoglobulin D (IgD) is a monomeric blood plasma antibody that may function as a control mechanism for the immune response. The function of this antibody is not clearly understood, but it may be involved in the determination of which antibody type a plasma cell produces, because it is found on the surface of B-cells.

Immunoglobulin E (IgE) is also called P-K antibody (in honor of the two scientists, Prausnitz and Küstner, who first identified it). These monomeric antibodies are more abundant in persons with allergy, drug sensitivity, or *anaphylactic shock.* Most often, these antibodies are bound to target cells (basophils in blood and mast cells in certain other tissues) and cause the rash or hives seen in various allergies, the runny nose and eyes of hay fever, asthma, the local reaction in immediate hypersensitivity skin tests, and shock reaction of *anaphylaxis.*

Immunoglobulin G (IgG) is the most abundant of the antibodies in blood. It binds (attaches) to antigens in blood plasma as well as in lymph and intercellular fluids. Invading bacteria are more easily destroyed by phagocytes or lysed by complement when antibodies are attached to them because the antibodies also may attach to the phagocytes or complement. A substance (such as an antibody) that enhances phagocytosis of a particle to which it is bound is known as an *opsonin.*

Immunoglobulin G is the smallest of the antibodies. It can cross the placenta and move freely in intercellular fluids and blood. Like all monomeric immunoglobulin molecules, it consists of four polypeptide chains (chains of amino acids): two identical short (light) chains and two longer (heavy) chains held together by disulfide (— S — S —) bonds with attached carbohydrate (Fig. 9 – 11). As just described, the molecule is referred to as a monomer. The antibody molecule is bivalent; that is, it has two sites that can bind specifically to the antigenic determinant that stimulated its production. These sites are called antigen-binding sites. The F_C region of this macromolecule can bind nonspecifically to complement, phagocytes, or tissue cells.

Immunoglobulin M (IgM) is the largest of the gamma globulins. It is a pentamer, consisting of five monomers held together by a J-chain. It can potentially combine with a total of 10 identical antigenic determinants. The IgM antibodies are the first antibodies formed in response to infections, especially those caused by gram-negative bacteria. Immunoglobulin M agglutinates bacteria, activates complement, and enhances phagocytosis. Red blood cell agglutinin antibodies and heterophile antibodies are also IgM antibodies.

Monoclonal Antibodies Purified antibodies against specific antigens are produced by scientists by an innovative technique in which a single plasma cell

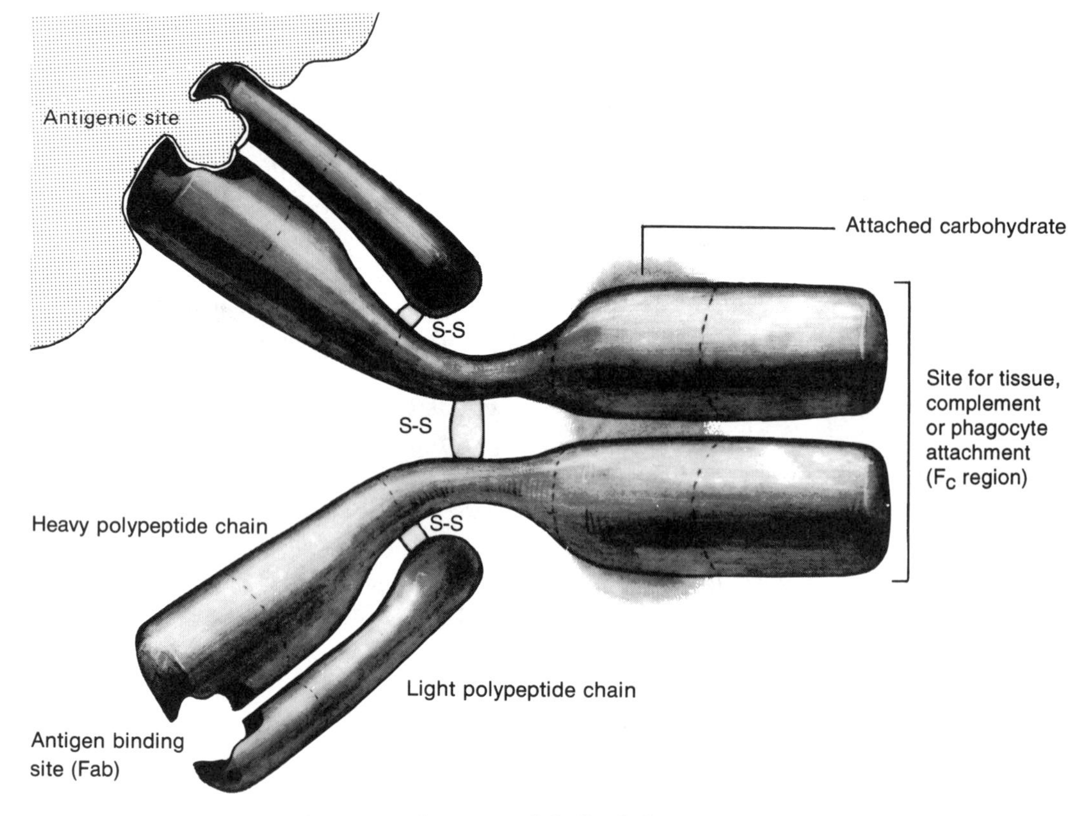

FIGURE 9 – 11
Basic structure of immunoglobulin IgG.

that produces only one specific type of antibody is fused with a rapidly dividing tumor cell. The new long-lived, antibody-producing cell is called a *hybridoma.* Hybridomas are capable of producing large amounts of specific antibodies called *monoclonal antibodies.* These antibodies are widely used, especially in developing new immunodiagnostic laboratory tests. Because they are uniform, highly specific, and can be produced in large quantities, monoclonal antibodies can be used routinely to bind with the antigens of particular pathogens in laboratory tests and in allergy and disease testing. Many new procedures are continuously being developed. Monoclonal antibodies may even be used for passive immunization to protect against certain diseases or to neutralize the toxins of botulism, tetanus, and perhaps snake and insect bites.

HYPERSENSITIVITY

Hypersensitivity is the defensive immune response gone awry, *i.e.,* being overly sensitive. Sometimes, instead of protecting a person, antibodies irritate and damage certain cells in the body. This can be compared to the person who builds a fire in the living room to warm the house and burns it down.

There are several different types of hypersensitivity reactions. Some involve various antibodies and others do not. All depend on the presence of an antigen and T-cells sensitized to that antigen.

Hypersensitivity reactions are divided into two general categories, immediate and delayed, depending on the nature of the immune reaction and the time required for an observable reaction to occur. An immediate reaction occurs from within a few minutes to 24 hours. There are several types of immediate reactions. Type I hypersensitivity reactions (also known as anaphylactic reactions) include the classic allergic responses of hay fever, asthma, and hives due to food allergies; allergic responses to insect stings; drug allergies; and anaphylactic shock. These reactions all involve IgE antibodies and the release of chemical mediators (especially histamine) from mast cells and basophils. Type II hypersensitivity reactions include the cytotoxic reactions seen in blood transfusion and Rh incompatibility reactions, and in myasthenia gravis, involving IgG or IgM antibodies and complement. Type III hypersensitivity reactions are immune complex reactions as in serum sickness and autoimmune diseases (systemic lupus erythematosus [SLE], rheumatoid arthritis) involving IgG or IgM antibodies, complement, and neutrophils. Delayed hypersensitivity, which is usually observed after 24 hours, is designated as type IV hypersensitivity. It is also referred to as cell-mediated immunity, such as occurs in tuberculin and fungal skin tests, contact dermatitis, and transplantation rejection. Delayed hypersensitivity does not involve antibodies but does depend on sensitized or primed T-cells that secrete lymphokines in response to a second exposure to the sensitizing antigen (Fig. 9 – 12).

The Allergic Response Immediate hypersensitivity (type I) is probably the most commonly observed type of hypersensitivity because approximately 80% of the American population is allergic to something. Persons who are prone to allergies (*atopic persons*) produce IgE (sometimes called reagin) antibodies when they are exposed to *allergens* (antigens that cause allergic reactions). The IgE molecules bind to the surface of basophils and mast cells by their F_C regions. The type and severity of an allergic reaction depend on a combination of factors, including the nature of the antigen, the amount of antigen entering the body, the route by which it enters, the length of time between exposures to

	IMMEDIATE			DELAYED
Type	**I**	**II**	**III**	**IV**
	Allergy, Anaphylaxis	Cytolysis, Blood type reactions	Auto-immunity	TB skin test, Transplantation rejection

FIGURE 9 – 12
Types of hypersensitivity immune responses.

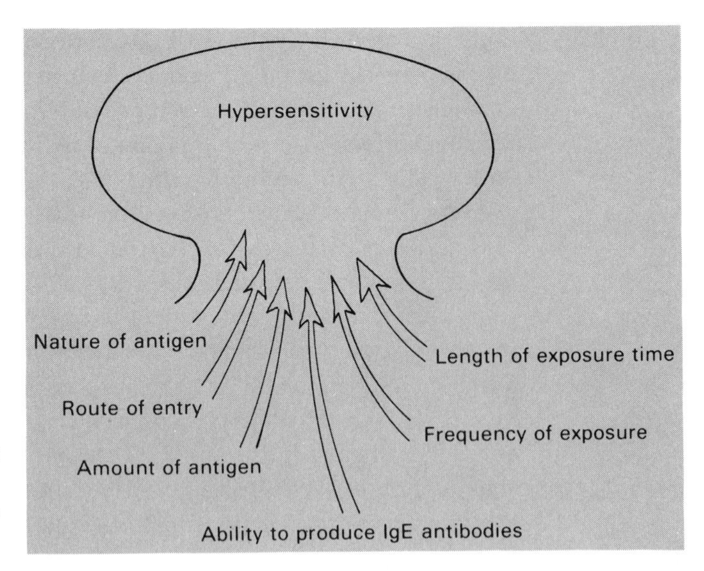

FIGURE 9 – 13
Factors in development of hypersensitivity.

the antigen, the person's ability to produce IgE antibodies, and the site of IgE attachment (Fig. 9 – 13).

The allergic reaction results from the presence of IgE antibodies bound to basophils in the blood or to mast cells in connective tissues, following the person's first exposure to the allergen. When the cell-bound IgE binds with the allergen during a subsequent exposure to the allergen, the sensitized cells respond by releasing irritating and damaging substances (chemical mediators). These mediators of the allergic response include *histamine*, prostaglandins, serotonin, bradykinin, slow-reacting substance of anaphylaxis (SRS-A), leukotrienes, and chemicals that attract eosinophils (eosinophilotactic agents).

Localized anaphylaxis Hay fever, asthma, and hives are localized allergic anaphylactic diseases. The symptoms depend on how the allergen enters the body and the sites of IgE attachment. If the allergen (pollens, dust, fungal spores) is inhaled and deposited on the mucous membranes of the respiratory tract, the IgE antibodies that are formed attach to mast cells in that area. Subsequent exposure to those inhaled allergens allows them to bind to the attached IgE, causing the mast cells to release large amounts of histamine. This substance initiates the classic symptoms of hay fever. Antihistamines function by binding to and, thus, blocking sites where histamine binds. Antihistamines are not as effective in treating asthma, however, because the mediators of this lower respiratory allergy include chemical mediators in addition to histamine. Allergens entering through the digestive tract (food and drugs) can also sensitize the host, and subsequent exposure may result in hives, vomiting, and diarrhea.

Systemic anaphylactic shock The severe allergic reaction called systemic anaphylactic shock occurs following an injection of an allergen to which the host has been sensitized. Frequently, these allergens are drugs or insect venom. With penicillin, the drug serves as a hapten, first binding to host blood proteins and then stimulating production of IgE, which binds to (sensitizes) mast cells and circulating basophils. Subsequent injections of large doses of penicillin into the sensitized host may cause the release of large amounts of histamine and other chemical mediators of allergy into the circulatory system.

The shock reaction usually occurs immediately (within 20 minutes) after reexposure to the allergen. The first symptoms are flushing of the skin with itching, headache, facial swelling, and difficulty in breathing; this is followed by falling blood pressure, nausea, vomiting, abdominal cramps, and urination (caused by smooth muscle contractions). In many cases, acute respiratory distress, unconsciousness, and death may follow shortly. Swift treatment with epinephrine (adrenaline) and antihistamine usually stops the reaction.

Health-care professionals must take particular care to ask patients if they have any allergies or sensitivities. In particular, those people with allergies to penicillin and other drugs and to insect stings should wear Medic-Alert tags so that they do not receive improper treatment during a medical crisis.

Anaphylactic reactions can be prevented by avoiding known allergens. In some cases, skin tests are used to identify the offending allergens. Then, desensitization may be accomplished by injecting small doses of allergen, repeatedly, several days apart. This treatment may be effective by causing the production of increased amounts of circulating IgG antibodies instead of IgE. In theory, the IgG should bind with the allergen and block its attachment to the cell-bound IgE. Such circulating IgG molecules are called *blocking antibodies.* This preventive measure usually works better for inhaled allergens than for allergens that are ingested or injected.

Autoimmune Diseases An *autoimmune disease* results when a person's immune system no longer recognizes certain body tissues as "self" and attempts to destroy those tissues as being "non-self" or "foreign." This may occur with certain tissues that are not exposed to the immune system during fetal development so that they are not recognized as self. Such tissues may include the lens of the eye, the brain and spinal cord, and sperm. Subsequent exposure to this tissue (by surgery or injury) may allow antibodies (IgG or IgM) to be formed, which together with complement could cause destruction of these tissues, resulting in blindness, allergic encephalitis, or sterility.

In some cases, immune complexes may cause destruction of heart, thyroid gland, or kidney tissues. Immunoglobulin G and IgM antibodies produced in response to group A, β-hemolytic streptococcal infection may bind with streptococcal antigens; the resultant immune complexes become deposited in heart tissue or the glomeruli of the kidney. Rheumatic fever and glomerulonephritis are serious complications (sequelae) of untreated or inadequately treated "strep" throat and other streptococcal infections.

Serum sickness is also a cross-reacting antibody immune reaction in which antibodies formed to globular proteins in horse serum (used for antitoxin treatments) may also bind with similar proteins in the patient's blood. The formation of these immune complexes (antigen + antibody + complement) causes the symptoms of fever, rash, kidney malfunction, and joint lesions of serum sickness.

It is believed that certain drugs and viruses may alter the antigens on host cells, thus inducing the formation of autoantibodies or sensitized T-cells to react against these altered tissue cells. This may be the underlying basis of multiple sclerosis, kuru, Hashimoto's thyroiditis, rheumatoid arthritis, SLE, and other autoimmune diseases.

Cell-Mediated Hypersensitivity Delayed hypersensitivity or type IV hypersensitivity is characterized by sensitized T-cells that secrete active substances in the absence of antibodies. As an example, when antigens enter the body, they may be phagocytized by macrophages and then the antigens sensitize T-cells. When these primed T-cells are subsequently exposed to the antigen, they secrete a group of chemicals called lymphokines. Lymphokines are cytokines that are produced by lymphocytes. The lymphokines then destroy the antigen directly, attract additional lymphocytes and macrophages to the area, and increase phagocytic activity. The action of T-cells, lymphokines, and macrophages causes the localized reaction seen in tuberculosis and fungal skin tests. The beginning of the reaction may be seen after a few hours, and the inflammation, fever, redness, and swelling remain for several days. A similar reaction occurs in contact dermatitis (poison oak and poison ivy).

The rejection of transplanted tissues containing foreign histological (tissue) antigens appears to occur in a similar manner, except that lymphokines as well as antibodies cause the rejection of the transplant.

Immunodiagnostic Procedures

By means of various clinical and research laboratory procedures, it has become possible to diagnose many infectious diseases by detecting either antigens or antibodies in clinical specimens, including body fluids and tissue specimens. Such procedures are collectively referred to as *immunodiagnostic procedures*—diagnostic procedures that use the principles of immunology. Any given immunodiagnostic procedure detects either antigens (direct evidence of the pathogen's presence in the patient) or antibodies (the patient's response to the presence of the pathogen). The antigens being detected are molecules (antigenic determinants) that may be part of the capsid or envelope of a virus or

INSIGHT
Specimen Quality and Clinical Relevance

Microbiology laboratory results are clinically relevant if they provide the physician with useful information that can be used to diagnose infectious diseases, monitor their progress, and guide therapy. To provide clinically relevant information, the microbiology laboratory must receive high-quality clinical specimens and the quality of the results can be no better than the quality of the specimen. If a poor-quality specimen is submitted to the laboratory, it is likely that the results obtained using that specimen will not be clinically relevant. In fact, results obtained from poor-quality specimens might very well be harmful to the patient.

What constitutes a high-quality clinical specimen? The best-quality specimen is one that has been selected, collected, and transported properly. First, it must be an appropriate specimen that was collected in a manner that will minimize its contamination with indigenous microflora. The specimen must also be transported to the laboratory in the proper manner—rapidly, if necessary; on ice, if necessary; anaerobically, if necessary; with the proper preservative, if necessary; etc. A specimen labeled "sputum" must contain sputum—not merely saliva. A urine specimen submitted for culture must be a clean-catch, midstream specimen. Care must be taken to adequately disinfect the phlebotomy site when blood is drawn for culture to minimize the chance of contamination of the specimen with indigenous skin flora.

To ensure that specimens are high quality, the laboratory must take the time to educate health-care workers about what constitutes an appropriate specimen for the diagnosis of each infectious disease. It is the laboratory's responsibility to publish and distribute a manual containing instructions for the proper selection, collection, and transport of specimens. Only in this way will the highest quality of service be assured and only then will the microbiology laboratory's results be clinically relevant.

bacterial capsules, cell walls, flagella, or toxins. As is true for any clinical specimens, specimens for immunodiagnostic procedures must be of the highest possible quality.

Serology, from the word serum, primarily refers to the application of immunodiagnostic procedures to serum and other body fluids. Serologic techniques are used to diagnose diseases, to type blood and select blood for transfusions, to identify microorganisms, to type tissue specimens for transplantation purposes, and to detect allergy or hypersensitivity. The most common *in vitro* serologic tests are *agglutination, precipitin,* lysis by complement, fluorescent antibody technique, radioimmunoassay, and enzyme-immunoassays. Sometimes it is necessary to use *opsonization,* capsular swelling, or immobilization tests.

These tests are all performed in the laboratory to determine if a patient's specimen contains either antibodies or antigens. The presence of specific antibodies to a particular pathogen would indicate one of three possibilities: (1) that the patient *currently* has the disease caused by the pathogen, (2) that the patient *had* the disease at some time in the past, or (3) that the patient had been vaccinated at some time in the past to protect him or her from developing the disease. Other tests are performed *in vivo* (in the living person), such as skin tests, in which antigens are injected within or beneath the skin (intradermally

TABLE 9 – 3.
Serologic Tests to Detect Antibodies

Reaction In Vitro	**Reagents** Antigen	Antibody	Other	**Results** +	−
Agglutination	Red blood cells or bacteria	Patient's serum		Clumping	No clumping
Precipitin	Toxins, hormones, proteins	Patient's serum	Agar or solution	Precipitate	No precipitate
Lysis by Complement	Cells, bacteria	Patient's serum	Complement	Lysis	No lysis
Fluorescent Antibody Technique	Pathogen	Patient's serum	Fluorescein-tagged rabbit antiserum	Fluorescent pathogen	No fluorescence
Opsonization	Bacteria	Patient's serum	Phagocytes	More bacteria in phagocytes	Fewer bacteria in phagocytes

continued

TABLE 9 – 3. (Continued)

Reaction In Vitro	**Reagents** Antigen	Antibody	Other	**Results** +	−
Capsular Swelling (Quellung Reaction)	Encapsulated bacteria 	Patient's serum 		Capsule appears to swell 	No appearance of swelling 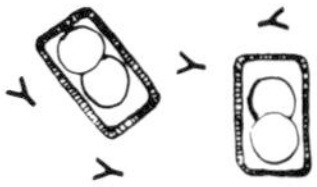
Immobilization	Motile bacteria 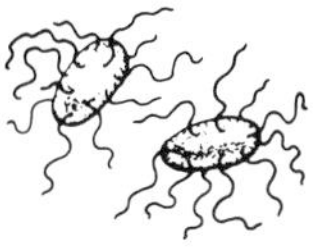	Patient's serum 		No motility 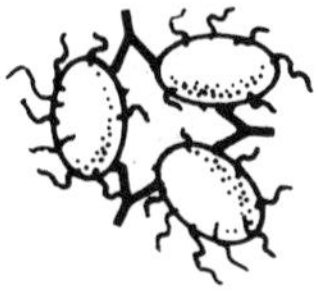	Motility 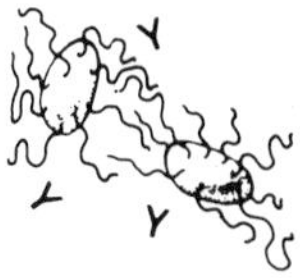
Radio-immunoassay	Patient's serum 	Radioactively-tagged rabbit antiserum 		High radioactivity 	Low or no radioactivity 
Enzyme-linked assay	Test microbe 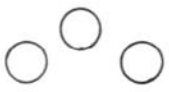	Patient's serum 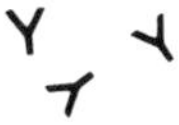	Enzyme linked antibody  +Substrate 	Color change 	No color change 

or subcutaneously) to detect the presence of antibodies or a cell-mediated immune reaction. Swelling and reddening at the site of the injection indicates that the patient either currently has the disease or has been exposed to the antigen at some time in the past, in which case the patient might be immune or allergic. Serologic tests are summarized in Table 9 – 3 and a description of them follows.

Agglutination Tests *Agglutination tests* are most often used to detect the presence and amount of humoral (serum) antibodies to particulate antigens, such as those on red blood cells or bacteria. As the word suggests, particulate antigens are antigenic sites on particles, such as cells or latex particles. The clumping of cells (such as red blood cells) or latex particles is referred to as *agglutination.*

The agglutination technique is used in the Venereal Disease Research Laboratory (VDRL) and rapid plasma reagin (RPR) tests for syphilis, the Widal test for typhoid fever, the Weil–Felix test for rickettsial diseases, for blood typing, and in various pregnancy tests. It also has many other applications.

Hemagglutination (the agglutination of red blood cells) is caused by certain viruses. Antibodies to these viruses would inhibit the hemagglutination reaction and could be used to positively identify the virus. This adaptation of the agglutination technique is called hemagglutination inhibition and is frequently used to identify influenza viruses.

Precipitin Test In one type of precipitin test, the patient's serum is used to demonstrate the presence of antibodies to soluble proteins or polysaccharide antigens, such as exotoxins or blood proteins. In the immunodiffusion test, antigens and antibodies are placed in wells in an agar plate and allowed to diffuse toward each other. At the point where the correct concentrations of antigen and antibody are reached (the zone of optimal proportions), a *precipitate* is formed that appears as a white line in the agar (Fig. 9 – 14). Precipitin tests can be used to detect either antigens or antibodies in patients' specimens. Precipitin tests are useful in detecting antibodies to the exotoxins of tetanus, diphtheria, and scarlet fever. They are also used to identify various proteins in blood. An interesting application of the precipitin test is its use in criminology laboratories as a means to determine whether a particular blood stain is from animal or human blood. Here the blood proteins are extracted and then identified by the antibodies that bind specifically with them.

Other precipitin-type tests (*e.g.,* immunoelectrophoresis) are performed by separating the proteins in human serum (by electrophoresis), adding known antibodies, and observing the precipitin lines (Fig. 9 – 15). This powerful test is often used to determine complement and antibody deficiencies in individuals.

Lysis by Complement When IgG and IgM antibodies are bound to antigens, the antibodies become activated and react with complement. This reaction, in turn, activates complement. If the antigens are on cells, such as red blood cells, or on pathogens, such as *Vibrio cholerae,* the activated complement will lyse (destroy) the cells. Lysis by complement can be used to identify the presence of specific antibodies to a pathogen in a patient's serum, as in cholera.

Fluorescent Antibody Technique Fluorescent dyes (called fluorochromes), which are easily seen glowing under an ultraviolet microscope, can be attached to the antibodies from the patient's serum, which will adhere to the pathogen.

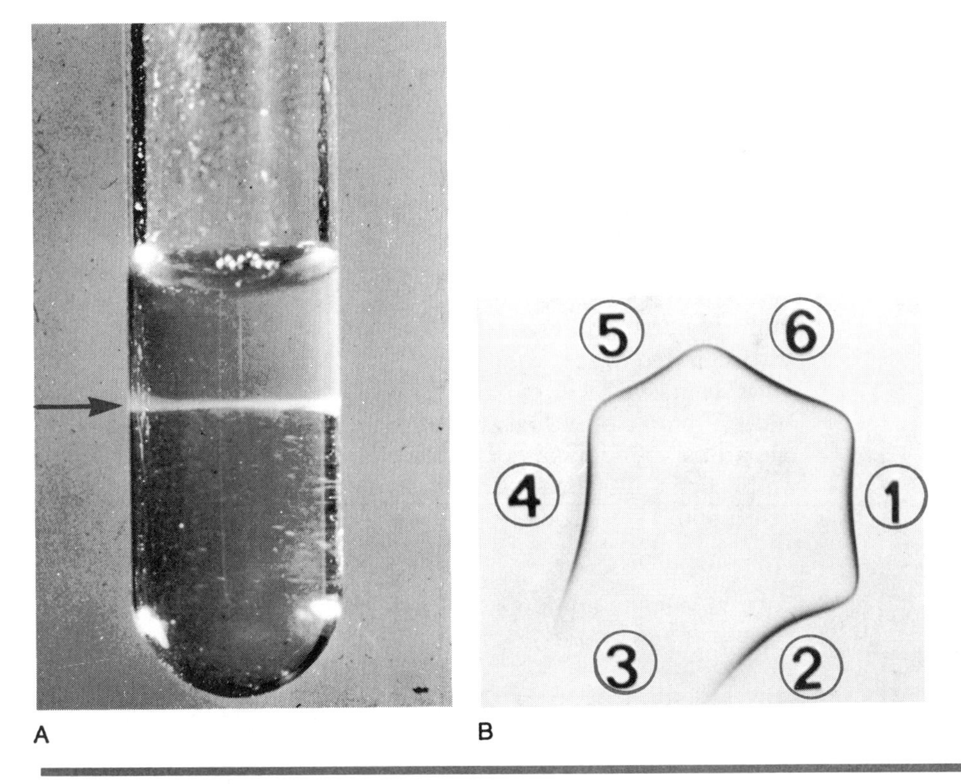

FIGURE 9 – 14
Precipitin ring test. (*A*) A line of precipitation forms where the antigen and antiserum meet (arrow). (Volk WA, Wheeler MF: Basic Microbiology, 3rd ed. Philadelphia, JB Lippincott, 1973) (*B*) Wells 1 through 6 contain different antigenic fractions of *Yersinia pestis,* the etiologic agent of plague. (Courtesy of Dr. Tornebene)

The pathogen might be *Treponema pallidum* (the etiologic agent of syphilis) or *Streptococcus pyogenes* (the etiologic agent of "strep" throat). The pathogen is beautifully illuminated or outlined by the fluorescent dye when it is examined with the ultraviolet microscope if the patient's serum contains the specific antibodies for those pathogens (Fig. 9 – 16). More often, the fluorescent dye is attached to rabbit antibodies to human gamma globulin, which will then outline the pathogen on the slide if the patient's gamma globulin contains antibodies to that pathogen. This test is often used for positive identification of Group a β-hemolytic *S. pyogenes, T. pallidum, Neisseria meningitidis, Salmonella typhi, Haemophilus influenzae,* rabies virus, and many other pathogens.

Radioimmunoassay Radioimmunoassay (RIA) is a sensitive, versatile technique using radioactively labeled antigen or antibody. It is frequently used to

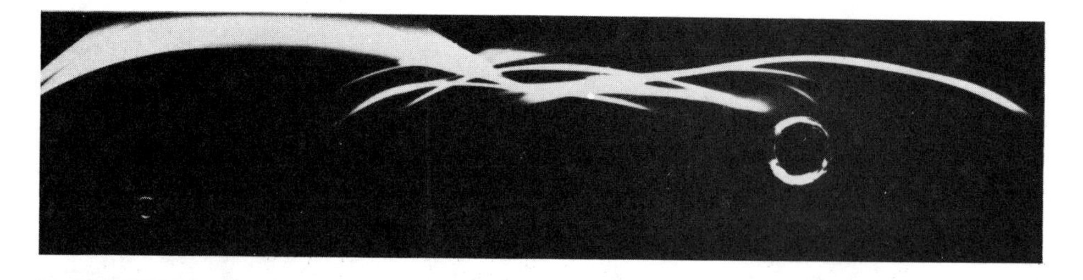

FIGURE 9 – 15
Immunoelectrophoresis: whole human serum was placed in the well (circle) and subjected to electrophoresis. After 2 hours, the electric current was turned off and anti-whole human serum was placed in a trough along the length of the slide. The precipitin lines were formed where the antibodies and the antigens (separated components of whole human serum) diffused together in optimal proportions. (Volk WA, et al: Essentials of Medical Microbiology. 5th ed. Philadelphia, JB Lippincott, 1991)

determine small amounts of drugs, hormones, or antigens, such as hepatitis B antigen in blood donor serum.

Enzyme-Linked Immunosorbent Assays (ELISA or EIA) These enzyme-linked assays are sensitive techniques that use an enzyme–antibody–antigen combination adsorbed onto the sides of a test well. If the patient's specimen contains antibodies against or antigens of the pathogen, the linkage is formed; when the substrate for the enzyme is added, a color change develops, which indicates a positive test result. If the patient does not have the antigen or the antibody sought, the enzyme does not bind and no color change is seen following addition of the substrate. These techniques are commonly used to test for AIDS, rubella virus (the etiologic agent of German measles), and certain drugs present in serum.

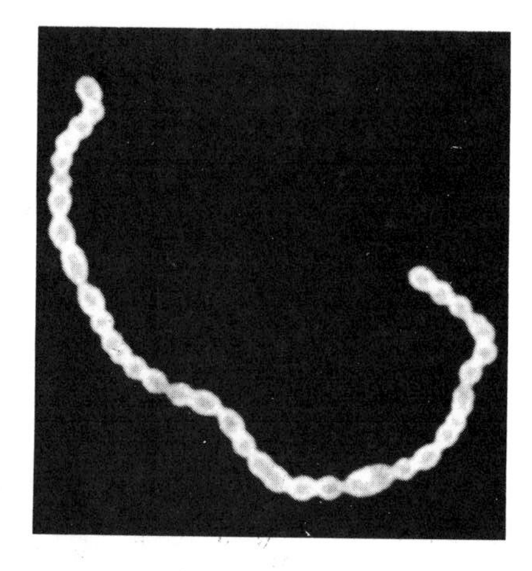

FIGURE 9 – 16
Streptococcus pyogenes stained with fluorescent antibody (original magnification, × 200). (Volk WA, Wheeler MF: Basic Microbiology, 4th ed. Philadelphia, JB Lippincott, 1984)

Opsonization The opsonization test is based on the fact that phagocytes can engulf more bacteria if the specific antibodies for those bacteria are present in a patient's specimen. A control slide with bacteria and phagocytes is observed, and the number of bacteria inside each phagocyte is counted. This number is compared with the number of bacteria engulfed by phagocytes in the presence of the patient's serum. The presence of antibodies on the surface of the bacteria makes them more readily phagocytised, a process known as *opsonization.* This test is seldom used in a clinical laboratory.

Capsular Swelling (Quellung Reaction) When serum containing antibodies to encapsulated bacteria is added to those bacteria and observed microscopically, the capsules appear to swell. The apparent swelling is frequently referred to as the Quellung reaction (see Figure 7 – 4). Various capsular types of *Streptococcus pneumoniae* and other encapsulated bacteria can be identified by this test.

Immobilization The immobilization test can be used if the pathogen is motile (flagellated). When a patient's serum is added to motile bacteria, the bacteria are immobilized if specific antibodies to the bacterial flagella are present. Identification of cholera and *Salmonella* and *Shigella* infections is more reliable using this technique; however, it is rarely used in clinical laboratories because better tests are readily available.

Toxin Neutralization The neutralization of toxins usually is demonstrated in laboratory animals by *in vivo* tests that demonstrate that specific antitoxins (antibodies) in the patient's serum neutralize the effect of a toxin.

The animal test for diphtheria A mouse injected with diphtheritoxin will die. But if the toxin is mixed with the serum of a patient who is immune to diphtheria and then the combined toxin and serum is injected into a mouse, the mouse will live.

The Schick test Diphtheria toxin is injected under the skin (subcutaneously). The site becomes reddened if the person is susceptible to diphtheria. No redness is seen if the person is immune, because antibodies are present that neutralize the toxin and, thereby, prevent the redness.

The Dick test The scarlet fever exotoxin is injected intracutaneously; redness occurs at the site in a susceptible person, but not in one who is immune.

The Schultz–Charlton test The rash of scarlet fever can be positively identified by means of the Schultz–Charlton test. An antitoxin to the scarlet fever toxin is injected into an area of the rash. If the rash clears (due to toxin neutralization), the patient has scarlet fever; if it does not clear, the rash is due to other causes.

Use of Serologic Tests Serologic reactions can provide information that is useful in the diagnosis of many infectious diseases. Most serologic tests are extremely valuable because they can be performed quickly and easily with a high degree of specificity and sensitivity. However, some of the newer tests may not be widely used because of their cost.

SUMMARY

The nonspecific defenses of the human body are those general mechanisms that serve to protect it from harmful foreign substances. The first line of defense is the mechanical and chemical barrier against foreign invaders provided by the skin, the mucous linings of the body openings, and the defenses in the digestive, respiratory, and urogenital systems. The chemical and physical processes involved in inflammation and phagocytosis make up the second line of defense.

The human body's third line of defense against the invasion of foreign microorganisms is the immune response. Immunology is the science of the immune responses in the body.

An antigen is any high-molecular-weight foreign protein or polysaccharide that causes antibodies to be formed and that binds specifically with that antibody. An antibody is a protein produced by lymphocytes in response to a foreign antigen, and it usually attaches only to that particular antigen.

There are several types of acquired immunity, including two types of active acquired immunity and two types of passive acquired immunity.

All antibodies have specific structures and functions. There are five classes of antibodies or immunoglobulins—IgG, IgM, IgA, IgD, and IgE. Immunoglobulin G protects against disease, attaches to phagocytes and tissues, fixes complement, crosses the placental barrier, and causes certain immunological diseases. Immunoglobulin M, a pentamer, protects against early infection, is bactericidal to gram-negative bacteria, and fixes (attaches to and activates) complement. Secretory IgA, a dimer found in tears, saliva, colostrum, and other secretions, protects the mucous membranes against infections. Immunoglobulin E molecules are the tissue-bound antibodies of allergy, drug sensitivity, anaphylaxis, and hypersensitivity. The function of IgD is not understood, but it is found on the surface of B-cells. Immunoglobulins G, D, and E are monomeric.

An individual may develop hypersensitivity, in which case the defense immune response has gone awry, and instead of protecting the individual, the antibodies irritate and—in some cases—destroy certain cells in the body. There are two major types of hypersensitivity: immediate and delayed.

The formation of antibodies against antigens of one's own tissues and the destruction of these tissues is called an autoimmune response. An example of this type of response is rheumatoid arthritis.

Clinical and research tests have been developed to indicate the presence of antigens and antibodies in blood and other clinical specimens. This study of antigen–antibody reactions is called serology. These tests include agglutination, precipitin, lysis by complement, fluorescent antibody technique, opsonization, capsular swelling, immobilization, and toxin neutralization.

PROBLEMS AND QUESTIONS

1. What factors contribute to natural resistance to disease?
2. List some nonspecific defenses of the skin, respiratory system, digestive system, and urogenital tract.
3. List some of the blood proteins that aid in the destruction of invading microorganisms.
4. Describe the process of phagocytosis.
5. What are the four main symptoms of inflammation?
6. What are the causes of the symptoms of inflammation?
7. What is the immune response? Is it always a protective mechanism?
8. What types of substances are effective antigens?
9. Where in the body are antibodies formed?
10. When are antibodies formed?
11. What is the anamnestic response?
12. What is the main difference between active and passive acquired immunity?
13. How are antibodies transferred in passive immunity?
14. List five classes of immunoglobulins. Where are each of these found?
15. What are the main differences among the four types of hypersensitivity?
16. What is an allergen? Give some examples.
17. Give an example of the delayed hypersensitivity reaction.
18. Cite an example of an autoimmune disease.
19. List seven *in vitro* antigen-antibody tests and when they are used.
20. List four *in vivo* antigen–antibody tests and when they are used.

Self Test

After you have read Chapter 9, examined the objectives, reviewed the chapter outline, studied the new terms, and answered the problems and questions above, complete the following self test.

Matching Exercises

Complete each statement from the list of words provided.

RESISTANCE AGAINST PATHOGENS

bile	lysozyme	nonspecific
interferon	specific	innate
species	digestive enzymes	indigenous microflora
complement		

1. Complement, interferon, and phagocytes are some of the body's ________________ defenses.
2. ________________ is secreted by the liver, stored in the gallbladder, and released into the small intestine, where it lowers the surface tension of particles, such as bacteria, to make them more digestible.
3. An enzyme found in nasal secretions that lyses the cell walls of certain bacteria is ________________.
4. When antibodies are formed that bind with the specific antigens that caused their formation, the body is activating its ________________ resistance against foreign substances.
5. The usually harmless microorganisms that reside on the skin and in the mucous membranes of many body systems are the ________________.
6. ________________ is a protein that is secreted by cells infected by viruses and serves to prevent the surrounding cells from producing viruses.
7. A certain species of animals resistant to a disease found in other species is said to have ________________ resistance.
8. The digestive tract contains many ________________, which frequently digest certain bacteria.
9. The natural, inherited defense mechanisms that give certain individuals or species some resistance to certain diseases are ________________ resistance.
10. A complex group of proteins found in blood plasma that aids in the inactivation and destruction of bacteria is the ________________ system.

Immunology

immunology
agammaglobulinemia
antitoxins
IgM
allergen
complement
immunocompetent
anamnestic response
hypogammaglobulinemia
immunoglobulins
anaphylactic response
IgA
plasma
antibodies
IgE
primary response
antigen
IgG
serum

1. When blood clots, the remaining liquid is called ________________.
2. When a person is not able to produce antibodies, he or she has an abnormality known as ________________.
3. The smallest but most abundant of the antibodies found in the serum is ________________.
4. Any foreign material that can stimulate the production of antibodies is an ________________.
5. When a person's immune system is functioning properly, the person is said to be ________________.
6. ________________ is the antibody found in allergic diseases that is usually bound to cells or tissues.

7. An antigen that causes an allergic reaction is an ________________.
8. The antibodies that are formed in response to toxin or toxoid antigens are ________________.
9. When small lymphocytes are initially stimulated by antigens (such as vaccines) to develop into plasma cells to produce antibodies, this process is the ________________.
10. The fluid that remains when cells are removed from blood is called ________________.
11. The largest serum antibody and the first to be formed against pathogens is ________________.
12. If an individual produces less than the normal amount of antibodies, the abnormality is ________________.
13. A group of serum proteins that are activated by specific antigen–antibody combinations is ________________.
14. When memory lymphocytes are restimulated by antigens to produce the same antibodies again (as when a booster vaccine is given), the ________________ has occurred.
15. The secretory antibody found in tears, saliva, and colostrum is ________________.
16. The protein molecules that are produced by B-cells in response to the presence of antigens are called ________________.
17. When the anamnestic response occurs as a serious allergic shock reaction, it is called a / an ________________.

IMMUNE RESPONSE

allergy
artificial active acquired immunity
immediate
delayed
autoimmune
artificial passive acquired immunity
natural active acquired immunity
natural passive acquired immunity

1. Hay fever, asthma, and anaphylactic shock are examples of ________________ hypersensitivity.
2. ________________ results when a person normally produces antibodies to a disease.
3. Skin rashes caused by ________________ are examples of ________________ hypersensitivity.
4. When people are vaccinated so that they produce antibodies to the antigens in a vaccine, they develop ________________.
5. The rejection of an organ transplant, such as a kidney transplant, is an example of ________________ hypersensitivity.
6. When antitoxin is given to a person who has tetanus, the individual has been given ________________.
7. When people's bodies form antibodies against their own tissues and destroy their own organs, such as the heart, kidney, or thyroid, we say that they have a / an ________________ disease.
8. A newborn has temporary maternal antibodies to protect him or her from disease; we say he or she has ________________.

ANTIGEN–ANTIBODY REACTIONS

lysis by complement
fluorescent antibody
capsular swelling
agglutination
opsonization
precipitation
lysis by complement
toxin neutralization
immobilization

1. When complement is activated by antibodies attached to cellular antigens, the cells are destroyed as a result of ________________.
2. When antibodies inactivate an exotoxin, ________________ occurs.
3. The Schick and Dick tests are examples of ________________ reactions.
4. Blood typing is a good example of a/an ________________ test.
5. The ________________ test can only be used with motile pathogens.
6. The Quellung test or ________________ test can only be used on encapsulated pathogens.
7. The VDRL and RPR tests for syphilis are ________________ tests.
8. ________________ is used to indicate the presence of antibodies to soluble antigens, proteins, or toxins.
9. An increase in phagocytosis due to the attachment of specific antibodies to certain pathogens is a test called ________________.
10. Pathogens such as streptococci are often identified by attaching a fluorescent dye to the specific antibody that will bind with them; this is called the ________________ technique.

True or False (T or F)

___ 1. Phagocytes easily ingest and digest all types of bacteria.
___ 2. The symptoms of inflammation are heat, swelling, redness, pain, and sometimes loss of function.
___ 3. The main purpose of the inflammatory response is to destroy bacteria.
___ 4. The ciliated epithelial cells lining the postnasal membranes engulf and digest bacteria.
___ 5. There are indigenous microflora in the healthy bladder.
___ 6. Periodic flushing of urine through the urethra helps prevent pathogens from invading the urinary system.
___ 7. Interferon defends the body by lysing the cell wall of gram-negative bacteria.
___ 8. Most phagocytes are also leukocytes.
___ 9. The reticuloendothelial system is a system of tissues found in the spleen, liver, lymph nodes, blood vessels, and intestines.
___ 10. Antigens may be proteins or polysaccharides or both.
___ 11. Foreign proteins are stronger antigens than foreign polysaccharides.
___ 12. An effective vaccine must always contain living pathogens.

___ 13. A good vaccine stimulates the body to produce protective antibodies.

___ 14. Attenuated microorganisms are living, weakened, mutated pathogens that no longer cause disease but cause the production of antibodies that can protect against that disease.

___ 15. Colostrum is a good source of antibodies to protect newborns.

___ 16. Passive immunity is always temporary.

___ 17. The tuberculosis skin test reaction is an example of immediate hypersensitivity.

___ 18. Rheumatic fever is an autoimmune disease that may follow β-hemolytic streptococcal infections.

___ 19. *In vitro* means the tests are performed in the living animal.

___ 20. Precipitins, opsonins, and agglutinins are all antibodies.

___ 21. An electron microscope must be used to observe the fluorescent antibodies attached to a pathogen.

___ 22. Antibodies and antibiotics are two words for the same thing.

___ 23. A series of *in vitro* antigen–antibody tests on a sick person that show an increasing antibody titer indicates that the patient's body is fighting the infection.

___ 24. An exotoxin is liberated only after the cell is destroyed or dead.

___ 25. Antitoxins are often used in the treatment of diphtheria, botulism, and tetanus.

___ 26. Anaphylactic shock kills humans chiefly by severe spasms of the smooth muscles of the body.

Multiple Choice

1. The body's first line of defense is
- **a.** unbroken skin
- **b.** antibody molecules
- **c.** antigen molecules
- **d.** T-cells
- **e.** phagocytic cells

2. What stimulates leukocytes to migrate to an injured area of the body?
- **a.** phagocytosis
- **b.** chemotaxis
- **c.** leukotrienes
- **d.** leukocytes do not migrate to an injured area
- **e.** prostaglandins

3. Interferon, an antiviral substance, has the following properties:
- **a.** it is cell-specific, not virus-specific
- **b.** it is a small protein molecule
- **c.** it can inhibit certain viral infections
- **d.** all of the above
- **e.** none of the above

4. The reticuloendothelial system functions as follows:
- **a.** removes bacteria and other particulate matter from circulating fluid
- **b.** cells migrate to the areas of injury and engulf foreign substances
- **c.** in inflammatory response, contributes fibrin
- **d.** all of the above
- **e.** none of the above

5. Mechanical and chemical defenses against infection include
- **a.** intact skin and mucous membranes
- **b.** fatty acids secreted by the sebaceous glands
- **c.** lysozyme in tears and other secretion
- **d.** ciliary action in respiratory tract
- **e.** all of the above

6. Phagocytes function by
- **a.** destroying all bacteria they encounter
- **b.** producing complement
- **c.** engulfing foreign material
- **d.** providing the first line of defense

7. If no urinary tract infection is present, urine drawn during a catheterization should contain
- **a.** erythrocytes
- **b.** *Enterobacter aerogenes*
- **c.** *Candida albicans*
- **d.** none of the above

8. Pyogenic bacteria, such as *Pseudomonas,* cause
- **a.** greater heat in the area of inflammation than other bacteria
- **b.** pus formation
- **c.** vasoconstriction
- **d.** destruction of phagocytes

9. The humoral response is
- **a.** a nonspecific defense
- **b.** initiated by mast cells
- **c.** the production of antibodies
- **d.** possible only against bacteria

10. Antibodies are produced by
- **a.** blood cells
- **b.** B-lymphocytes
- **c.** antigens
- **d.** complement

11. Immunity following actual illness is usually
- **a.** temporary natural passive acquired immunity
- **b.** permanent artificial active acquired immunity
- **c.** permanent natural passive acquired immunity
- **d.** temporary natural active acquired immunity

12. Naturally acquired passive immunity would result from
- **a.** subclinical disease
- **b.** colostrum
- **c.** injection of antitoxin
- **d.** all of the above

13. The vaccines for typhus, measles, and mumps are
- **a.** toxoids
- **b.** killed pathogens
- **c.** extracts of pathogens
- **d.** attenuated living organisms

14. The newborn's antibodies are
- **a.** IgG
- **b.** IgA
- **c.** IgM
- **d.** there are no antibodies present at birth

15. Botulism would be treated with
- **a.** hyperimmune gammaglobulin
- **b.** botulism toxoid
- **c.** antitoxin
- **d.** autogenous vaccine

16. The Schultz–Charlton test involves
- **a.** the injection of antitoxin into a test animal
- **b.** the injection of antitoxin into the patient
- **c.** the injection of toxin into the patient
- **d.** the combining of patient serum with toxin and complement

17. An immunocompetent adult human being can respond immunologically to

a. protein antigens only
b. polysaccharide antigens only
c. most substances recognized as self
d. nonhuman substances only
e. many thousands of different antigenic determinants

CHAPTER 10

Major Infectious Diseases of Humans

OBJECTIVES

After studying this chapter, you should be able to

- Name the major organs that might become infected in each body system
- List the most common members of the indigenous microflora usually found in the various body systems
- Outline the causative agent, reservoir, mode of transmission, pathogenesis, treatment, and control measures for the major infectious diseases of each body system
- For each body system, list some examples of diseases that are caused by bacteria, viruses, fungi, and protozoa

CHAPTER OUTLINE

Skin Infections
Healthy Skin
Opportunists
Major Microbial Opportunists
Viral Infections of the Skin
Bacterial Infections of the Skin
Fungal Infections of the Skin
Burn and Wound Infections

Eye Infections
Viral Eye Infections
Bacterial Eye Infections

Infectious Diseases of the Mouth
Indigenous Microflora and Oral Diseases
Bacterial Infections of the Oral Cavity

Ear Infections
Viral and Bacterial Ear Infections

Infectious Diseases of the Respiratory System
Nonspecific Respiratory Infections
Viral Respiratory Infections
Specific Bacterial Respiratory Infections
Protozoan Respiratory Infections

Infectious Diseases of the Gastrointestinal Tract
Viral Gastrointestinal Infections
Bacterial Gastrointestinal Infections
Bacterial Foodborne Intoxications, Food Poisoning
Protozoan Gastrointestinal Diseases

Infectious Diseases of the Urogenital Tract
Sexually Transmitted Diseases
Urinary Tract Infections
Viral Urogenital Infections
Bacterial Urogenital Infections
Protozoan Urogenital Infections
Other Sexually Transmitted Diseases

Infectious Diseases of the Circulatory System
Viral Lymphatic and Cardiovascular Infections
Rickettsial Cardiovascular Infections
Bacterial Cardiovascular Infections
Protozoan Cardiovascular Infections

Infectious Diseases of the Nervous System
Viral Nervous System Infections
Bacterial Nervous System Infections

NEW TERMS

Arbovirus
Botulinal toxin
Cervicitis
Choleragin
Cystitis
Cytotoxins
Dermatophytes
Encephalitis
Encephalomyelitis
Epididymitis
Gingivitis
Immunocompetent person
Immunosuppressed person
Lymphadenitis
Lymphadenopathy
Lymphocytosis
Malaise
Mastitis
Meninges (sing. *meninx*)
Meningitis
Meningoencephalitis
Myelitis
Myocarditis
Necrotoxin
Nephritis
Oncogenic
Oophoritis
Orchitis
Parotitis
Pericarditis
Periodontitis
Proctitis
Prophylaxis
Prostatitis
Pyelonephritis
Salpingitis
Sebum
Tetanospasmin
Tinea infections
Toxemia
Ureteritis
Urethritis

In previous chapters you have studied the disease process, how pathogens cause disease, modes of transmission, how the body attempts to defend itself, and the chemotherapeutic agents and vaccines used to cure and prevent certain infectious diseases. This chapter summarizes the major infectious diseases of the skin, eyes, ears, oral cavity, respiratory system, gastrointestinal tract, urogenital areas, cardiovascular system, and nervous system. Some infections involve several body systems simultaneously or the pathogen may move from one area of the body to another. The disease-causing microbe may be an opportunistic member of the indigenous microflora, but usually the causative agent is transmitted to the recipient from a reservoir of infection (see Chapter 7).

The early signs and symptoms of many diseases are flu-like; usually slight fever, headache, fatigue, *malaise* (a feeling of bodily discomfort), gastrointestinal upset, or sneezing and coughing develops in the patient. Only specific signs and symptoms characteristic of the major diseases are mentioned here.

SKIN INFECTIONS

Healthy Skin

As seen in Figure 10 – 1, the structure of skin is not simple. Healthy, intact skin serves as a formidable protective barrier for the underlying tissues. Vast numbers of microbes survive on and within the epidermal layers, in pores, and in hair follicles (see Table 6 – 1). Their growth is controlled by the (1) amount of moisture present, (2) pH, (3) temperature, (4) salinity of perspiration, (5) chemical wastes such as urea and fatty acids, and (6) other microbes present that secrete fatty acids and antimicrobic substances. Proper hygienic cleanliness and washing serves to flush away dead epithelial cells, many transient and resident microbes, and the odorous organic materials present in perspiration, *sebum* (sebaceous gland secretions), and microbial secretions.

Once the skin barrier is broken (*e.g.*, by wounds, surgery, or burns), the opportunists may infect underlying tissues, invade capillaries and lymph, and be carried by blood, lymph, or phagocytes to many regions of the body.

Opportunists

Most microbes that colonize the skin are harmless, but several genera of aerobes and anaerobes may cause infections when the ecological balance of the skin environment changes chemically, physically, or microbiologically (Fig. 10 – 2).

Throughout this chapter, the number of cases reported to the CDC is as of 2/6/95, and does not include any 1994 cases reported after that date.

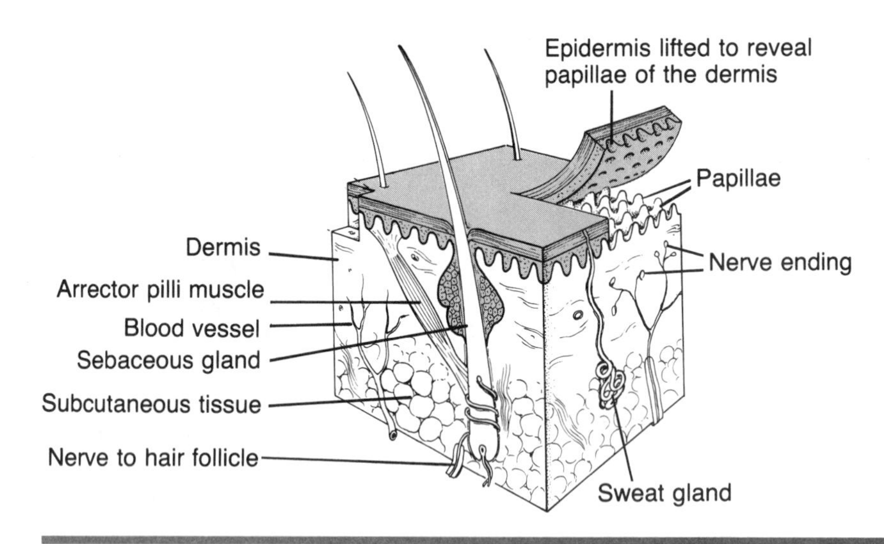

FIGURE 10 – 1
Cross-section of the skin. (Lindberg JB, Hunter ML, Kruszewski AZ: Introduction to Person-Centered Nursing. Philadelphia, JB Lippincott, 1988)

Major Microbial Opportunists

Gram-Positive Cocci *Staphylococcus epidermidis, Staphylococcus aureus, Micrococcus* species, and *Streptococcus* species are facultative anaerobes that may invade through breaks in the skin to cause local, deep, or systemic infections. Many of them produce invasive enzymes and damaging exotoxins capable of causing serious diseases, such as the toxic shock syndrome (TSS) caused by *S. aureus.* A total of 183 U.S. cases of TSS were reported to the Centers for Disease Control and Prevention (CDC) during 1994.

Gram-Positive Bacilli These pleomorphic rods, frequently referred to as diphtheroids, include *Corynebacterium, Brevibacterium,* and *Propionibacterium* species, which frequently cause hair follicle and sweat gland infections.

Gram-Negative Bacilli In moist areas, armpits, perineum, and between toes, *Pseudomonas* and some enteric rods may be found. Microbes can grow profusely on the organic compounds in perspiration and sebum, producing malodorous fatty acids. Many deodorants contain antimicrobial agents that inhibit growth of these bacteria.

Viral Infections of the Skin

CHICKENPOX AND SHINGLES

Characteristics Chickenpox (also known as varicella) is a respiratory infection and generalized viremia with local vesicular lesions on the skin of the face,

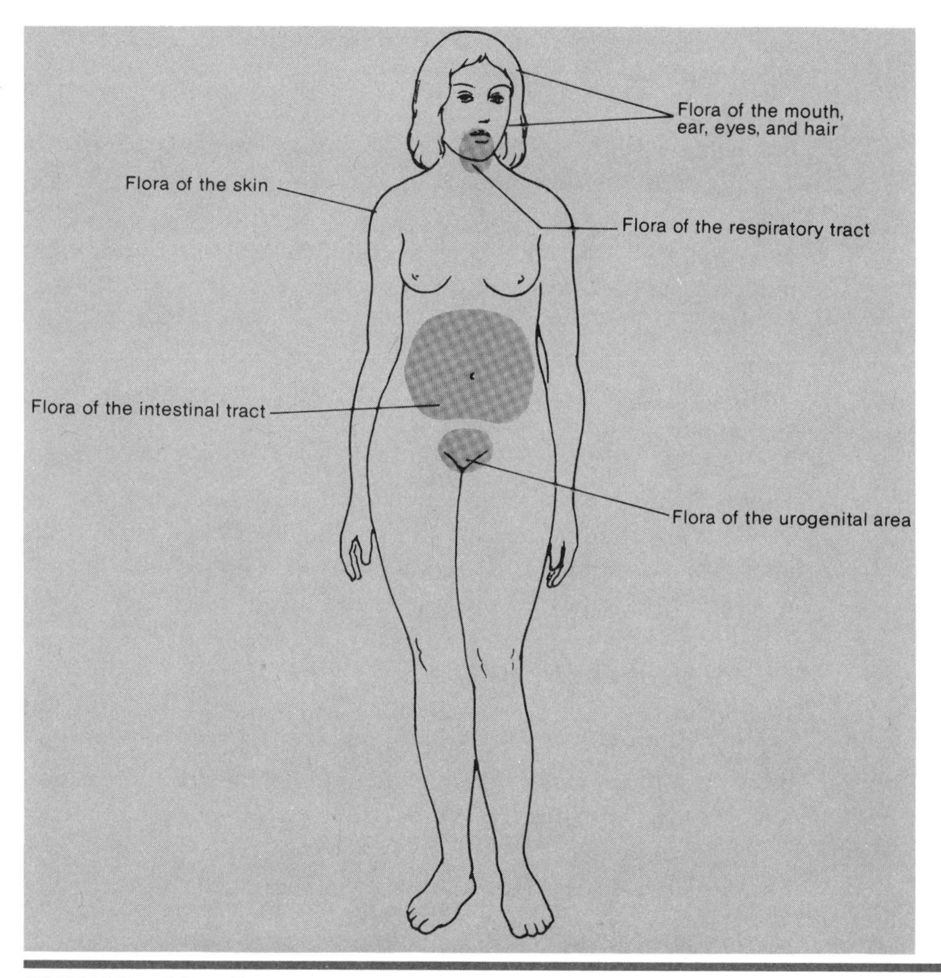

FIGURE 10 – 2
Areas where most of the indigenous microflora reside: skin, hair, mouth, ears, eyes, throat, nose, intestinal tract, and urogenital tract

thorax, and back that become encrusted. Vesicles also form in mucous membranes. Although usually a mild, self-limiting disease, it can be severely damaging to a fetus. Reye's syndrome (a severe encephalomyelitis with liver damage) may follow clinical chickenpox if aspirin is given to children younger than 16 years of age. Secondary bacterial infections (*e.g.*, pneumonia, otitis media, and bacteremia) frequently occur. Shingles (also known as herpes zoster) is a reactivation of varicella provirus in adults; it is an inflammation of sensory ganglia of cutaneous sensory nerves, producing rash and pain.

Pathogens Varicella-zoster virus (VZ) virus; a herpes virus.

Reservoir Humans.

Transmission Inhalation of virus via contaminated droplets or direct contact with vesicles or contaminated articles.

Incubation Period Usually 13–17 days; contagious from 1 to 2 days before onset of rash to 5 days after first crop of vesicles.

Epidemiology Usually occurs in winter and early spring; an estimated 2 million cases each year. Long-lasting immunity usually follows childhood disease.

Control Isolation for 1 week after eruption of vesicles or until vesicles become dry.

Usual Treatment An attenuated varicella vaccine is now available. Varicella-zoster immune globulin (VZIG) is effective if given within 96 hours after exposure; antiviral chemotherapy (acyclovir or vidarabine) may be effective. Antibiotics may be prescribed to prevent secondary bacterial infections.

MEASLES, "HARD MEASLES," RUBEOLA

Characteristics Upper respiratory infection, high fever, coughing, light sensitivity, Koplik's spots in mouth, red blotchy skin rash. Complications include bronchitis, pneumonia, otitis media, encephalitis. Autoimmune subacute sclerosing panencephalitis (SSPE) may follow a latent period.

Pathogen Measles (rubeola) virus; a paramyxovirus.

Reservoir Humans.

Transmission Inhalation of droplets during contact with children.

Incubation period Usually 14 days until rash appears; contagious until 4 days after appearance of rash.

Epidemiology In 1994, 895 U.S. cases were reported to the CDC. Vaccination produces immunity in more than 95% of recipients.

Control Immunization with live, attenuated measles virus in measles-mumps-rubella (MMR) vaccine.

Usual treatment Bed rest, fluids, preventive nursing care to prevent complications and secondary infections; antiviral chemotherapy and antibiotics to prevent bacterial infections may be helpful.

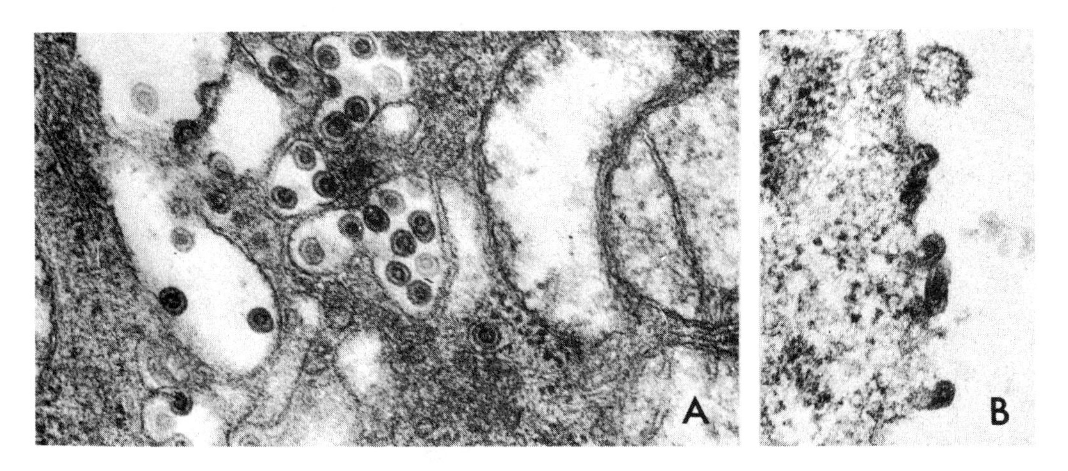

FIGURE 10 – 3
Development of rubella virus in the surface and cytoplasmic membranes of infected cell cultures. (*A*) Viral particles budding from cytoplasmic membranes into vacuoles and cytoplasm. Numerous mature virions are present within vacuoles (original magnification, ×60,000). (*B*) Viral particles budding from the surface of an infected cell (original magnification, ×60,000). (Oshiro LS et al: J Gen Virol 5:205)

GERMAN MEASLES, RUBELLA

Characteristics Rash, flat pink spots spreading from face, tender lymph nodes, low-grade fever. This is a milder disease than "hard measles;" it may be subclinical. During the first trimester of pregnancy, it may cause congenital rubella syndrome in the fetus. Encephalitis is a rare complication.

Pathogen Rubella virus; a togavirus (Fig. 10 – 3).

Reservoir Humans.

Transmission Inhalation of droplets during direct contact with infected individuals; occasionally via contaminated articles.

Incubation period 14–21 days.

Epidemiology In 1994, 209 U.S. cases were reported to the CDC. Endemic worldwide. Fewer epidemics in immunized communities.

Control Immunization with live, attenuated rubella virus in MMR vaccine, especially for young women before pregnancy. Serologic testing is performed to determine immunity. Isolation for 7 days after onset of rash.

Usual treatment No chemotherapy available.

WARTS, CONDYLOMA ACUMINATA

Characteristics Fibrous superficial papules of the skin (See Fig. 10 – 19) and mucous membranes, including genitalia.

Pathogen Human papillomavirus, of the papovavirus group.

Reservoir Humans.

Transmission Direct contact and sexual intercourse, fomites, spread by scratching.

Incubation period 1–20 months.

Epidemiology Worldwide occurrence.

Control Destroyed by freezing, acids, electrosurgery, or laser therapy.

Usual Treatment Podophyllum chemotherapy for persistent epidermal warts or idoxuridine for genital warts. Interferon may prove to be effective for genital and laryngeal warts.

Bacterial Infections of the Skin

IMPETIGO, PYODERMA

Characteristics Skin lesions forming pustules, then amber crusts (Fig. 10 – 4), itching, peeling of skin.

Pathogens *S. aureus* or *Streptococcus pyogenes* or both; identified by culture, biochemical, and immunologic techniques.

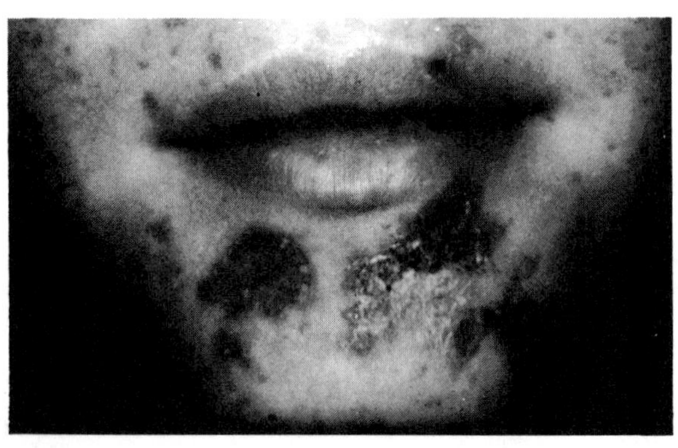

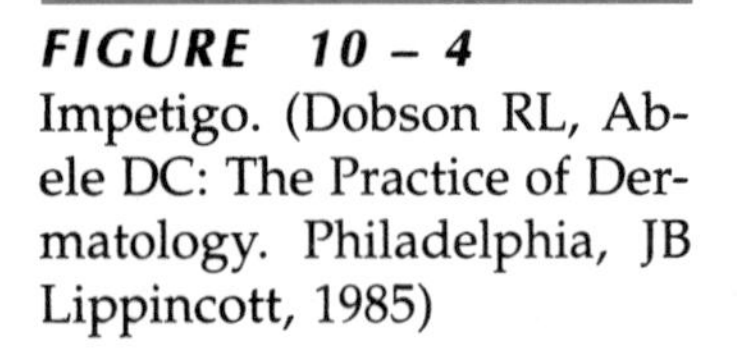

FIGURE 10 – 4
Impetigo. (Dobson RL, Abele DC: The Practice of Dermatology. Philadelphia, JB Lippincott, 1985)

Pathogenicity Spreads through skin by producing hyaluronidase. Staphylococcal toxins cause scalded skin syndrome and peeling.

Reservoir Humans.

Transmission By direct contact, through broken skin, scratching.

Incubation period 4–10 days.

Control Disinfectant skin lotions.

Usual treatment Topical disinfectants and antibiotics, such as penicillin, erythromycin, cephalosporin, clindamycin, vancomycin.

SCARLET FEVER, SCARLATINA, ERYSIPELAS

Characteristics Skin rash.

Pathogen *S. pyogenes* (see Fig. 9 – 15); group A, β-hemolytic, gram-positive cocci that form chains; identified by culture, biochemical, and immunologic techniques.

Pathogenicity The result of erythrogenic toxins; a secondary complication of "strep" throat (or "strep" pharyngitis) in susceptible individuals.

Reservoir Humans.

Transmission Person-to-person, by secretions and close contact.

Incubation period 1–3 days.

Control Isolation until after 24 hours of antibiotic therapy.

Usual treatment Penicillin, erythromycin, cephalosporins, clindamycin.

FOLLICULITIS, HAIR FOLLICLE INFECTIONS, FURUNCLES (BOILS), CARBUNCLES

Characteristics Hair follicle pustules, small or large.

Pathogens *S. aureus* (gram-positive cocci in clusters), *Pseudomonas aeruginosa* (gram-negative bacilli).

Reservoir Humans, especially the skin (*S. aureus*), and contaminated hot tubs, whirlpool baths, swimming pools, and luffa sponges (*P. aeruginosa*).

Transmission Direct contact, water or fomites, poor personal hygiene, skin abrasions, cuts, contaminated hot tubs, whirlpool baths, swimming pools, and luffa sponges. *S. aureus* may become systemic and cause recurring boils and carbuncles.

Incubation period Variable, usually 4–10 days.

Control Good personal hygiene, properly treated hot tubs and pools; allow luffa sponges to dry well between uses.

Usual treatment Topical bacitracin or polymyxin B for *S. aureus;* gentamicin or carbenicillin systemically for *Pseudomonas.*

ACNE

Characteristics Inflammatory pustules, cysts, and papules; infections of sebaceous glands in hair follicles on face, chest, and back.

Pathogens Interaction of *Propionibacterium acnes, S. aureus,* and *Corynebacterium* spp.

Reservoir Humans.

Transmission Direct contact with susceptible individuals; usually occurs during puberty because of hormonal changes.

Incubation period Variable, usually 4–10 days.

Control Cleanliness, good personal hygiene.

Usual treatment Locally applied benzoyl peroxide, salicylic acid, or sulfur; systemic antibiotics such as tetracycline, minocycline (Minocin®), and Cleacin-T® or drug therapy with isotretinoin (Accutane®) for severe cases, and Retin A® for moderate control.

ANTHRAX, WOOLSORTER'S DISEASE

Characteristics Cutaneous blackened lesions called eschars caused by *necrotoxin* (a substance that kills cells); pneumonia and / or systemic infection may develop.

Pathogen *Bacillus anthracis,* a gram-positive, aerobic, endospore-forming bacillus.

Pathogenesis Endospore germinates; bacteria produces necrotoxins.

Reservoir Spores in soil and on hides and skins of infected animals.

Transmission From animal hair, wool, and hides by (1) direct contact, endospores into skin abrasions; (2) inhalation of endospores; (3) ingestion of endospores.

Incubation period 3–5 days.

Epidemiology No U.S. cases were reported to the CDC in 1994.

Control Avoid contact with infected animals and soils contaminated by spores from infected animals; vaccination of animals and humans who work with animals and animal products.

Usual treatment Penicillin, tetracycline, erythromycin, chloramphenicol.

LEPROSY, HANSEN'S DISEASE

Characteristics (1) Neural, tuberculoid form: lesions on skin and peripheral nerves, loss of sensation; (2) cutaneous, lepromatous form: progressive disfiguring nodules in skin; invades throughout the body.

Pathogen *Mycobacterium leprae,* an acid-fast bacillus, cultured only in laboratory animals (armadillos, monkeys, and mouse footpads).

Reservoir Humans.

Transmission (1) Prolonged exposure through skin or mucous membranes; (2) via droplets to respiratory tract.

Epidemiology In 1994, 111 U.S. cases were reported to the CDC. Worldwide prevalence.

Control Early detection and isolation until treatment is effective.

Usual treatment Prolonged treatment with rifampin or with dapsone and clofazimine.

Fungal Infections of the Skin

Many fungi cause skin lesions and may enter the body through the skin, such as *Sporothrix* and *Candida,* but the most common are the *dermatophytes,* the fungi that cause superficial fungal infections of the skin, hair, and nails (called ringworm or *tinea infections*). The cell-mediated immune responses to fungi result in inflammation and limitation of the spread of the fungi.

DERMATOMYCOSIS, TINEA, RINGWORM, DERMATOPHYTOSIS

Characteristics Fungal lesions on skin (tinea corporis), scalp (tinea capitis), groin (tinea cruris or jock itch), foot (tinea pedis or athlete's foot), and nails (tinea unguium).

Pathogens Various species of *Microsporum, Epidermophyton,* and *Trichophyton;* filamentous fungi.

Reservoir Humans, animals, and soil.

Transmission Direct or indirect contact with fungal spores from lesions or fomites, such as shower stalls, toilet articles, and athletic supporters. Spores enter through breaks in skin and moist areas and germinate into filamentous growths.

Incubation period 4–14 days.

Control Keep susceptible areas clean and dry.

Usual treatment Topical application of miconazole (Lotrimin®), clotrimazole, and other antifungal agents; oral griseofulvin for scalp and nail infections.

Burn and Wound Infections

When the protective skin barrier is broken by burns, wounds, and surgical procedures, many of the indigenous microflora and environmental bacteria can invade and cause local or deep-tissue infections. They may also become systemic and produce exotoxins, causing severe damage to the individual.

BURNS

Many patients die after being burned severely due to a loss of body fluids and the presence of toxic microbial invaders including *S. aureus, S. pyogenes, P. aeruginosa,* and many fungi. These organisms usually grow aerobically in the burned area. They can produce many exotoxins, such as *cytotoxins* (which damage cells), necrotoxins, and neurotoxins (which affect the nervous system), and may ultimately cause the death of the patient.

Burn victims must be treated in an aseptic environment in which health-care personnel are gloved, gowned, and masked. The burns may be left open to the air to speed healing or covered with artificial skin or film to reduce contamination and loss of fluids. Some antimicrobial topical agents (silver sulfadiazine, for example) may be used to reduce the possibility of infections during the prolonged healing period.

WOUND AND SURGICAL INFECTIONS

Most gunshot, stab, puncture, bite, and abrasive wounds are contaminated during the wounding process. Microbes introduced are frequently anaerobes from dust or dirt and indigenous microflora that grow rapidly deep within the wound; other facultative anaerobes flourish on the surface of the wound, secreting enzymes that enable them to invade the blood and lymph. Bacterial and fungal spores may also be introduced, causing local and deep-seated infections.

Any traumatic injury must be opened and thoroughly cleansed to remove debris and to inhibit the growth of microbes; then it is usually covered lightly to prevent further contamination.

Staphylococcus (see Figs. 1 – 8 and 1 – 9) and *Streptococcus* (see Fig. 9 – 15) species are the most frequent bacteria that cause focal infections. They may invade by secreting hyaluronidase and other spreading factors to cause severe toxic systemic infections of many sites, including the brain, spinal cord, and bone. *P. aeruginosa* from soil and feces is notorious for causing deep, antibiotic-resistant infections. This gram-negative bacillus produces protease enzymes that enable it to move through tissues and an exotoxin that inhibits protein synthesis. Antimicrobial susceptibility tests are essential to ensure that the appropriate drug is administered.

Anaerobic *Clostridium* species are of grave concern in puncture wounds. Endospores of *Clostridium tetani* (Fig. 10 – 28) may be introduced from soil or fecal contamination. When these spores germinate into vegetative bacteria, neurotoxins are produced that cause involuntary muscle spasms and respiratory failure. The availability of vaccines and antitoxins has greatly reduced deaths due to tetanus in developed countries.

Clostridium perfringens, the major causative agent of gas gangrene, is not only of concern in accidental wounds but in surgical sites as well. The contaminating spores germinate, and the vegetative pathogens produce many invasive enzymes and exotoxins, resulting in necrotic areas with gas present. Rapid and extensive destruction of muscle tissue may necessitate amputation of an infected extremity.

Fungal infections, such as sporotrichosis, are caused by the introduction of soil fungi and spores that invade cutaneous and lymphatic tissues through wounds that are sometimes as small as a thorn prick. This slowly progressing disease is usually localized and self-limiting, but severe cases may be treated with amphotericin or potassium iodide.

INFECTIONS ASSOCIATED WITH BITES

Human and animal bites introduce oral microflora into the wound. Because the organisms from human bites are adapted to humans, they are more likely to produce infections than those of animal sources. Every bite should be thoroughly cleansed, disinfected, and left open to the air to prevent microbial growth. Bite wounds should not be closed tightly with stitches and bandages because this procedure would encourage the growth of anaerobic bacteria.

Human Bites Severe infections may result from anaerobic growth of *Bacteroides* and *Actinomyces,* as well as facultative *Streptococcus* and *Staphylococcus* species and many other oral microbes. There is even some concern about the oral introduction of some viruses, such as human immunodeficiency virus (HIV), into the bloodstream through bites by infected individuals.

Animal Bites In general, bites of animals may introduce microbes, such as staphylococci and streptococci, which cause local infections in humans. The bacteria *Pasteurella multocida* and *Pasteurella haemolytica* are frequently found in dog-bite and cat-bite wounds and may cause severe local infections and pasteurellosis. Rat-bite fever, caused by *Streptobacillus moniliformis* or *Spirillum minor,* is naturally carried by rats, mice, cats, squirrels, and weasels. The rabies virus is transmitted by the bites of infected dogs, foxes, coyotes, wolves, cats, prairie dogs, bats, skunks, raccoons, and other rabid animals.

Insect Bites Many other disease-causing microbes are introduced through the skin to the circulatory or nervous systems by bites of arthropod vectors (ticks, fleas, lice, mites, and mosquitoes); these diseases include encephalitis, Colorado tick fever, Rocky Mountain spotted fever, typhus, plague, tularemia, malaria, and Lyme disease; these are discussed elsewhere in this chapter.

EYE INFECTIONS

The eye consists of tissues similar to and contiguous with the skin. The anatomy and structure of the eye are illustrated in Figure 10 – 5.

The external surface of the eye is lubricated, cleansed, and protected by tears, mucus, and sebum. Thus, continual production of tears and the presence of lysozyme and other antimicrobial substances found in tears greatly reduce the numbers of indigenous microflora organisms found on the eye surfaces.

Infections of the eye caused by bacteria (including chlamydias) and viruses should be differentiated from allergic manifestations and conjunctivitis by microscopic examination of the exudate (oozing pus), culture of pathogens, or immunodiagnostic procedures (*e.g.,* fluorescent antibody and enzyme-linked immunosorbent assay [ELISA]).

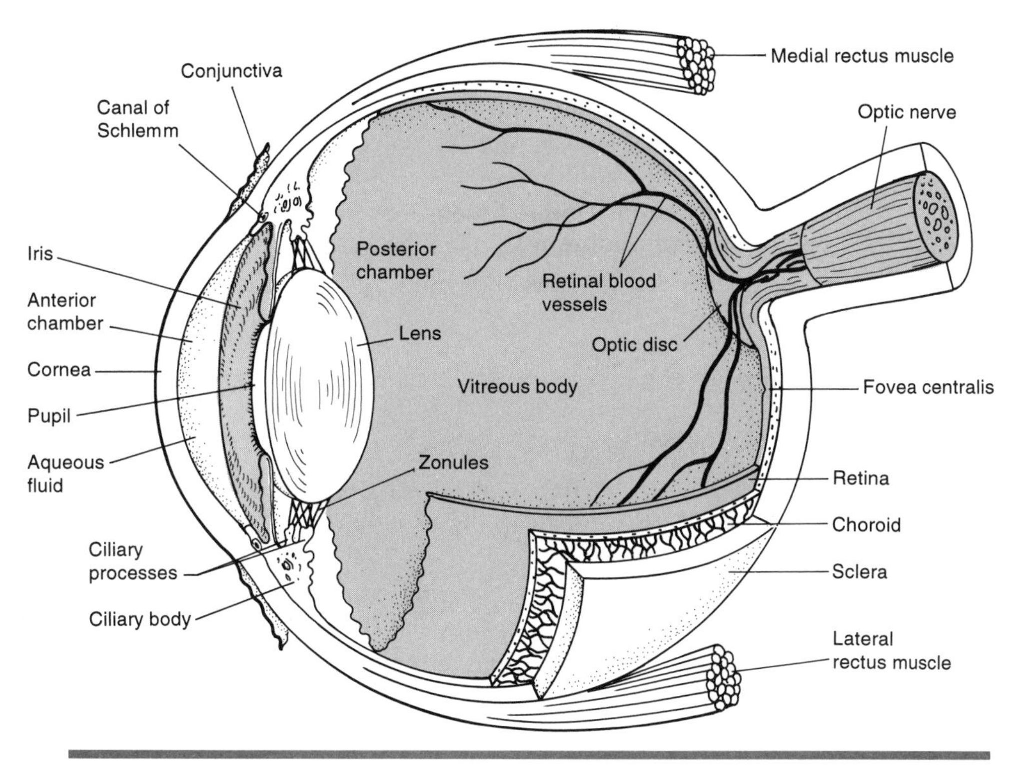

FIGURE 10 – 5
Anatomy of the eye. (Scherer JC: Introductory Medical-Surgical Nursing, 6th ed. Philadelphia, JB Lippincott, 1991)

Viral Eye Infections

VIRAL CONJUNCTIVITIS, ACUTE HEMORRHAGIC CONJUNCTIVITIS (AHC), HERPES CONJUNCTIVITIS

Characteristics Viral invasion of eyelids, conjunctiva, and cornea; usually self-limiting in 1–3 weeks. Subconjunctival hemorrhages caused by adenovirus or enterovirus. Herpes lesions and occasional blindness from herpesvirus infections.

Pathogens Picornaviruses (enterovirus 70), coxsackievirus, adenoviruses, herpes simplex types 1 and 2.

Reservoir Humans.

Transmission Person-to-person, direct or indirect contact with infected eye secretions. Herpes is transmitted through saliva or genital secretions.

Incubation period Herpes, 2–12 days; others, 1–3 days.

Control Personal hygiene, drainage / secretion precautions, strict asepsis in eye clinics.

Usual treatment Herpes lesions may be treated with idoxuridine or adenine arabinoside (Vidarabine®, Vira-A®, or Ara-A®) ophthalmic ointments. There is no recommended treatment for other self-limiting viral eye infections.

Bacterial Eye Infections

BACTERIAL CONJUNCTIVITIS, PINK EYE

Characteristics Irritation, reddening of conjunctiva, edema of eyelids, mucopurulent discharge, sensitivity to light. Highly contagious.

Pathogens *Haemophilus aegyptius, Streptococcus pneumoniae,* other streptococci, staphylococci, *Haemophilus influenzae, Moraxella lacunata, P. aeruginosa, Corynebacterium diphtheriae* may be causative agents.

Reservoir Humans.

Transmission Person-to-person, through eye and respiratory discharges, fingers, clothing, eye makeup, eye medications, ophthalmic instruments, and contact lens-wetting and lens-cleaning agents.

Incubation period 1–3 days, depending on causative agent.

Control Personal hygiene, drainage / secretion precautions, sterilization of fomites.

Usual treatment Ophthalmic tetracycline, erythromycin, gentamicin, or sulfonamide depending on the susceptibility of the pathogen.

CHLAMYDIAL CONJUNCTIVITIS, INCLUSION CONJUNCTIVITIS, PARATRACHOMA

Characteristics In neonates, acute conjunctivitis with mucopurulent discharge; may result in mild scarring of conjunctivae and cornea. May be concurrent with chlamydial nasopharyngitis or pneumonia. In adults, may be concurrent with nongonococcal urethritis or cervicitis.

Pathogens *Chlamydia trachomatis,* immunotypes B and D through K.

Reservoir Humans.

Transmission Nasopharynx, rectal and genital secretions to eye, by fingers, sexual contact; to newborns, by infected birth canal; nonchlorinated swimming pools, by genitourinary exudates; spread by flies to eyes.

Control Identification and treatment of chlamydial genital infections of expectant parents; drainage/secretion precautions; chlorination of swimming pools; erythromycin or tetracycline ophthalmic ointments in eyes of newborns. Chlamydias are not susceptible to silver nitrate or penicillin.

Usual treatment Erythromycin orally for infected pregnant women and infected neonates; topical tetracycline, erythromycin, or sulfonamide for eye infections.

TRACHOMA, CHLAMYDIA KERATOCONJUNCTIVITIS

Characteristics Highly contagious, acute or chronic conjunctival inflammation, resulting in scarring of cornea and conjunctiva, deformation of eyelids, and blindness.

Pathogens *C. trachomatis,* immunotypes C, A, and B.

Reservoir Humans.

Transmission Direct contact with infected ocular or nasal secretions or contaminated articles; sexually transmitted via infected cervix or urethra; to newborns through the birth canal; spread by flies.

Incubation period 5–12 days.

Control Personal hygiene, improved sanitation and living conditions; drainage/secretion precautions of hospitalized patients.

Usual treatment Topical ophthalmic tetracycline or erythromycin, oral tetracycline or erythromycin.

GONOCOCCAL CONJUNCTIVITIS, OPHTHALMIA NEONATORUM

Characteristics Acute redness and swelling of conjunctiva, purulent discharge (Fig. 10 – 6); corneal ulcers, perforation, and blindness, if untreated.

Pathogens *Neisseria gonorrhoeae;* gram-negative, kidney bean-shaped diplococci; identified by Gram-stained smears, culture, biochemical and immunologic tests.

Reservoir Humans.

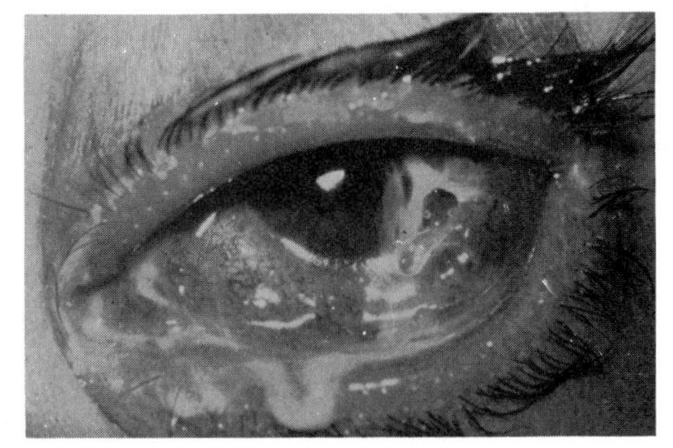

FIGURE 10 – 6
Purulent conjunctivitis caused by *Neisseria gonorrhoeae.* (Moffett HL: Clinical Microbiology, 2nd ed. Philadelphia, JB Lippincott, 1980)

Transmission Infection of cervix, birth canal, and secretions. To newborn, during passage through infected birth canal. To adults, through infected genital secretions.

Incubation period 1–5 days.

Control Instill 1% silver nitrate solution in the eyes of neonates. Ophthalmic erythromycin or tetracycline may be used to prevent chlamydial and gonococcal eye infections.

Usual treatment Penicillin or oral cefotaxime for penicillin-resistant gonococci (penicillinase-producing *N. gonorrhoeae,* PPNG).

INFECTIOUS DISEASES OF THE MOUTH

The oral cavity is a complex ecosystem suitable for growth and interrelationships of many types of microorganisms. The actual indigenous microflora of the mouth varies greatly from person to person.

When the anatomy of the mouth (Fig. 10 – 7) is studied, it is obvious that there are many areas where bacteria can attach, colonize, and proliferate, even in the presence of the normal defenses. In the healthy mouth, saliva secreted by salivary and mucous glands helps control the growth of opportunistic oral flora. Saliva contains enzymes, immunoglobulins (IgA), and buffers to control the near-neutral pH and continually flushes microbes and food particles through the mouth. Other antimicrobial secretions and phagocytes are found in the mucus that coats the oral surfaces. The hard, complex, calcium tooth enamel, bathed in protective saliva, usually resists damage by oral microbes; however, if the ecological balance is upset or is not properly maintained, oral disease may result.

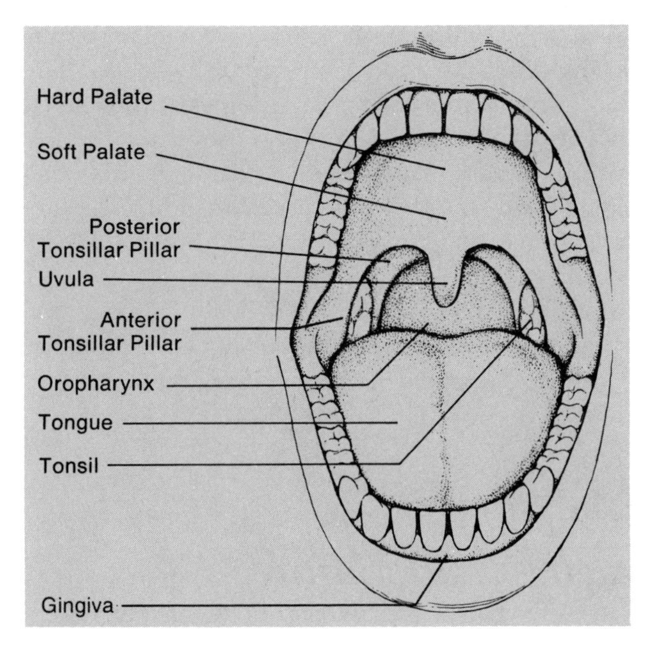

FIGURE 10 – 7
Anatomy of the mouth

Indigenous Microflora and Oral Disease

The indigenous microflora of the mouth have recently been shown to include more than 1,000 different types of bacteria, both aerobes and anaerobes. Only a fraction of these oral microbes have been completely identified. Some of them are beneficial because they produce secretions that are antagonistic to other bacteria. Although several species of *Streptococcus* (*S. salivarius, S. mitis, S. sanguis,* and *S. mutans*) and *Actinomyces* species often interact to protect the oral surfaces, they are involved in oral disease in other circumstances.

The anaerobic condition produced by oxidation-reduction interactions of the oral flora organisms allow certain genera of anaerobic bacteria (*Bacteroides, Fusobacterium, Actinomyces,* and *Treponema*) to become involved in the production of oral diseases. The coating that forms on unclean teeth, called dental plaque, is a coaggregation of bacteria and their products. Many of these microorganisms produce a slime layer or glycocalyx that enables them to attach firmly and cause damage to the tooth enamel. Certain carbohydrates are metabolized by streptococci (especially *S. mutans*) and lactobacilli, producing lactic acid, which encourages other bacteria to grow and become involved in the decay process. Calculus, a complex calcium phosphate material produced by some bacteria (*e.g., Actinomyces*), is a hard plaque deposit that may be produced when tooth enamel is dissolved by lactic acid-producing bacteria.

The progressive microbial activities involving formation of dental plaque, dental caries (decay), *gingivitis* (infection of the gingiva or gums), and *periodon-*

titis (infection of subgingival tissue) result from the unique microbial population, reduced host defenses, improper diet, and poor dental hygiene. These diseases are the consequence of at least four microbial activities, including (1) formation of dextran from sugars by streptococci, (2) acid production by lactobacilli, (3) deposition of calculus by *Actinomyces,* and (4) release of inflammatory substances (endotoxin) by *Bacteroides* species. This combination of circumstances damages the teeth, soft tissues (gingiva), alveolar bone, and the periodontal fibers attaching teeth to bone.

Oral disease can be prevented by maintaining good health, proper oral hygiene (brushing and flossing), an adequate diet without sugars, and regular fluoride treatments to help control the microbial population and to prevent damaging bacterial interactions.

Bacterial Infections of the Oral Cavity

DENTAL CARIES, GINGIVITIS, PERIODONTITIS

Characteristics Damage to teeth, gums, and alveolar bone, resulting from complex interaction of bacteria with dietary sugars.

Pathogens *Streptococcus mutans* (Fig. 10 – 8), *S. mitis, S. salivarius,* and *S. sanguis, Actinomyces naeslundii, A. viscosus, Lactobacillus, Bacteroides, Actinobacillus, Fusobacterium, Capnocytophaga.*

Pathogenicity Lactic acid and endotoxin production.

Reservoir Human oral cavity.

Transmission Direct contact.

Incubation period Variable.

Control Good oral hygiene that includes tooth brushing and flossing, adequate diet without sucrose, and fluoride in water, toothpaste, and other preparations.

Usual treatment Cleaning, filling decayed tooth areas, fluoride paste, and other treatments by dentists.

ACUTE NECROTIZING ULCERATIVE GINGIVITIS (ANUG), VINCENT'S ANGINA, TRENCH MOUTH

Characteristics A noncontagious periodontitis with gingival ulcers. It is associated with lack of oral hygiene, nutritional deficiency, debilitating disease, stress, and at least two synergistic microorganisms.

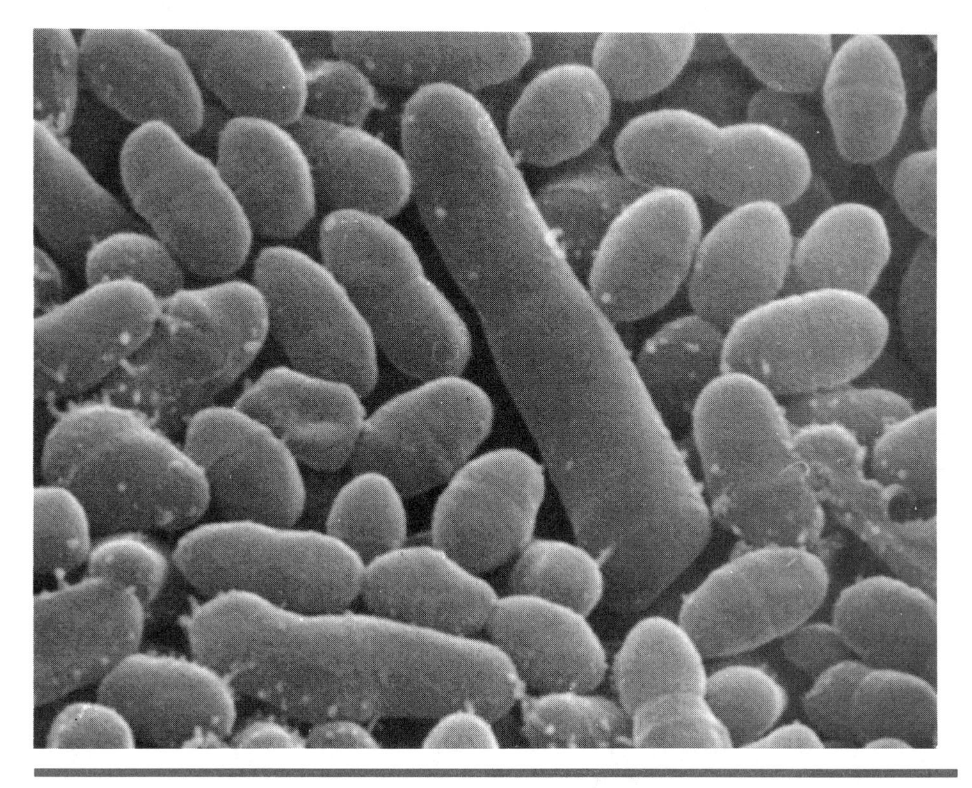

FIGURE 10 – 8
Electron micrograph of *Streptococcus mutans* in dental plaque. (Photo by Joan Foster)

Pathogens *Bacteroides* and *Fusobacterium* spp. (anaerobic, gram-negative bacilli); *Actinomyces* spp. (anaerobic, gram-positive bacilli); *Borrelia* and *Treponema* spp. (anaerobic, gram-variable, spiral-shaped bacteria); sometimes *Capnocytophaga, Eikenella,* and other bacteria may occur in lesions.

Pathogenicity Lactic acid and endotoxin production.

Reservoir Human mouth.

Transmission Food, direct contact, lack of oral hygiene.

Incubation period Variable.

Control Gentle local debridement, oral hygiene, adequate nutrition, high fluid intake, and rest.

Usual treatment Penicillin, erythromycin, tetracycline, or metronidazole.

EAR INFECTIONS

When the anatomy of the ear (Fig. 10 – 9) is studied, one observes that there are only three pathways for pathogens to enter: (1) through the eustachian (auditory) tube, from the throat and nasopharynx; (2) from the external ear; and (3) through the blood or lymph. Usually, the bacteria are trapped in the middle ear when a bacterial infection in the throat and nasopharynx causes the eustachian tube to close. The result is an anaerobic condition in the middle ear, allowing anaerobes to grow and cause pressure on the tympanic membrane (eardrum). Swollen lymphoid (adenoid) tissues, viral infections, and allergies may also close the eustachian tube, especially in young children.

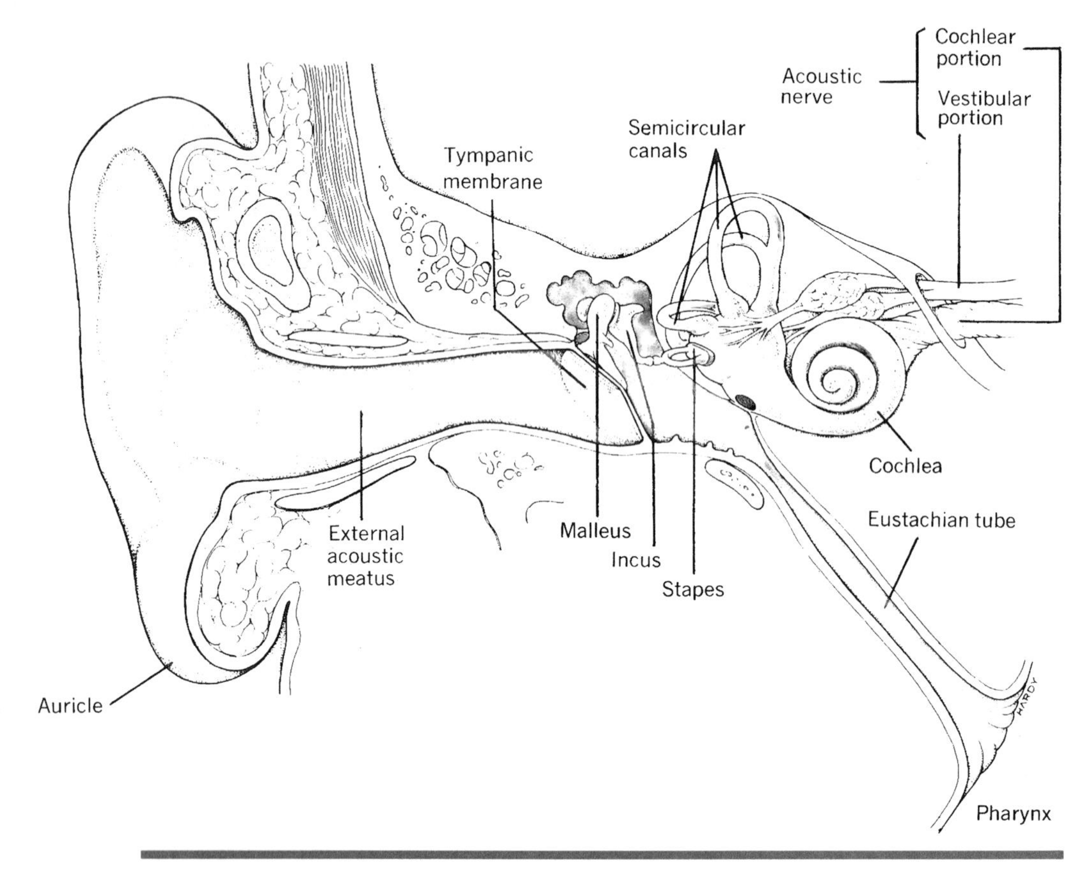

FIGURE 10 – 9
Anatomy of the ear. (Chaffee EE, Lytle IM: Basic Physiology and Anatomy, 4th ed. Philadelphia, JB Lippincott, 1980)

Viral and Bacterial Ear Infections

OTITIS MEDIA, MIDDLE EAR INFECTION

Characteristics Pressure in middle ear, flattening of the tympanic membrane, pain, fever, production of exudate, perforation and scarring of eardrum, and, if untreated, loss of hearing. Most common in children younger than 8 years of age.

Pathogens Bacteria include *H. influenzae, S. pneumoniae, S. pyogenes, S. aureus;* viruses include measles virus, parainfluenza virus, and respiratory syncytial virus (RSV).

Pathogenicity Bacterial exotoxins and viral destruction of cells.

Reservoir Human respiratory nasopharynx.

Transmission Complication following colds and nose and throat infections; direct contact.

Incubation period Variable.

Control For *H. influenzae,* vaccinate to prevent meningitis; rifampin *prophylaxis* (preventive treatment) is also used for prevention of meningitis. For streptococci, search for and treat carriers.

Usual treatment Ampicillin for *H. influenzae,* amoxicillin or penicillin for streptococci, erythromycin for penicillin-resistant strains. Enhance body defenses for viral infections. Insertion of a tube through the tympanic membrane by a physician to relieve pressure and anaerobic conditions.

OTITIS EXTERNA, EAR CANAL INFECTIONS, SWIMMER'S EAR

Characteristics Infection of ear canal with itching, pain, discharge, tenderness, redness, swelling, loss of hearing.

Pathogens *Escherichia coli, P. aeruginosa, Proteus vulgaris, S. aureus;* rarely by a fungus, such as *Aspergillus.*

Reservoir Contaminated water trapped in ear canal; also normal opportunistic flora.

Transmission Contaminated swimming pool or bath water; articles inserted in ear canal for cleaning out debris and wax.

Incubation period 1–3 days.

Control Prevent contaminated water from entering and being trapped by wax in external ear canal. Allow physician to clean wax and debris from ears.

Usual treatment Remove infected debris. Topical treatment with neomycin and polymyxin B for gram-negative rods; 1% hydrocortisone for swelling.

INFECTIOUS DISEASES OF THE RESPIRATORY SYSTEM

To simplify discussion of the functions, defenses, and diseases of the respiratory system (Fig. 10 – 10), it is often separated into the upper respiratory tract (URT)—consisting of the nose and pharynx (throat)—and the lower respiratory tract (LRT)—including the larynx, trachea, bronchial tubes, and alveoli. The most common diseases are those of the upper respiratory tract (*e.g.*, colds and sore throats), which may predispose the patient to more serious infections such as sinusitis, otitis media, bronchitis, and pneumonia.

Typical indigenous microflora found in the mucosa of the nose and throat include bacterial species of *Streptococcus, Staphylococcus, Haemophilus, Corynebacterium, Neisseria, Bacteroides, Branhamella, Fusobacterium,* and *Actinomyces.* Many of these microorganisms may cause opportunistic diseases of the respiratory tract.

Nonspecific Respiratory Infections

PNEUMONIA

Characteristics An acute nonspecific infection of the alveolar spaces and tissue of the lung, with fever, cough, acute chest pain, and respiratory distress; identified by chest x-rays. Pneumonia is usually a secondary infection that follows a primary viral respiratory infection.

Pathogens May be caused by gram-positive or gram-negative bacteria, mycoplasmas, viruses, fungi, or protozoa; bacteria causing pneumonia include *S. pneumoniae* (see Fig. 7 – 4), other streptococci, *S. aureus* (see Figs. 1 – 8 and 1 – 9), *H. influenzae, Neisseria meningitidis, P. aeruginosa, Yersinia pestis, Klebsiella pneumoniae,* and other gram-negative enteric bacilli. *Rickettsia, Chlamydia,* and *Mycoplasma* spp. may also be causative agents. Many viral respiratory infections may result in pneumonia. Fungi such as *Histoplasma, Coccidioides, Candida, Cryptococcus, Blastomyces,* and *Pneumocystis carinii* (considered a protozoan by some microbiologists) may be etiologic agents, especially in immunocompromised individuals.

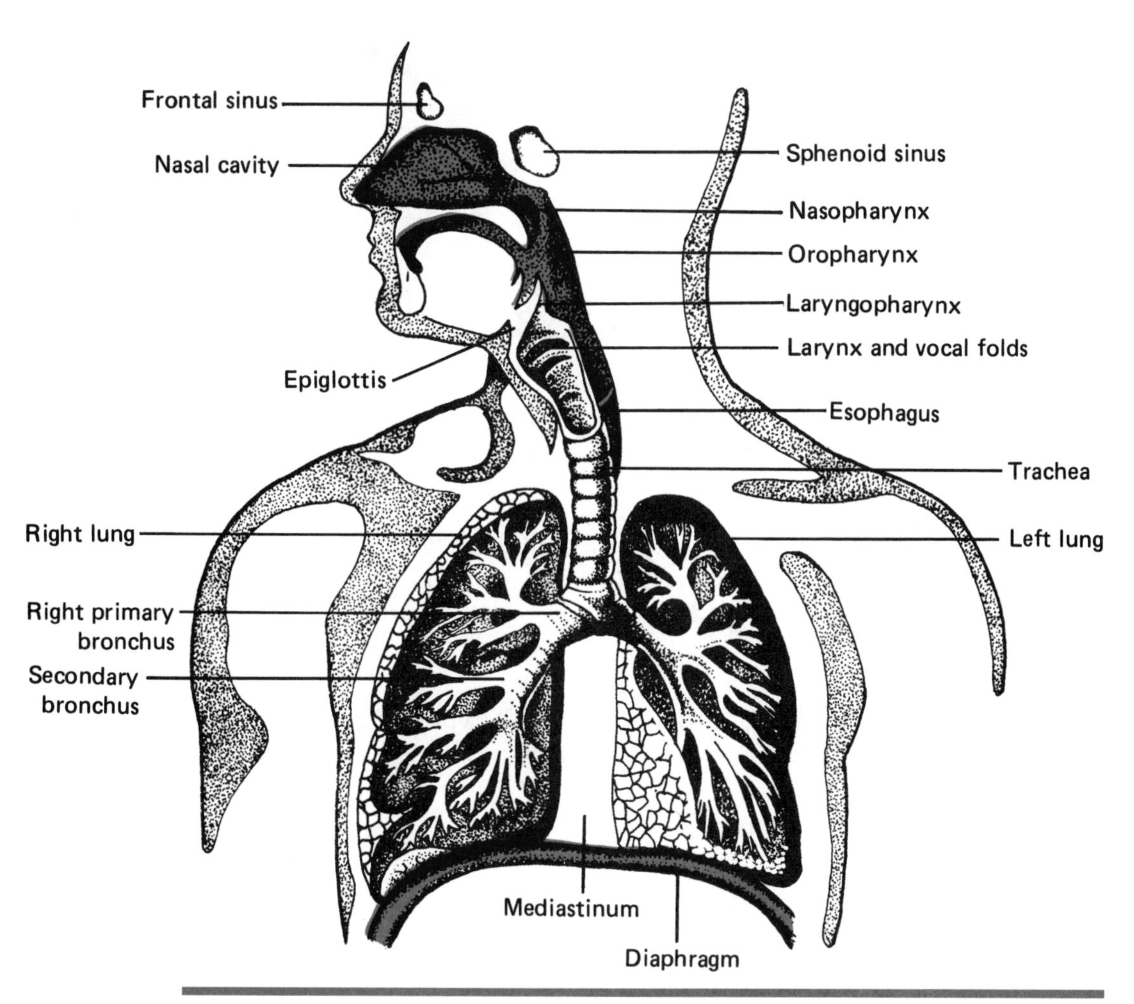

FIGURE 10 – 10
The respiratory system. (Chaffee EE, Lytle IM: Basic Physiology and Anatomy, 4th ed. Philadelphia, JB Lippincott, 1980)

Reservoir Humans.

Transmission Direct contact, respiratory secretions, hands, and fomites.

Incubation period Depends on pathogen involved.

Control Depends on pathogen involved.

Usual treatment Antibiotic therapy depends on the causal agent identified in sputum or other LRT specimens.

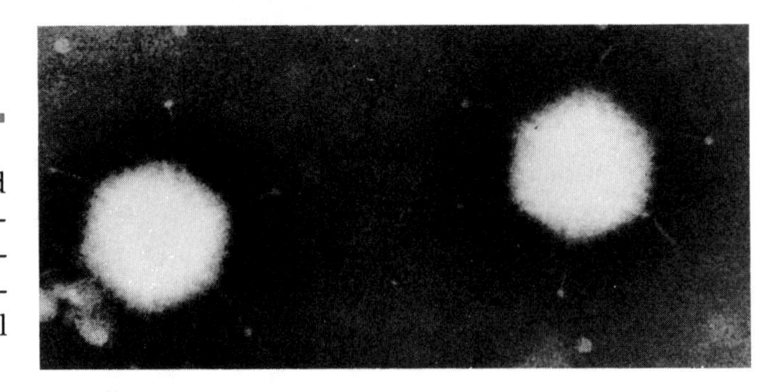

FIGURE 10 – 11
Electron micrograph of purified type 5 adenovirus particles embedded in sodium silicotungstate (original magnification, ×350,000). (Valentine RC, Pereira HG: J Mol Biol 13:13, 1965)

Viral Respiratory Infections

COMMON COLD, ACUTE VIRAL RHINITIS, ACUTE CORYZA, UPPER RESPIRATORY INFECTION (URI)

Characteristics Coryza (profuse discharge from nostrils), sneezing, sore throat, bronchiolitis, bronchitis. Secondary bacterial infections frequently follow.

Pathogens Many (more than 100) types of rhinovirus, adenovirus (Fig. 10 – 11), coronavirus, (RSV), influenza, parainfluenza, and other viruses.

Reservoir Human respiratory system.

Transmission Respiratory secretions by way of hands, direct contact, airborne droplets, soiled articles.

Incubation period 1–3 days.

Epidemiology Most common in fall, winter, and spring.

Control Sanitary disposal of oral and nasal discharges. Disinfect eating and drinking utensils. Avoid contact with infected individuals. Oral live adenovirus vaccine is effective in the military for specific adenovirus epidemics.

Usual treatment New antiviral agents. Antibiotics should be reserved for bacterial complications (sinusitis, otitis media, tonsillitis, bronchitis, pneumonia) because they are ineffective against viruses.

CROUP, ACUTE LARYNGOTRACHEOBRONCHITIS

Characteristics Acute viral inflammation of respiratory tract with subglottic swelling, respiratory distress with a high-pitched sound during inspiration.

Pathogens Usually parainfluenza virus, less often RSV, influenza A or B, or other viruses and *Mycoplasma pneumoniae* (a bacterium).

Reservoir Humans.

Transmission Direct oral contact, droplets, fomites.

Incubation period Variable.

Epidemiology Usually seen in susceptible children 3 months to 3 years old; symptoms are most dramatic at night, lasting 3–4 days.

Control Usually self-limiting and not contagious.

Usual treatment Vaporizers, humidifiers, hot shower steam may relieve symptoms. Hospitalization with oxygenation may be necessary in severe cases. Antibiotics are rarely indicated, but amantadine or other antiviral agents may help.

INFLUENZA, FLU

Characteristics A specific acute viral respiratory infection with fever, chills, headache, cough, nasal drainage; sometimes causing bronchitis, pneumonia, and death in severe cases. Nausea, vomiting, and diarrhea are rare.

Pathogens Influenza virus, types A, B, and C; type A is usually associated with pandemics and widespread epidemics.

Reservoir Humans.

Transmission Respiratory secretions, direct contact, fomites, hands.

Incubation period 24–72 hours.

Epidemiology Pandemics occurred in 1889, 1918, 1957, and 1968. Type A epidemics occur in the U.S. almost every year, whereas type B epidemics occur every 2–3 years.

Control Yearly immunization. Prophylactic amantadine or rimantadine may be used against type A for high-risk patients. Good personal hygiene and avoiding crowds during epidemics may prevent infection.

Usual treatment Amantadine reduces symptoms if given early in type A disease.

Specific Bacterial Respiratory Infections

DIPHTHERIA

Characteristics An acute contagious respiratory disease, with fibrinous pharyngeal pseudomembrane, causing myocardial and neural tissue damage.

Pathogens *Corynebacterium diphtheriae,* pleomorphic, gram-positive bacilli that group in palisade arrangements. Toxigenic strains are infected with a corynebacteriophage, resulting in exotoxin (diptheritoxin) production that causes the heart and nerve damage. Identified by culture and biochemical techniques.

Reservoir Humans.

Transmission Airborne droplets, direct contact, contaminated fomites, raw milk.

Incubation period 2–5 days.

Epidemiology One U.S. case was reported to the CDC in 1994. Diphtheria occurs during colder months among nonimmunized children and adults. The disease has disappeared in areas with effective immunization programs.

Control Community programs of immunization of all infants and children with diphtheria-tetanus-pertussis (DTP) vaccine; give diphtheria-tetanus (DT) booster vaccines at appropriate intervals. Patient isolation, quarantine, and disinfection of all fomites.

Usual treatment Administer antitoxin in known and strongly suspected cases, with erythromycin or penicillin in confirmed cases. Treat carriers with erythromycin or penicillin.

LEGIONELLOSIS, LEGIONNAIRES' DISEASE, PONTIAC FEVER

Characteristics An acute bacterial pneumonia with headache, high fever, dry cough, chills, diarrhea, pleural and abdominal pain.

Pathogens *Legionella pneumophila,* a gram-negative bacillus, requiring special medium to grow *in vitro* (see Fig. 8 – 2); additional *Legionella* species have been identified. Identification by culture, fluorescent antibody, and ELISA tests.

Reservoir Soil, dust, and environmental water sources; lakes, creeks, hot-water and air-conditioning systems, shower heads, ultrasonic nebulizers, tap water, and water distillation systems.

Transmission Airborne from water or dust, not person-to-person.

Incubation period 2–10 days.

Epidemiology In 1994, 1,535 U.S. cases reported to the CDC.

Control Search for environmental sources of infection; periodic superchlorination or superheating of water supply.

Usual treatment Erythromycin with rifampin.

STREPTOCOCCAL PHARYNGITIS, "STREP" THROAT

Characteristics An acute bacterial infection of the throat with fever and pain; inflammation of pharynx and tonsils; white patches of pus on pharyngeal epithelium. Rheumatic fever and glomerulonephritis complication may result in heart and kidney damage because of immune complex deposition.

Pathogens *S. pyogenes* (see Fig. 9 – 15), β-hemolytic, gram-positive cocci, catalase-negative (Fig. 10 – 12); strains that produce erythrogenic toxin cause scarlet fever. Identified by throat culture on blood agar plate, by susceptibility to bacitracin disks, by fluorescent antibody serologic techniques, and other immunodiagnostic procedures.

Reservoir Humans.

Transmission Person-to-person, by direct contact, usually hands, aerosol droplets, secretions from patients, and carriers.

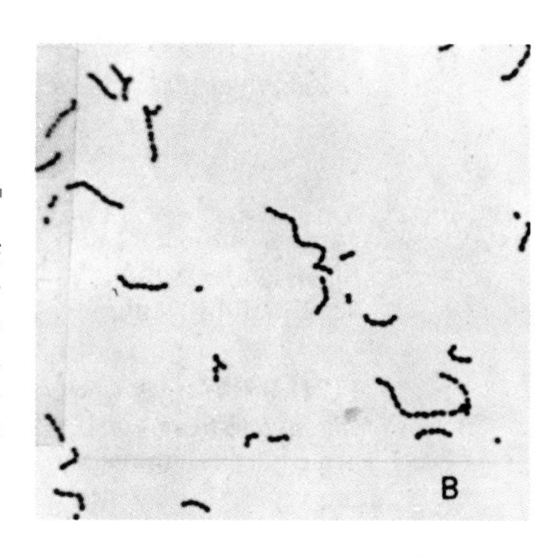

FIGURE 10 – 12
Chain formation characteristic of streptococci in liquid media. Smear made from 24-hr culture in serum broth (original magnification, ×1000). (Davis BD, et al: Microbiology, 4th ed. Philadelphia, Harper & Row, 1990)

Incubation period 1–3 days.

Epidemiology Over 200,000 cases per year in the United States, mostly among children (5–14 years of age); 3% develop rheumatic fever.

Control Throat cultures followed by antibiotic treatment; personal hygiene and cleanliness. No protective immunity following recovery; no effective vaccines.

Usual treatment Penicillin or erythromycin to prevent ear infections, rheumatic fever, glomerulonephritis, and other complications.

TUBERCULOSIS

Characteristics An acute or chronic mycobacterial infection of the pulmonary tract; may invade lymph nodes to cause systemic disease. Infected patients show a positive hypersensitivity skin test, and pulmonary tubercles may be seen on chest x-rays.

Pathogens Primarily *Mycobacterium tuberculosis* (a slow-growing, acid-fast, gram-positive bacillus); occasionally other *Mycobacterium* spp. Identified by culture techniques.

Reservoir Humans. In diseased cattle rarely.

Transmission Airborne droplets, prolonged direct contact, milk, and contact with infected cattle.

Incubation period 4–12 weeks.

Epidemiology In the United States, more than 20,000 new cases per year; 22,152 cases were reported to the CDC in 1994.

Control Tuberculin testing of humans and cattle. Chest x-ray and prompt treatment of tuberculosis skin test-positive individuals. BCG (Bacillus of Calmette & Guérin) vaccination or preventive treatment of close contacts of infected individuals.

Usual treatment A combination of antimicrobial drugs such as isoniazid (INH) with rifampin, streptomycin, ethambutol, or parazinamide is given for 9–12 months. Multi–drug-resistant strains are increasingly common.

WHOOPING COUGH, PERTUSSIS

Characteristics An acute bacterial childhood (usually) infection. The initial catarrhal (inflammatory) stage produces mild symptoms resembling the common cold. The second paroxysmal (symptomatic) stage is marked by uncontrollable sieges of coughing in an attempt to expel the thick mucus in the trachea and bronchi. Pneumonia may be a complication.

Pathogens *Bordetella pertussis,* a small, encapsulated, nonmotile, gram-negative coccobacillus that produces endotoxins and exotoxins. Identified by culture, biochemical and fluorescent antibody serologic techniques.

Reservoir Human respiratory tract.

Transmission Airborne via droplets from coughing.

Incubation period Usually 7–10 days; up to 21 days.

Epidemiology In 1994, 3,590 U.S. cases were reported to the CDC.

Control Vaccination of all young children with DTP (D for diphtheria, T for tetanus, P for pertussis) containing diphtheria and tetanus toxoids and killed *B. pertussis.* An acellular vaccine containing inactivated bacterial components is also available.

Usual treatment Erythromycin, tetracycline, or chloramphenicol.

Protozoan Respiratory Infections

PNEUMOCYSTOSIS, P. CARINII PNEUMONIA (PCP), INTERSTITIAL PLASMA-CELL PNEUMONIA

Characteristics A subacute pulmonary disease found in chronically ill children, *immunosuppressed* patients (patients whose immune systems are not functioning properly), and persons with AIDS. Asymptomatic in *immunocompetent* people (people whose immune systems are functioning properly). Has been the most common cause of death in AIDS patients. Patients show respiratory insufficiency with cyanosis. Pulmonary infiltration of intra-alveolar tissue with frothy exudate. Usually fatal in untreated patients.

Pathogen *Pneumocystis carinii* (Fig. 10 – 13) has both protozoan and fungal properties; thus, it is considered a protozoan by some microbiologists and a fungus by others.

Reservoir Humans.

Transmission Direct contact, transfer of pulmonary secretions from infected to susceptible persons. Perhaps airborne.

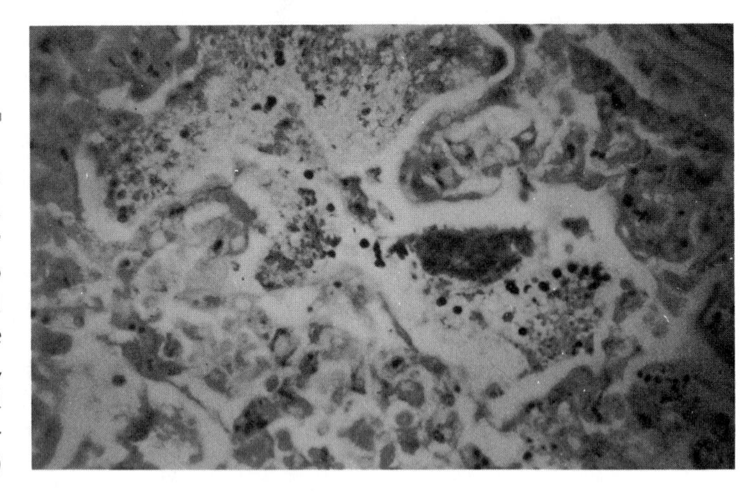

FIGURE 10 – 13
Pneumocystis carinii appears as dark oval bodies when stained with a special silver stain. Foamy material is also seen throughout the several alveoli in this section of the lung (original magnification, ×250). (Moffett HL: Clinical Microbiology, 2nd ed. Philadelphia, JB Lippincott, 1980)

Incubation period 1–2 months or longer.

Control Prophylaxis treatment of immunosuppressed patients with co-trimoxazole. Careful disinfection of respiratory therapy equipment.

Usual treatment Co-trimoxazole or pentamidine.

INFECTIOUS DISEASES OF THE GASTROINTESTINAL TRACT

The digestive tract consists of a long tube with many expanded areas designed for the digestion of food, the absorption of nutrients, and the elimination of undigested materials (Fig. 10 – 14). Transient and resident microbes continuously enter and leave the gastrointestinal tract. Most of the microorganisms ingested with food are destroyed in the stomach and duodenum by the low pH and are inhibited from growing in the lower intestines by the resident microflora. They are then flushed from the colon during defecation, along with large numbers of indigenous microbes.

The largest number of resident microorganisms are found in the lower small intestine and colon. There, the availability of nutrients and moisture at a constant 37° C allows many facultative and obligate anaerobes to thrive. Most of these intestinal microbes are obligate anaerobes—many *Bacteroides* and *Fusobacterium* species, with fewer *Clostridium* and *Veillonella* species. Facultative anaerobes are less abundant, but are better understood because they are easier to isolate and cultivate. Included in this group are the coliforms, the indicator organisms of fecal contamination of water supplies. These enteric bacilli include *Escherichia, Enterobacter, Proteus,* and *Klebsiella* species.

Many viruses, most of which are harmless, are also found in fecal material;

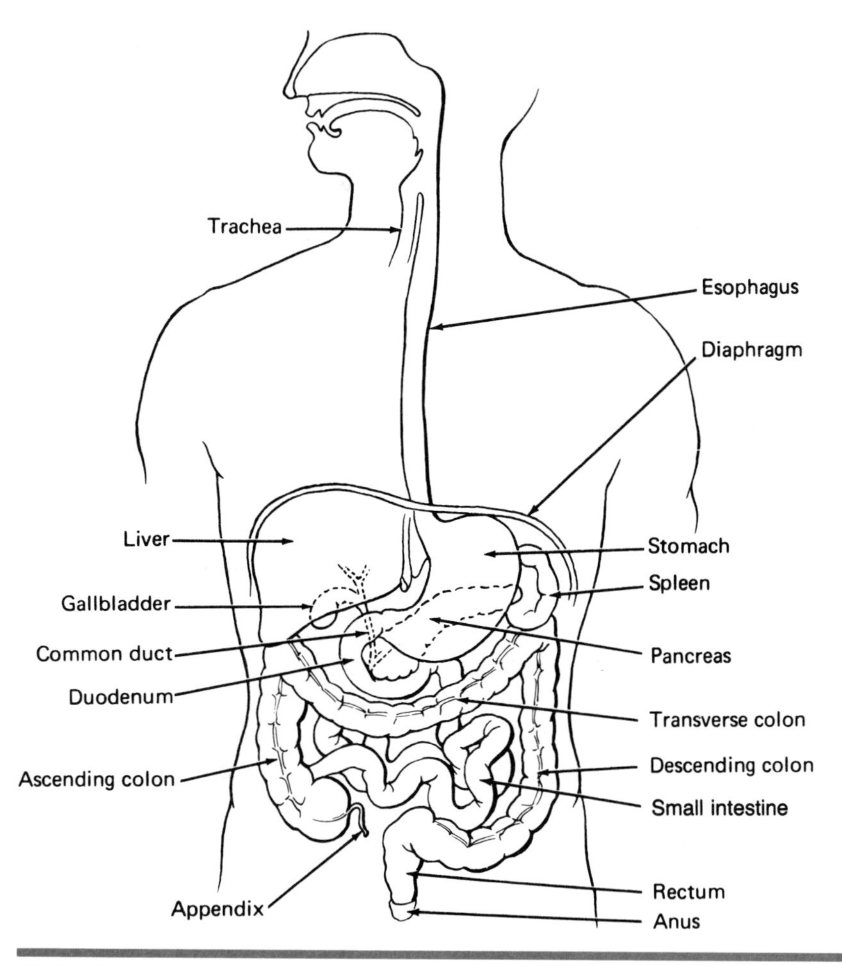

FIGURE 10 – 14
Anatomy of the digestive system. (Scherer JC: Introductory Medical-Surgical Nursing, 5th ed. Philadelphia, JB Lippincott, 1991)

however, some of these viruses can cause gastroenteritis under certain conditions. Some may be ingested with contaminated food, such as hepatitis A virus in fecally contaminated raw oysters.

Viral Gastrointestinal Infections

VIRAL GASTROENTERITIS, VIRAL DIARRHEA, EPIDEMIC ACUTE INFECTIOUS NONBACTERIAL GASTROENTEROPATHY

Characteristics A self-limiting viral infection of the lining of the gastrointestinal tract with nausea, vomiting, diarrhea, abdominal pain, headache, malaise, and low-grade fever. Although sometimes referred to as stomach flu or 24-hour flu, the term flu should be reserved for influenza, which is a respiratory infection.

Pathogens Norwalk virus, adenoviruses, echoviruses, rotavirus, coxsackieviruses, poliovirus, and others. Identified by cell culture and serologic techniques.

Reservoir Humans.

Transmission Fecal-oral route, fecal contamination of food, water, and hands.

Incubation period 24–48 hours.

Control Isolation of patient, enteric precautions.

Usual treatment Replacement of fluids and electrolytes. Amantadine occasionally reduces symptoms.

VIRAL HEPATITIS

Viral hepatitis includes a group of diseases involving inflammation of the liver caused by several viruses: hepatitis A virus (HAV), hepatitis B virus (HBV), hepatitis C virus (HCV), hepatitis D virus (HDV), and hepatitis E virus (HEV).

VIRAL HEPATITIS A, TYPE A HEPATITIS, INFECTIOUS HEPATITIS, EPIDEMIC HEPATITIS

Characteristics Liver inflammation with fever, nausea, abdominal discomfort, usually jaundice. Prolonged convalescence.

Pathogen Hepatitis A virus (HAV), a picornavirus, similar to an enterovirus; hepatitis C virus (HCV) may cause similar hepatitis symptoms. Serologic and radioimmunologic identification techniques.

Reservoir Humans.

Transmission Fecal-oral route, person-to-person; fecal contamination of water, food, milk, undercooked shellfish.

Incubation period 15–50 days, average 28–30 days.

Epidemiology In 1994, 23,507 U.S. cases were reported to the CDC.

Control Good sanitation practices, personal hygiene. Passive immunization of exposed individuals with hyperimmune gamma globulin. Vaccination for high-risk persons. Isolation of infected patients, enteric precautions.

Usual treatment None. Immune serum may reduce symptoms and speed recovery.

VIRAL HEPATITIS B, TYPE B HEPATITIS, SERUM HEPATITIS, AUSTRALIA ANTIGEN HEPATITIS

Characteristics Viral infection and liver damage with anorexia, rash, and jaundice. May lead to chronic hepatitis or cirrhosis of the liver in immunodeficient patients.

Pathogen Hepatitis B virus (HBV), a DNA virus with several major antigenic subtypes; may also involve coinfection with hepatitis D virus (HDV), an RNA virus. Hepatitis C virus causes similar blood–liver infections when transferred via transfusions. Identified by serologic tests.

Reservoir Humans.

Transmission Person-to-person, by saliva, semen, and other body fluids; parenteral route, through contaminated syringes, needles, dialysis, and other equipment serving as vectors.

Incubation period 45–108 days, average 60–90 days.

Epidemiology In 1994, 11,402 U.S. cases were reported to the CDC.

Control Vaccination against HBV (recommended for all health-care personnel). Active and passive immunization of exposed individuals. Isolation of infected persons with blood and body fluid precautions. Sterilization of all equipment.

Usual treatment Immune globulin (IG), hyperimmune gamma globulin (HBIG), or antiviral chemotherapy when available.

Bacterial Gastrointestinal Infections

CAMPYLOBACTER *GASTROENTERITIS*

Characteristics An acute enteric disease with diarrhea, nausea, vomiting, fever, malaise, abdominal pain, usually self-limiting, lasting 1–4 days.

Pathogens *Campylobacter jejuni* and *C. coli* of many serotypes. Rigid, curved, gram-negative bacteria. Identified by culture, biochemical and serologic techniques.

Reservoir Animals including cattle, sheep, swine, poultry, other birds, rodents, cats, dogs, other pets.

Transmission Ingestion of bacteria in food, milk, water. Contact with infected pets, wild animals, and infected infants.

Incubation period 1–10 days, average 3–5 days.

Control Thorough cooking of foods, especially poultry; pasteurize milk; chlorinate water supplies. Hand-washing after animal contact. Isolation of hospitalized patients, enteric precautions.

Usual treatment Replacement of fluids and electrolytes. Antibiotic therapy with erythromycin, tetracycline, or aminoglycosides.

CHOLERA

Characteristics An acute, enteric diarrheal disease with watery stools, vomiting, rapid dehydration, loss of blood volume, shock; death frequently results if untreated.

Pathogens *Vibrio cholerae* type 01, E1 Tor strain of *V. cholerae,* or other serotypes. A gram-negative curved (comma-shaped) bacillus that secretes an enterotoxin (a toxin that adversely affects cells in the intestinal tract) called *choleragin.* Identified by culture, biochemical and serologic techniques.

Reservoir Humans and environmental reservoirs.

Transmission Fecal-oral route; vomitus and feces-contaminated water and foods; soiled hands; flies; raw or undercooked seafoods from contaminated waters; sometimes carriers.

Incubation period 1–5 days.

Epidemiology Occurs worldwide. Only 39 U.S. cases were reported to the CDC in 1994. U.S. cases most often involve people who ingested oysters from the coastal areas of the Gulf of Mexico or seafood while traveling in Central and South America.

Control Isolation of hospitalized patients, enteric precautions. Concurrent disinfection of feces, vomitus, hands, linen, and fomites. Adequate sewage processing and water treatment. Fly control. Vaccination in some countries. Prophylactic tetracycline for exposed families; doxycycline or furazolidone treatment.

Usual treatment Prompt fluid and electrolyte replacement therapy. Tetracycline, co-trimoxazole, furazolidone, and other chemotherapy to shorten duration of disease.

ENTEROVIRULENT ESCHERICHIA COLI *DIARRHEA, TRAVELER'S DIARRHEA*

Characteristics Watery diarrhea with or without mucus or blood, vomiting, abdominal cramping.

Pathogens There are at least five different types of enterovirulent *E. coli* (different serotypes of *E. coli*); gram-negative bacilli. Some serotypes produce enterotoxin, others invade colon mucosa. Identified by culture, biochemical and serologic tests (see Fig. 2 – 9).

Reservoir Infected humans, carriers, animals.

INSIGHT
"Microbes In The News"—E. coli *0157:H7*

Escherichia coli is a gram-negative bacillus that lives in the intestinal tract of virtually everyone. As long as these "garden variety" strains of *E. coli* remain in the intestinal tract, they cause no harm and, in fact, even do some good by producing certain vitamins. However, because they are opportunistic pathogens, they do possess the ability to cause harm when they gain access to certain parts of the body (*e.g.*, urinary bladder, bloodstream, wounds) where they do not belong. Intestinal strains of *E. coli* are the number one cause of urinary tract infections, septicemia, and nosocomial infections.

There are other strains (or serotypes) of *E. coli* that are not members of our indigenous flora and are enteric pathogens whenever they are ingested. These are referred to as *enterovirulent E. coli,* and there are at least five known types.

1. *Enterotoxigenic E. coli* (ETEC) are a leading cause of infant diarrhea and mortality in developing countries and the leading cause of traveler's diarrhea. These strains produce enterotoxins that cause a profuse, watery diarrhea with no blood or mucus, abdominal cramping, vomiting, and dehydration. ETEC possess fimbriae (pili) that enable the bacteria to adhere to the intestinal wall.

2. *Enteroinvasive E. coli* (EIEC) are able to invade and proliferate within intestinal cells, killing the infected cells. EIEC cause dysentery with blood and mucus present in the patient's stool specimens.

3. *Enteropathogenic E. coli* (EPEC) primarily causes infant diarrhea, mucus in stools, fever, and dehydration. These strains destroy the microvilli of the small intestine.

4. The so-called *enteroaggregative E. coli* (EaggEC) strains are not well understood, but they are known to cause infant diarrhea in developing countries.

5. *Enterohemorrhagic E. coli* (EHEC), also known as verocytotoxic *E. coli* (VTEC) has been the subject of media attention and congressional inquiry. The serotype known as O157:H7 has caused a number of epidemics that have been associated with eating undercooked, contaminated hamburger meat. The first hamburger-associated epidemic occurred in 1982. EHEC strains cause hemorrhagic (bloody) diarrhea and a severe, sometimes fatal, urinary tract condition known as hemolytic uremic syndrome (HUS) in children. Hamburger meat has not been the only source of EHEC strains; other sources have included drinking water, lake water, and apple cider.

Transmission Fecal-oral route, fecally contaminated water, food, hands, fomites, or direct contact.

Incubation period 12–72 hours.

Control Proper sewage disposal, water treatment, hand-washing practices. Enteric precautions for hospitalized patients.

Usual treatment Fluid and electrolyte replacement therapy. Ampicillin or co-trimoxazole treatment for severe cases and carriers.

SALMONELLOSIS AND TYPHOID FEVER

Characteristics A gastroenteritis, inflammation of stomach and intestines with abdominal pain, headache, nausea, usually vomiting and diarrhea. Typhoid fever, the most severe form, is a bacteremia, characterized by constipation, intestinal hemorrhage, enlarged spleen and lymph nodes, rose spots on trunk, sustained fever.

Pathogens *Salmonella typhi* causes typhoid fever; *S. cholerae-suis, S. typhimurium,* and many other serologic types of *S. enteritidis* cause salmonellosis. Motile, gram-negative bacilli with endotoxin in their cell walls. *S. typhi* produces exotoxins and endotoxin. Identified by culture, biochemical and serologic techniques.

Reservoir Humans, animals, wild and domestic cattle, poultry, pigs, dogs, cats, turtles, fish, and many others.

Transmission Fecal-oral route; uncooked and undercooked feces-contaminated meat, eggs, and milk products; sewage, polluted water; hands and fingers of infected persons and carriers; flies and other flying insects.

Incubation period 1–3 weeks for typhoid fever; 1–3 days for salmonellosis.

Epidemiology In 1993, 41,641 U.S. cases of salmonellosis and in 1994, 410 U.S. cases of typhoid fever were reported to the CDC.

Control Isolation of hospitalized patients with enteric precautions. Water purification; effective sewage disposal. Pasteurization of milk; thorough cooking of foods of animal and egg origin. Proper hand-washing techniques. Control of flies. Clean food preparation area. Treatment of carriers and infected pets. Eliminate food preparation by infected persons and carriers.

Usual treatment Fluid and electrolyte replacement therapy. Ampicillin or amoxicillin with co-trimoxazole or chloramphenicol for resistant strains, quinolones, and cephalosporins.

SHIGELLOSIS, BACILLARY DYSENTERY

Characteristics An acute bacterial infection of the lining of the small and large intestine. Diarrhea with blood, mucus, and pus; nausea, vomiting, cramps, fever, sometimes *toxemia* (toxins in the blood) and convulsions.

Pathogens Several serotypes of *Shigella dysenteriae, S. flexneri, S. boydii,* and *S. soneii;* nonmotile, gram-negative bacilli. Plasmid associated with toxin production and virulence. Identified by culture, biochemical and serologic techniques.

Reservoir Humans, some primates.

Transmission Fecal-oral route from patients or carriers. Fecally contaminated hands and fingernails, contaminated food and milk, sewage-polluted water. Flies and cockroach vectors.

Incubation period 1–7 days, average 1–3 days.

Epidemiology In 1993, 32,198 U.S. cases were reported to the CDC.

Control Eliminate fecal contamination of food, milk, and water. Proper hand-washing and fomite cleaning. Enteric precautions with patients. Infected individuals should neither prepare nor serve food.

Usual treatment Fluid and electrolyte replacement therapy. Co-trimoxazole, ampicillin, tetracyclines, or nalidixic acid antibiotic therapy; chloramphenicol for resistant strains.

Bacterial Foodborne Intoxications, Food Poisoning

Foodborne intoxication or food poisoning refers to diseases resulting from the ingestion of food or water contaminated with bacterial exotoxins produced by such bacteria as *V. cholerae,* enterovirulent *E. coli, Clostridium botulinum, C. perfringens,* and *S. aureus* or, in some cases, by the bacteria themselves.

BOTULISM

Characteristics A neuromuscular disease caused by ingestion of food contaminated by *C. botulinum* spores, containing neurotoxins secreted by the pathogen. Infant botulism and wound botulism are produced by living exotoxin-producing bacteria in the gastrointestinal tract or wounds, respectively. Neurotoxins may cause nerve damage, visual difficulty, respiratory failure, flaccid paralysis of voluntary muscles, brain damage, coma, and death within 1 week if untreated.

Pathogen *C. botulinum,* a spore-forming, gram-positive, anaerobic bacillus that produces several exotoxins.

Reservoir Dust, soil, dirty foods, honey, improperly canned foods, neutral pH foods, lightly cured foods.

Transmission Ingestion of foods in which the pathogen has produced exotoxin (*botulinal toxin*); improperly cooked foods with near-neutral pH.

Incubation period 12–36 hours.

Epidemiology In 1994, only 142 U.S. cases were reported to the CDC, including 59 cases of foodborne botulism and 76 cases of infant botulism.

Control Careful washing, canning, processing, and cooking of food. Thorough cleaning of wounds. Never feed raw honey to babies or add it to their milk or foods.

Usual treatment Antitoxin and treatment of respiratory failure.

CLOSTRIDIUM PERFRINGENS *FOOD POISONING*

Characteristics A gastrointestinal toxemia with colic, diarrhea, nausea, rarely vomiting. Usually a mild disease of short duration, less than 24 hours, rarely fatal.

Pathogen *C. perfringens,* a gram-positive, spore-forming, enterotoxin-producing, anaerobic bacillus.

Reservoir Spores in soil, gastrointestinal tract of humans and animals.

Transmission Ingestion of food, contaminated by dirt or fecal material, containing vegetative forms or spores of *C. perfringens;* the enterotoxins are produced *in vivo.*

Incubation period 6–24 hours.

Control Cook food well, serve hot or cold. Train food processors in proper hand-washing and food-handling techniques.

Usual treatment None; usually a self-limiting disease.

STAPHYLOCOCCAL FOOD POISONING

Characteristics A gastroenteritis and toxemia with acute onset of cramps, vomiting, nausea, occasional diarrhea, subnormal body temperature and blood pressure resulting from ingested enterotoxins in foods.

Pathogens Enterotoxin-producing strains of *S. aureus* growing in foods.

Reservoir Humans—skin, abscesses, nasal secretions; bovine-contaminated milk.

Transmission Ingestion of *S. aureus*-contaminated foods containing staphylococcal enterotoxin (a type of exotoxin); particularly starchy foods and meats.

Incubation period 1–7 hours.

Control Training food preparers in proper personal hygiene, in cleanliness in kitchen areas, and in the need for refrigeration of starchy foods and meats. Prevent infected persons from preparing food.

Usual treatment Fluid replacement if needed.

Protozoan Gastrointestinal Diseases

PRIMARY AMEBIASIS, AMEBIC DYSENTERY

Characteristics A protozoan intestinal disease with dysentery, fever, chills, bloody or mucoid diarrhea or constipation, and colitis; may progress to liver, lungs, pericardium, brain, or other organs.

Pathogen *Entamoeba histolytica,* an ameba in the subphylum Sarcodina. Occurs in two stages: the cyst stage (the dormant, infective stage) and the motile, metabolically active, reproducing trophozoite stage (the ameba). Amebae may invade mucous membranes of the colon, forming abscesses. Diarrheal specimens contain cysts and trophozoites.

Reservoir Humans, usually asymptomatic carriers.

Transmission Fecally contaminated water and vegetables, flies on food. Soiled hands of infected food handlers. Oral-anal sexual contact.

Incubation period Variable, days to months.

Epidemiology Amebiasis causes an estimated 40,000-100,000 deaths per year, worldwide.

Control Sanitary sewage disposal; protection and treatment of water supplies. Personal hygiene and hand-washing before preparing and eating food. Halt the practice of fertilizing crops with human feces. Control flies around foods. Enteric precautions with hospitalized patients.

Usual treatment Metronidazole (Flagyl®) with iodoquinol. Fluid and electrolyte replacement.

GIARDIASIS

Characteristics A protozoan infection of the duodenum (the uppermost portion of the small intestine), with nausea, gas, abdominal pain, acute malodorous diarrhea, malabsorption, and damage to mucosal membranes; may be asymptomatic.

Pathogen *Giardia lamblia* (see Fig. 2 – 16B), a flagellate. Trophozoites attach to mucosal membranes. Cysts and sometimes trophozoites are expelled in feces.

Reservoir Humans; possibly beaver and other wild and domestic animals.

Transmission Fecal-oral route; ingestion of cysts in fecal-contaminated water or foods; person-to-person by soiled hands to mouth.

Incubation period 5–25 days or longer.

Control Proper treatment (including filtration) of public water supplies (routine chlorination does not destroy cysts); sanitary disposal of feces. Boil all emergency water supplies.

Usual treatment Quinacrine (Atabrine®) or metronidazole (Flagyl®).

INFECTIOUS DISEASES OF THE UROGENITAL TRACT

When the anatomy of the male and female urinary and reproductive systems are studied, it is easy to locate many areas where infections may occur. The urinary tract is usually protected from pathogens by the frequent flushing action of urination (Fig. 10 – 15). The acidity of normal urine also discourages growth of many microorganisms. Indigenous microflora are found at and near the outer opening (meatus) of the urethra of both males and females. Inhabitants of the distal urethra include species of *Bacteroides, Streptococcus, Mycobacterium, Lactobacillus,* and nonpathogenic *Neisseria,* as well as some gram-

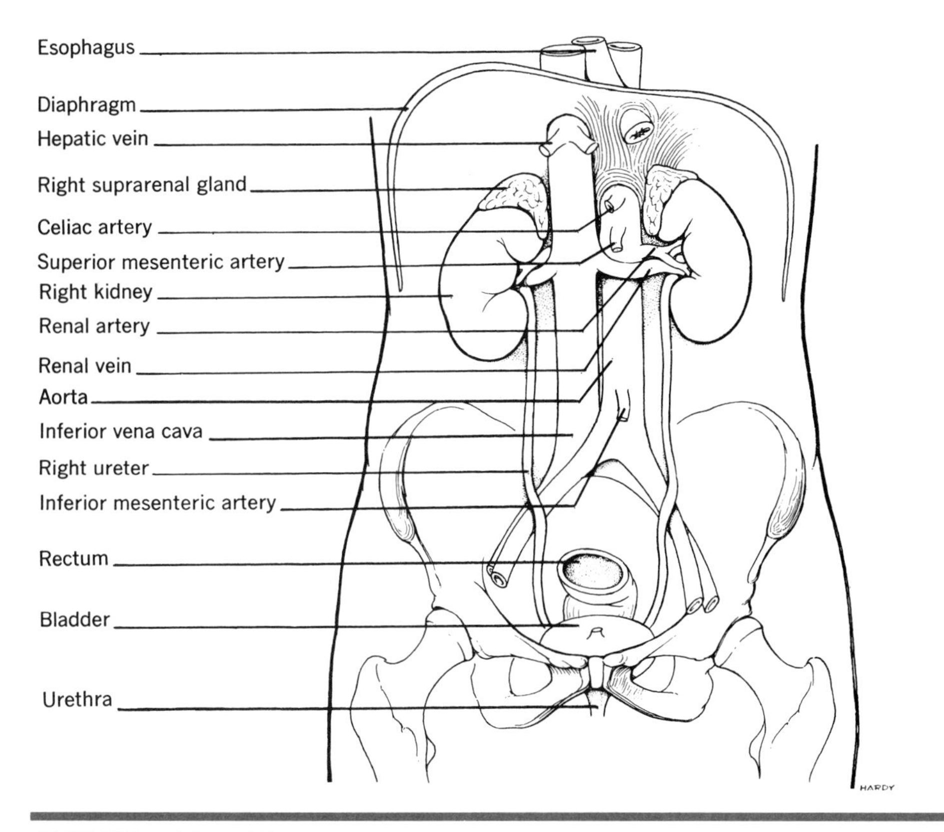

FIGURE 10 – 15
Urinary tract. (Chaffee EE, Lytle IM: Basic Physiology and Anatomy, 3rd ed. Philadelphia, JB Lippincott, 1973)

negative enteric bacteria. Additionally, the female genital area supports the growth of many other microorganisms. In the adult vaginal microflora, there are many species of *Lactobacillus, Staphylococcus, Streptococcus, Enterococcus, Neisseria, Clostridium, Actinomyces, Prevotella,* diphtheroids, enteric bacilli, and *Candida.* The balance among these microbes depends on the estrogen levels and pH of the site. The rest of the reproductive systems of both sexes should be free of microbial life (Fig. 10 – 16). Should any of these or other microorganisms invade further into the genitourinary tract, a variety of nonspecific infections may occur.

Sexually Transmitted Diseases

The term sexually transmitted disease (STD), formerly called venereal disease (VD), includes any of the infections transmitted by sexual activities. They are diseases of not only the reproductive and urinary tracts, but also of the skin, mucous membranes, blood, lymphatic and digestive systems, and many other body areas. Acquired immunodeficiency syndrome (AIDS), chlamydial and

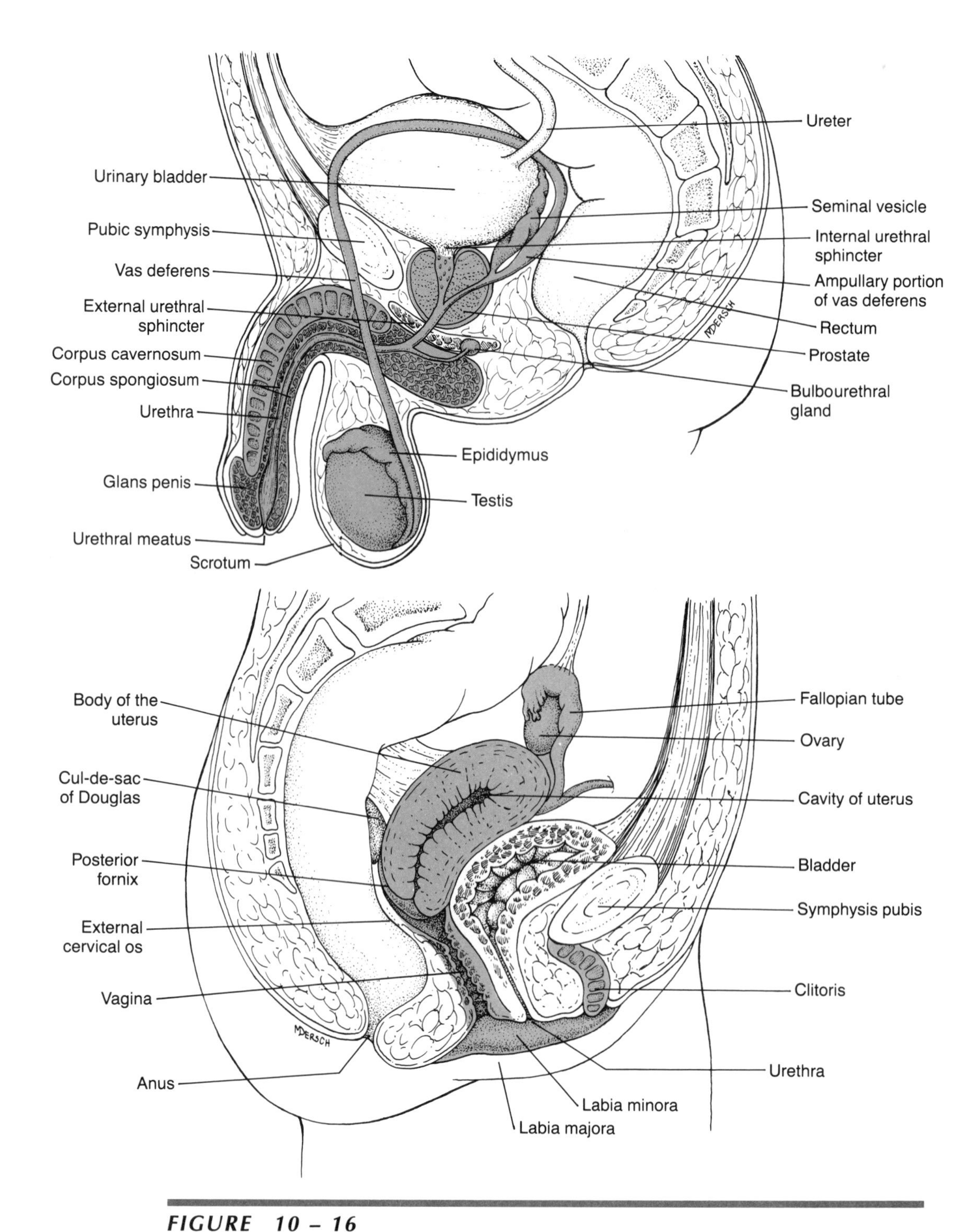

FIGURE 10 – 16

Reproductive systems. (*A*) Male; (*B*) Female. (Koniak D. Maternity Nursing: Family, Newborn, and Women's Health Care, 17th ed. Philadelphia, JB Lippincott, 1992) herpes infections, gonorrhea, and syphilis are the epidemic STDs of the 1990s. Because the AIDS virus (HIV) primarily causes damage to helper T-cells and, thus, inhibits antibody production, it is discussed later with diseases of the circulatory system. Because they can be transmitted by sexual activities, hepatitis B, amebiasis, and giardiasis can also be considered STDs, as well as many other diseases.

Urinary Tract Infections

URETHRITIS, CYSTITIS, URETERITIS, PROSTATITIS

Characteristics Inflammatory infections of the urethra (*urethritis*), the urinary bladder (*cystitis*), the ureters (*ureteritis*), or the prostate (*prostatitis*) may lead to kidney infections (*pyelonephritis*) and damage.

Pathogens Any of the normal flora introduced by poor personal hygiene, sexual intercourse, insertion of catheters, etc. The most common pathogens include *E. coli*, *Proteus* spp., *Pseudomonas* spp., and fecal streptococci (*Enterococcus* spp.). Nosocomial infections caused by *Serratia*, *Pseudomonas*, and *Klebsiella* spp. are common. Nonspecific urethritis is frequently caused by species of *Chlamydia*, *Ureaplasma*, and *Mycoplasma*, usually introduced by sexual contact. Gonococci may invade the urinary tract and the reproductive system.

Reservoir Humans.

Transmission Poor personal hygiene; sexual intercourse; introduction of catheters and other instruments.

Incubation period Variable, 1–15 days.

Control Personal cleanliness, particularly before intercourse; use of condoms. Sterilization of catheters, cystoscopes, and other instruments and proper disinfection of the urethral area before use of these instruments.

Usual treatment Appropriate antibiotic or chemotherapeutic agent for the identified pathogen.

Viral Urogenital Infections

GENITAL HERPES

Characteristics A herpes virus infection of the reproductive system producing herpetic lesions externally and / or internally with fever, headache, malaise.

FIGURE 10 – 17
Herpes simplex virus particles embedded in phosphotungstate. (*A*) Enveloped virion showing the thick envelope surrounding the nucleocapsid. (*B*) Naked viral particle; the structure of the capsomeres is plainly visible (original magnification, ×200,000). (Watson DH, et al: Virology 19:250, 1963)

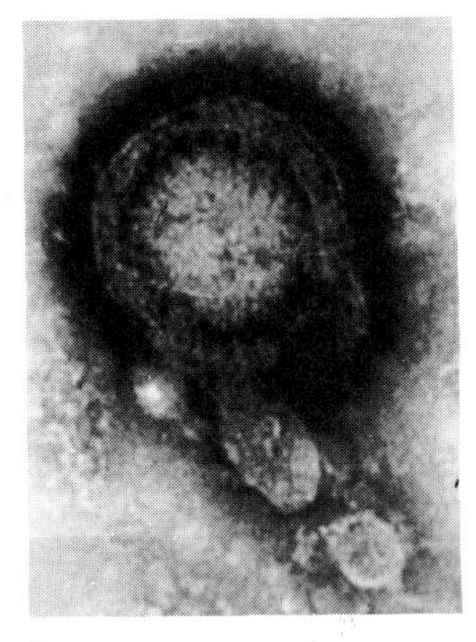
A

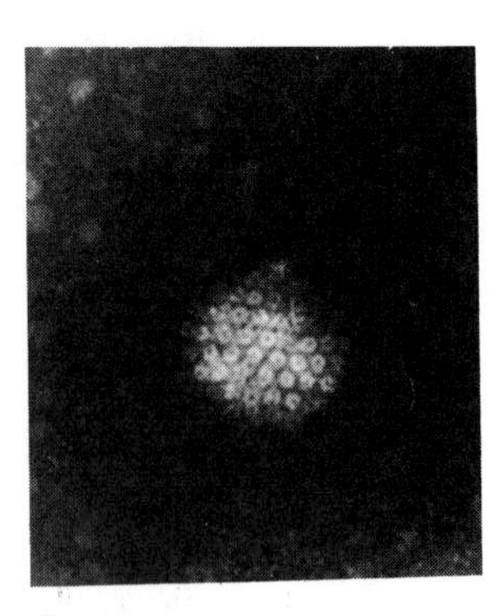
B

May become latent in nerve ganglia and asymptomatic. Progression of lesion: papule → vesicle → pustule → ulcer → crusting → healing.

Pathogen Herpes simplex virus, type 2 (HSV-2) usually; HSV-1 occasionally (Fig. 10 – 17).

Reservoir Humans.

Transmission Direct sexual contact; oral-genital or oral-anal contact during presence of lesions (Fig. 10 – 18). Mother-to-fetus or mother-to-neonate during pregnancy and birth or autoinoculation from another lesion.

Incubation period 2–28 days or longer; may remain as latent infection indefinitely.

Control Refrain from intercourse with person with herpetic lesions; use of condoms. Cesarean delivery of infected mothers before membranes rupture. Prophylactic oral acyclovir for women of reproductive age.

Usual treatment Acyclovir; intravenous, oral, or topical.

GENITAL WARTS, PAPILLOMA VENEREUM, CONDYLOMA ACUMINATA

Characteristics Viral disease with skin and mucous membrane lesions that are cauliflower-like, fleshy growths in genital areas, internally or externally (Fig. 10 – 19); laryngeal papillomas in babies of infected mothers. Some infections become malignant.

Pathogen Human papilloma virus (HPV) of genotypes 6, 11, 16, or 18; the latter two are more often involved in malignancies.

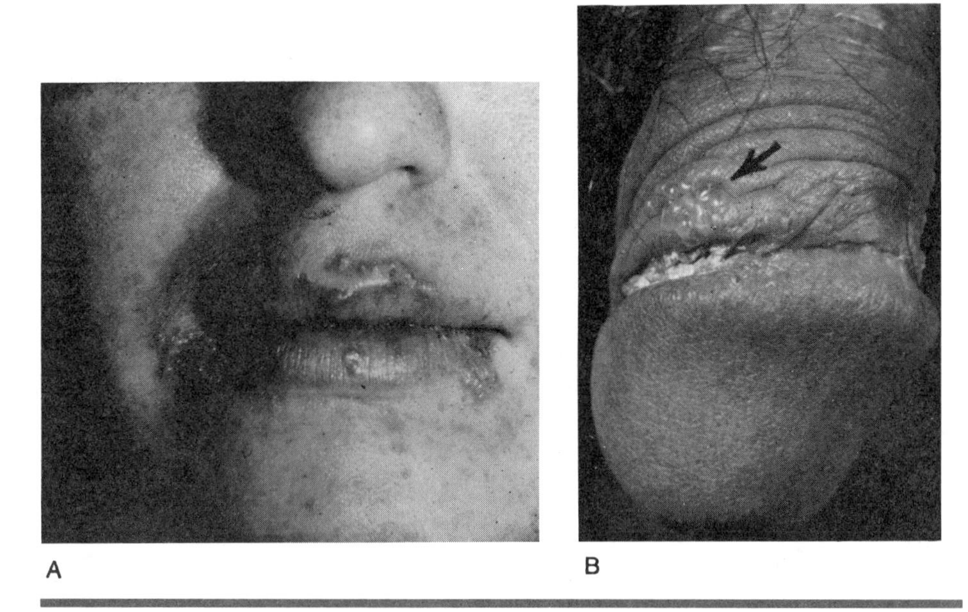

FIGURE 10 – 18
(*A*) Recurrent herpes simplex of the lip. (*B*) Recurrent herpes progenitalis. (Dobson RL, Abele DC: The Practice of Dermatology. Philadelphia, JB Lippincott, 1985)

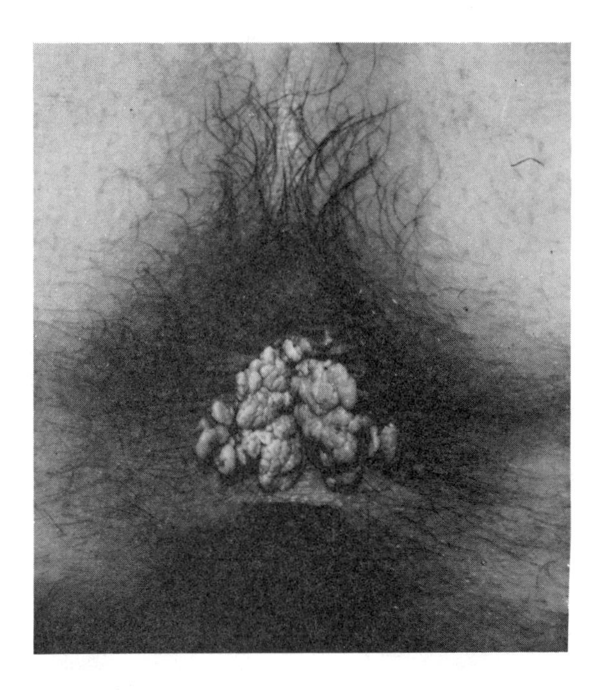

FIGURE 10 – 19
Anogenital warts (condylomata accuminata or moist warts). (Dobson RL, Abele DC: The Practice of Dermatology. Philadelphia, JB Lippincott, 1985)

Reservoir Humans.

Transmission Direct contact, usually sexual. Through breaks in skin or mucous membranes. From mother to neonate during birth.

Incubation period 1–8 months or longer.

Control Avoid direct contact with lesions. Use of condoms. Cesarean section delivery for infected mothers.

Usual treatment Local treatment with 25% podophyllin, 5-fluorouracil (5-FU), or other cytotoxic agents, but not in pregnant women. Cryosurgery with liquid nitrogen. Surgical removal, laser therapy, or electrocautery. Treatment repeated with recurrence.

Bacterial Urogenital Infections

CHLAMYDIAL INFECTIONS, GENITAL CHLAMYDIASIS

Characteristics The most frequent cause of nonspecific urethritis (NSU) and nongonococcal urethritis (NGU), *cervicitis,* vaginitis, *salpingitis, epididymitis,* and newborn conjunctivitis and pneumonia; many asymptomatic carriers.

Pathogen *C. trachomatis* serotypes D–K; tiny, obligately intracellular, gram-negative bacteria in two forms: elementary body and reticulate body. Concomitant infectious agents may include *Ureaplasma ureolyticum, Mycoplasma hominis,* and other sexually transmitted pathogens in 50% of cases. Identified by cell culture, staining, and immunodiagnostic techniques.

Reservoir Humans.

Transmission Direct sexual contact or mother-to-neonate during birth.

Incubation period 2–3 weeks, may be latent.

Control Personal cleanliness, sex education, use of condom. Prophylactic treatment of contacts and pregnant women with infection.

Usual treatment Tetracycline or doxycycline; erythromycin for pregnant women and resistant cases.

GONORRHEA

Characteristics A very common acute infectious disease of urethra, anus, vagina, cervix, fallopian tubes (salpingitis), and other reproductive organs.

Causes pelvic inflammatory disease (PID), a yellow purulent urethral discharge (most commonly seen in men). Frequently asymptomatic. Also, may result in rectal gonorrhea (*proctitis*), pharyngitis, conjunctivitis (see Fig. 10 – 6), and disseminated infections.

Pathogen *N. gonorrhoeae,* gram-negative diplococci; some strains have plasmids for β-lactamases (penicillinase-producing *N. gonorrhoeae,* or PPNG, strains). Identified by culture, Gram stain morphology, biochemical and immunodiagnostic techniques.

Reservoir Humans.

Transmission Direct mucous membrane-to-mucous membrane or sexual contact. Adult-to-child (may indicate sexual abuse). Mother-to-neonate during birth.

Incubation period 2–7 days or longer; frequently asymptomatic.

Epidemiology In 1994, 400,592 new U.S. cases were reported to the CDC.

Control Refrain from intercourse with infected partners; use of condoms. Vaginal and cervical cultures for pregnant women.

Usual treatment Ceftriaxone, doxycycline, ciprofloxacin; erythromycin for pregnant women.

SYPHILIS

Characteristics A treponemal disease that occurs in three stages: a primary lesion known as a chancre (Fig. 10 – 20A), secondary skin rash with fever and mucous membrane lesions (Fig. 10 – 20B), a long latent period, and then a tertiary stage with damage to central nervous system, cardiovascular system, visceral organs, bones, sense organs, and other sites.

Pathogen *Treponema pallidum,* a gram-variable spirochete, too thin to be seen with brightfield microscopy. Identified using darkfield microscopy and serologic tests.

Reservoir Humans.

Transmission Direct contact with lesions, body secretions, blood, semen, saliva, vaginal discharges; usually during sexual contact. Blood transfusions. Transplacentally from mother to fetus.

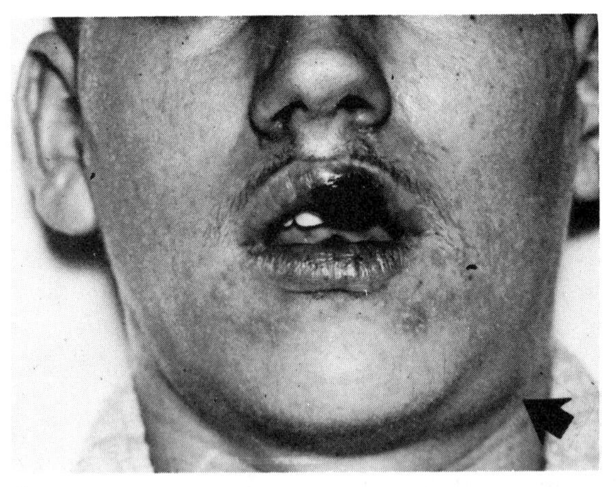

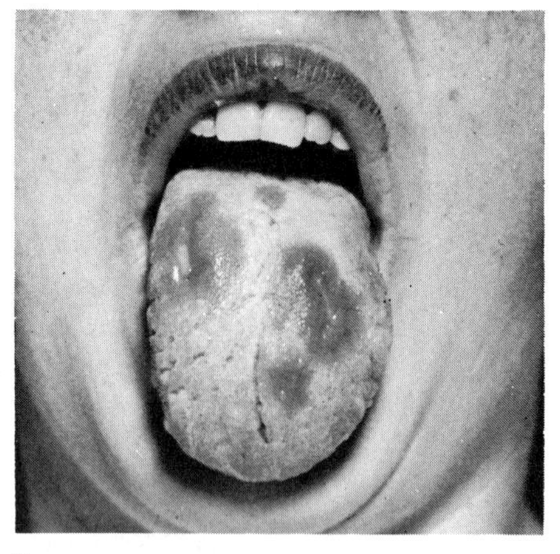

FIGURE 10 – 20
Syphilis. (*A*) Primary syphilis, showing a chancre of the lip with unilateral adenopathy (arrow). (*B*) Secondary syphilis, with mucous patches on the tongue. (Dobson RL, Abele DC: The Practice of Dermatology. Philadelphia, JB Lippincott, 1985)

Incubation period 10 days to several weeks.

Epidemiology In 1994, 21,306 U.S. cases were reported to the CDC.

Control VDRL or RPR tests with confirming serology for high-risk persons and pregnant women. Avoid sexual contact with infected persons; use of condoms.

Usual treatment Long-acting penicillin G (benzathine penicillin) or procaine penicillin; tetracycline or doxycycline for penicillin-sensitive persons.

Protozoan Urogenital Infections

TRICHOMONIASIS

Characteristics A protozoan disease causing vaginitis in women. May cause a profuse, thin, malodorous, yellowish discharge; often asymptomatic. In men, an infection of the urethra, prostate, or seminal vesicles; usually asymptomatic.

Pathogen *Trichomonas vaginalis,* a flagellate. Identified microscopically or immunologically, rarely by culture.

Reservoir Humans.

Transmission Direct mucous membrane contact, sexual intercourse.

Incubation period 4–20 days.

Control Avoid sexual relations with infected persons; use of condoms. Concurrent treatment of sexual partners.

Usual treatment Metronidazole (Flagyl®)

Other Sexually Transmitted Diseases

Many other pathogens may be sexually transmitted. Three, seen more often in parts of the world other than in the United States, are chancroid, granuloma inguinale, and lymphogranuloma venereum (LGV). Chancroid, caused by the gram-negative bacterium *Haemophilus ducreyi,* can be treated with erythromycin, ceftriaxone, or co-trimoxazole. Granuloma inguinale, a chronic infection caused by a gram-negative bacterium named *Calymmatobacterium granulomatis* (*Donovania granulomatis*), is treatable with tetracycline or co-trimoxazole. Lymphogranuloma venereum is a chlamydial infection involving the lymph nodes, rectum, and reproductive tract. It is caused by *Chlamydia trachomatis* types L-1, L-2, and L-3 (see Fig. 2 – 14) and treated with tetracycline, erythromycin, or sulfonamides. It should be noted that many STDs are transmitted simultaneously (*e.g.*, gonorrhea and syphilis together).

INFECTIOUS DISEASES OF THE CIRCULATORY SYSTEM

The circulatory system, consisting of the cardiovascular system and lymphatic system (Fig. 10 – 21), carries blood and lymph fluid throughout the body. Included in these fluids are many cells: erythrocytes, leukocytes (including lymphocytes), and platelets (thrombocytes). Most leukocytes function to protect the body from pathogens either by phagocytosis or antibody production.

Normally, the blood is sterile; it contains no resident microflora. When microbes invade the bloodstream, septicemia results if they are not quickly destroyed. A patient with septicemia experiences chills, fever, and prostration and has bacteria and/or their toxins in his or her bloodstream. Lymph occasionally picks up microorganisms from the intestine, lungs, and other areas, but these transient organisms are usually quickly engulfed by phagocytic cells in the liver and lymph nodes. Frequently, transient bacteremia (the temporary presence of bacteria in the blood) results from dental extractions, wounds,

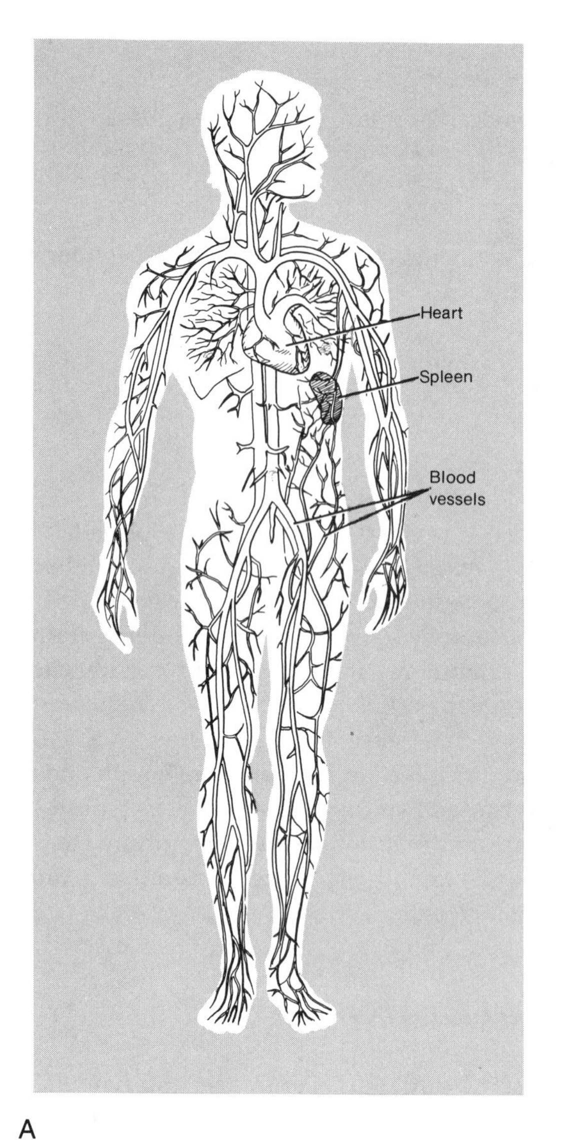

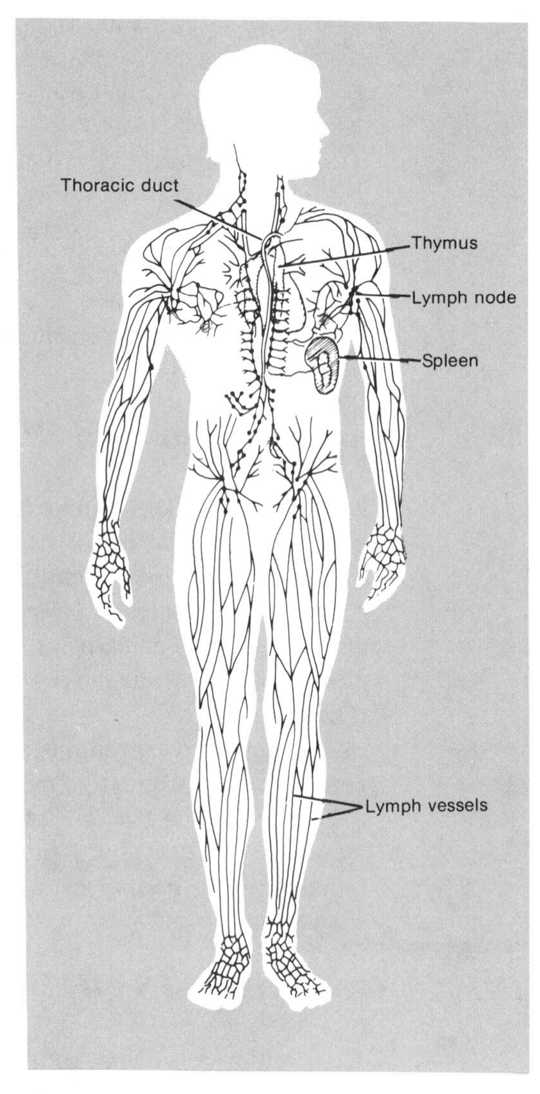

FIGURE 10 – 21
(*A*) The cardiovascular system. (*B*) The lymphatic system. (Tortora GJ, Anagnostakos NP: Anatomy and Physiology, 4th ed. Philadelphia, Harper & Row, 1985)

bites, and damage to the intestinal, respiratory, or reproductive tract mucosa. However, when pathogenic organisms are capable of resisting or overwhelming the phagocytes and other body defenses—or when the individual is immunosuppressed or is otherwise more susceptible than normal—a systemic disease may occur.

Viruses often invade the circulatory system and damage certain "target" cells, an example being HIV, which destroys helper T-cells, causing the immune system to malfunction. Some cardiovascular diseases are the result of toxemia, the production of damaging exotoxins secreted by bacteria and carried by the bloodstream to target areas. Toxic shock syndrome, scarlet fever, tetanus, and botulism are examples of such diseases. Certain fungal and protozoan cardiovascular diseases may cause damage by actually clogging the capillaries in various regions of the body.

Viral Lymphatic and Cardiovascular Infections

ACQUIRED IMMUNODEFICIENCY SYNDROME (AIDS)

Characteristics A viral infection that destroys helper T-cells (Fig. 10 – 22) of the immune system, causing immunosuppression and allowing secondary infections caused by viruses (CMV, herpes), protozoa (*Pneumocystis, Toxoplasma*), bacteria (mycobacteria), and / or fungi (*Candida*). The secondary infections become systemic and cause death. Kaposi's sarcoma (a previously rare type of cancer) is a frequent complication. Symptoms of AIDS include fever, fatigue,

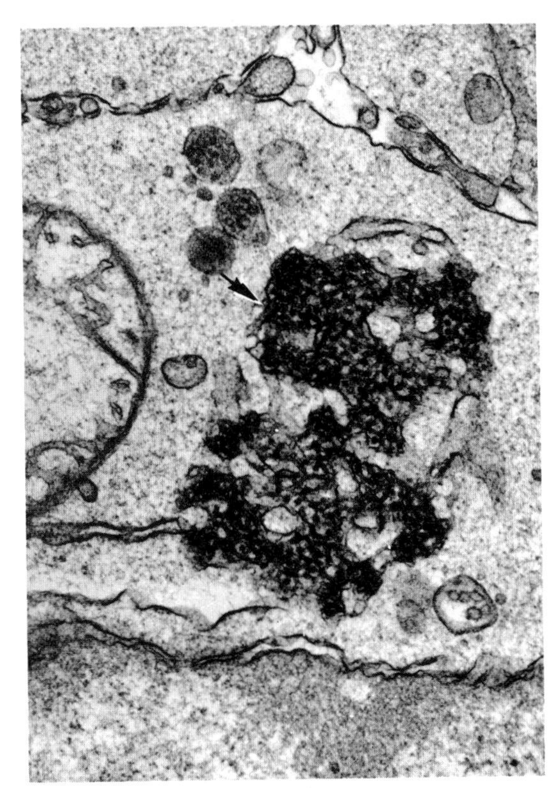

FIGURE 10 – 22
A large inclusion (arrow) within the cytoplasm of a peripheral blood mononuclear cell from a patient with AIDS (original magnification, ×40,000). (DeVita VT, Jr, Hellman S, Rosenberg SA: AIDS. Philadelphia, JB Lippincott, 1985)

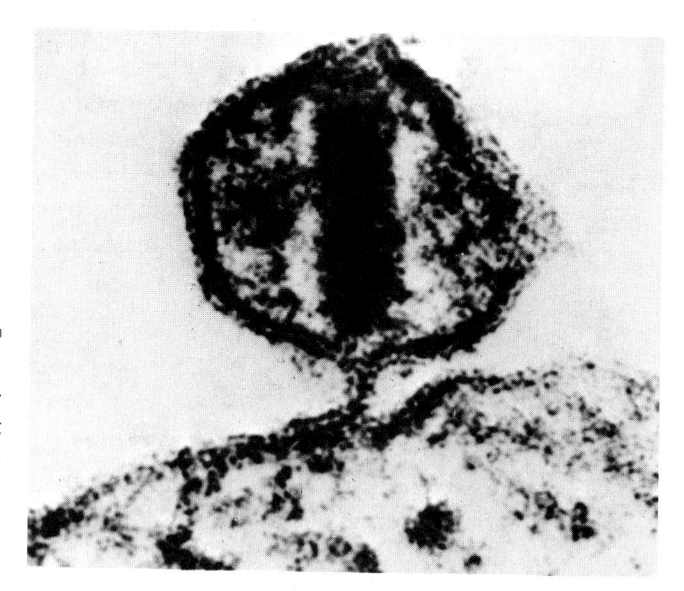

FIGURE 10 – 23
Extracellular HIV retrovirus particle from a lymphocyte of a patient with ARC. Note the dense cylindrical core. (DeVita VT, Jr, Hellman S, Rosenberg SA: AIDS. Philadelphia, JB Lippincott, 1985)

diarrhea, weight loss, enlarged lymph nodes (*lymphadenitis*), brain and spinal cord damage. May be asymptomatic as AIDS-related complex (ARC) for several to many years. Considered to be a universally fatal disease. Average time from initial infection to death, 10 years.

Pathogen Human immunodeficiency virus (HIV) (Fig. 10 – 23).

Reservoir Humans.

Transmission Direct sexual contact, homosexual or heterosexual. Contaminated intravenous needles and syringes. Blood transfusion with contaminated blood and blood products. Transplacental and mother-to-child transfer.

Incubation period 6 months to 5 years or longer.

Epidemiology Pandemic; 80,691 new U.S. cases were reported to the CDC in 1994 (see Fig. 7 – 9).

Control Avoid sexual contact with high-risk persons, *i.e.,* those having multiple sex partners, and those sharing IV drug paraphernalia (soak needles in 10% Clorox). Use of condoms. For hospitalized patients, practice universal body substance precautions.

Usual treatment Azidothymidine (AZT), other derived nucleotides such as dideoxycytidine; treatment of secondary infections.

COLORADO TICK FEVER

Characteristics An acute viral infection with fever, headache, fatigue, aching, and occasionally *encephalitis* or *myocarditis,* following tick bite; in western North America; usually self-limiting.

Pathogen An *arbovirus.*

Reservoir Tick-infested small mammals.

Transmission Bite of virus-infected tick to humans, from infected small mammals, ground squirrels, porcupines, or chipmunks.

Incubation period 4–5 days.

Control Control of ticks and infected rodent hosts; blood and body fluid precautions for hospitalized patients.

Usual treatment None, except rest and nutritious diet.

INFECTIOUS MONONUCLEOSIS, MONO, KISSING DISEASE

Characteristics An acute viral disease of B-lymphocytes, spleen, and liver with high fever, sore throat, fatigue, weakness, and generalized *lymphadenopathy* (diseased lymph nodes). Showing transformed B-cells. Usually self-limiting, may be asymptomatic and latent.

Pathogen Epstein–Barr virus (EBV), a human herpes virus type 4 that may be *oncogenic* (cancer causing). Identified by heterophile antibody and immunofluorescent antibody (IFA) tests.

Reservoir Humans.

Transmission Person-to-person, by direct contact with oral and nasal secretions, by exchange of saliva during kissing, and blood transfusion.

Incubation period 4–6 weeks.

Control Limit kissing contacts. Disinfect articles contaminated with nose and throat discharges. Hand-washing.

Usual treatment None, except rest and nutritious diet.

MUMPS, INFECTIOUS PAROTITIS

Characteristics An acute viral infection of salivary glands, usually the parotid glands (*parotitis*). Deafness, *meningoencephalitis, mastitis, nephritis,* thyroiditis, and *pericarditis* complications may occur. In adults, *orchitis* (inflammation of the testes) or *oophoritis* (inflammation of an ovary).

Pathogen Mumps virus, a paramyxovirus. Identified by serologic techniques.

Reservoir Humans.

Transmission Direct contact and fomites via respiratory secretions, aerosol droplets, saliva.

Incubation period 2–3 weeks, usually 18 days.

Epidemiology In 1994, 1,322 U.S. cases were reported to the CDC.

Control Attenuated MMR (measles-mumps-rubella) vaccine for children and nonimmune adults. Respiratory isolation for hospitalized patients.

Treatment No specific chemotherapy.

Rickettsial Cardiovascular Infections

ROCKY MOUNTAIN SPOTTED FEVER, TICKBORNE TYPHUS FEVER

Characteristics The most common rickettsial spotted fever; an infection of the vascular endothelial cells, with high fever, muscle pain, headache, chills, and maculopapular rash on extremities after third day. May result in death if untreated.

Pathogen *Rickettsia rickettsii,* a gram-negative bacterium; obligate intracellular parasite. Identified by IFA test and nonspecific Weil–Felix reaction.

Reservoir Infected ticks on dogs, rodents, and other animals.

Transmission Bite of infected tick, from animals to humans; tick parts or feces into break in skin.

Incubation period 2–14 days after tick bite.

Epidemiology In 1994, 441 U.S. cases were reported to the CDC.

Control Avoid tick-infested areas. Search for ticks on self and dogs after being in tick-infested areas. Avoid crushing ticks during removal.

Usual treatment Tetracycline or chloramphenicol.

ENDEMIC MURINE TYPHUS FEVER, FLEABORNE TYPHUS

Characteristics An acute febrile disease (similar to, but milder than, epidemic typhus) with fever, headache, and rash on body trunk.

Pathogen *Rickettsia typhi.*

Reservoir Infected rats with infective rat fleas.

Transmission Rats to fleas that defecate rickettsia into bite site and skin wounds of humans; rat→flea→human.

Incubation period 1–2 weeks.

Control Apply insecticide to rat-infested areas, then use rodent control measures. Avoid contact with rats.

Usual treatment Tetracycline, doxycycline, or chloramphenicol.

EPIDEMIC TYPHUS FEVER, LOUSEBORNE TYPHUS

Characteristics An acute rickettsial disease with very high prolonged fever, severe headache, rash appearing on trunk by sixth day; toxemia may involve kidneys, heart, spleen, and nerves.

Pathogen *Rickettsia prowazekii*. Identified by IFA tests.

Reservoir Humans.

Transmission Humans to body lice (Fig. 10 – 24) to humans.

Incubation period 1–2 weeks.

Control Use of insecticide to kill body lice. Improve personal cleanliness practices. Immunization of susceptible people.

Usual treatment Chloramphenicol or tetracycline.

Bacterial Cardiovascular Infections

SUBACUTE BACTERIAL ENDOCARDITIS (SBE)

Characteristics A bacterial infection of heart valves damaged by rheumatic fever, syphilis, atherosclerosis, congenital heart deformities. Bacterial masses

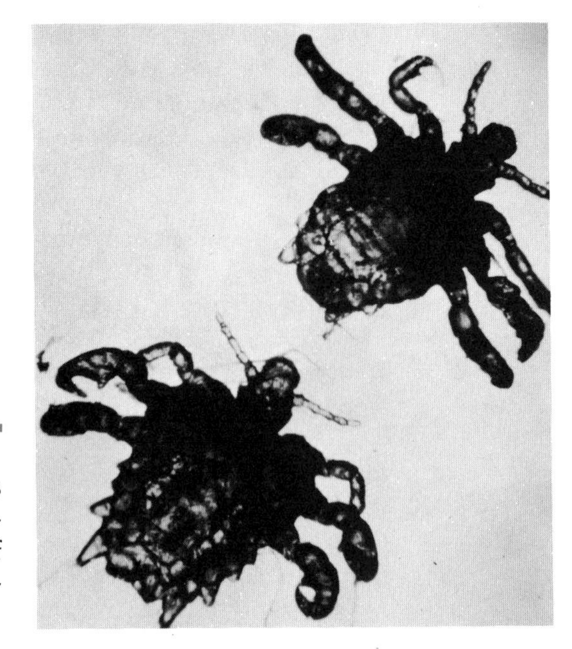

FIGURE 10 – 24
Pubic lice (*Phthirus pubis*) as seen with the ×7.5 lens of a microscope. (Sauer GC: Manual of Skin Diseases, 6th ed. Philadelphia, JB Lippincott, 1991)

with fibrin break away to form clots that clog small blood vessels in brain, heart, kidney, spleen, liver, or other organs. May cause death.

Pathogens Most often, species of non–group-A streptococci, gram-positive cocci of low virulence. Identified by blood culture techniques.

Reservoir Humans.

Transmission (1) Parenteral; (2) mechanical invasion of blood vessels; (3) surgery, catheterization, minor trauma, dental work.

Incubation period Variable.

Control Personal hygiene. Sterile and aseptic techniques by dentists, physicians, and nurses.

Treatment Bactericidal antibiotics specific for the pathogen identified.

LYME DISEASE

Characteristics A tickborne disease characterized by three stages: (1) early local stage, distinctive large skin lesions (usually at site of tick bite); (2) early disseminated systemic stage, skin blotches, malaise, fatigue, arthritis, carditis, *meningitis*, encephalitis; (3) prolonged arthritis, debilitating fatigue, chronic *encephalomyelitis*, neuropathy (numbness, loss of memory), facial palsy (paralysis). Isolates from skin lesions grow at 33° C on Barbour-Stoenner-Kelley (BSK) medium and may be identified by specific antibody tests.

Pathogen A spirochete, *Borrelia burgdorferi.*

Reservoir Carried by ticks that feed on mice, dogs, horses, cattle, deer, and humans. The deer tick, *Ixodes dammini,* serves as the vector in the Northeast and upper Midwest United States; *I. pacificus* is the vector in the Western United States; *I. ricinis* in Europe; and *I. persulatus* in Asia.

Transmission Large animals or rodents to ticks that bite humans and transmit the spirochetes during several hours of feeding. Congenital transmission is rare. No person-to-person transmission has been reported.

Incubation period From 3 to 33 days after tick bite; the first early local stage may be almost asymptomatic.

Epidemiology In 1994, 11,424 U.S. cases were reported to the CDC.

Control Avoid tick-infested areas and animals. Wear light-colored clothing covering legs and arms, closed tightly at feet and hands. Use tick repellant (diethyltoluamide [DEET] or permethrin) on pantlegs and sleeves. Check total body surface for ticks every 4 hours and remove attached ticks including mouth parts with tweezers. A vaccine is available for dogs but is not yet available for humans.

Usual treatment For adults, treat with tetracycline, doxycycline, or cephalosporin for 10–30 days; for children younger than 8 years of age, treat with amoxicillin for 10–30 days. Erythromycin or ceftriaxone may be used for those allergic to penicillin or tetracycline.

PLAGUE, BUBONIC PLAGUE (BLACK DEATH), SEPTICEMIC PLAGUE, PNEUMONIC PLAGUE

Characteristics An acute, often severe, infection. Bubonic plague is named for the enlarged inguinal lymph nodes (buboes) that develop. Pneumonic plague involves the lungs; may become disseminated throughout the body, resulting in death if untreated. Pneumonic plague is very contagious and usually fatal.

Pathogen *Yersinia pestis,* a nonmotile, bipolar-staining, gram-negative coccobacillus. Identified by its typical bipolar appearance, culture of bubo aspirate, sputum, or CSF, and IFA or ELISA tests.

Reservoir Wild rodents and fleas on rodents.

Transmission Wild rodents to fleas to humans, from rodents directly to humans (bubonic plague), or respiratory pathway from person-to-person (pneumonic plague).

Incubation period 2–6 days.

Epidemiology In 1994, only 14 U.S. cases were reported to the CDC.

Control Avoid contact with rodents and their fleas. Strict isolation of pneumonic patients.

Usual treatment Streptomycin, tetracycline, or chloramphenicol.

TULAREMIA, RABBIT FEVER

Characteristics An acute infection with primary local ulcer, lymph node, swelling; may be systemic, pneumonic, or gastrointestinal disease.

Pathogen *Francisella tularensis,* a small, gram-negative, pleomorphic bacillus. Identified by IFA and heterophile antibody tests.

Reservoir Wild animals, such as rabbits, muskrats, beaver; some domestic animals; hard ticks.

Transmission By handling or ingesting contaminated meat, by drinking contaminated water, or by bites of infected animals, flies, and ticks.

Incubation period 2–10 days.

Epidemiology In 1994, 85 U.S. cases were reported to the CDC.

Control Use impervious gloves when handling and dressing rabbits, cook wild meat well, avoid bites of insects and ticks in infected areas. Vaccinate high-risk persons.

Usual treatment Streptomycin, gentamicin, or tobramycin.

Protozoan Cardiovascular Infections

MALARIA

Characteristics A systemic sporozoan infection with fever, chills, sweating, headache, asymptomatic periods, rarely progressing to shock, renal and liver failure, and coma.

Pathogens *Plasmodium vivax* (the most common species), *P. falciparum* (the most deadly species), *P. malariae,* and *P. ovale;* sporozoans with a complex life cycle involving a female *Anopheles* mosquito and the liver and erythrocytes of the infected human. Identified in Giemsa-stained blood smears.

Reservoir Humans.

Transmission Bite of infected female *Anopheles* mosquito; also by blood transfusion or contaminated syringes.

Incubation period 12–30 days or longer.

Epidemiology In 1994, 1,065 U.S. cases were reported to the CDC; mostly imported cases. Causes an estimated 1-2 million deaths per year, worldwide.

Control Insecticides and draining of breeding areas to destroy anopheline mosquitoes. Use of chemosuppressive drugs: chloroquine, primaquine, Fansidar®, or Maloprim® (pyrimethamine and dapsone). Use of diethyltoluamide insect repellent on skin.

Usual treatment Chloroquine, Fansidar®, mefloquine or primaquine (for chloroquine-resistant cases), quinine.

TOXOPLASMOSIS

Characteristics A systemic sporozoan disease with fever, lymphadenopathy, *lymphocytosis,* rarely with central nervous system symptoms, pneumonia, myocarditis, rash, and death. Primary infection may be asymptomatic. Common in patients with AIDS.

Pathogen *Toxoplasma gondii,* an obligate intracellular sporozoan.

Reservoir Rodents, cattle, sheep, pigs, goats, chickens, birds; cats shed oocysts in fecal material.

Transmission Humans become infected by (1) eating infected raw or undercooked meat (usually pork or mutton) containing the cyst form of the pathogen or (2) ingesting oocysts shed in the feces of infected cats in contaminated food, water, or dust. Children may inhale or ingest oocysts from sandboxes containing cat feces. Transplacental infection occurs during pregnancy, which may cause severe damage to the fetus. Cats become infected by eating infected birds and rodents.

Incubation period 5–23 days.

Epidemiology Estimated that more than 2 million people per year are infected.

Control Cook meats thoroughly. Wear disposable latex gloves when cleaning cat litter box. Dispose of cat feces daily in toilet or by burying. Pregnant women should avoid cat litter boxes and cat feces-contaminated soil. Wash hands after handling cats. Vaccinate pet cats.

Usual treatment Pyrimethamine and sulfadiazine with folinic acid.

AFRICAN TRYPANOSOMIASIS (AFRICAN SLEEPING SICKNESS), AMERICAN TRYPANOSOMIASIS (CHAGAS' DISEASE)

Characteristics Systemic diseases caused by flagellated protozoa. Chancre at site of tsetse fly or reduviid bug bite; then intense headache, lymphadenopathy, insomnia, anemia, and rash. Sleeping sickness (the result of central nervous system involvement) and death in chronic cases of African trypanosomiasis. Chronic Chagas' disease symptoms include enlarged heart, enlarged esophagus (megaesophagus), and enlarged colon (megacolon).

Pathogen *Trypanosoma brucei gambiense* causes most cases of sleeping sickness. *T. brucei rhodesiense* causes a more rapidly fatal form of African trypanosomiasis. *T. cruzi* causes American trypanosomiasis.

Reservoir Humans and wild animals.

Transmission To humans by bite of tsetse fly (African trypanosomiasis) or reduviid bugs (American trypanosomiasis) that have previously ingested the blood of infected humans or animals.

Incubation period 3–21 days or longer.

Epidemiology African trypanosomiasis causes an estimated 20,000 deaths per year.

Control Control of tsetse flies and reduviid bugs; treatment of infected humans.

Usual treatment Suramin, pentamidine, or melarsoprol. None for Chagas' disease.

INFECTIOUS DISEASES OF THE NERVOUS SYSTEM

The central nervous system (CNS) (Fig. 10 – 25) is well protected; it is encased in bone, covered with three membranes (called the *meninges*), bathed and cushioned in cerebrospinal fluid (CSF), and nourished by capillaries. These capil-

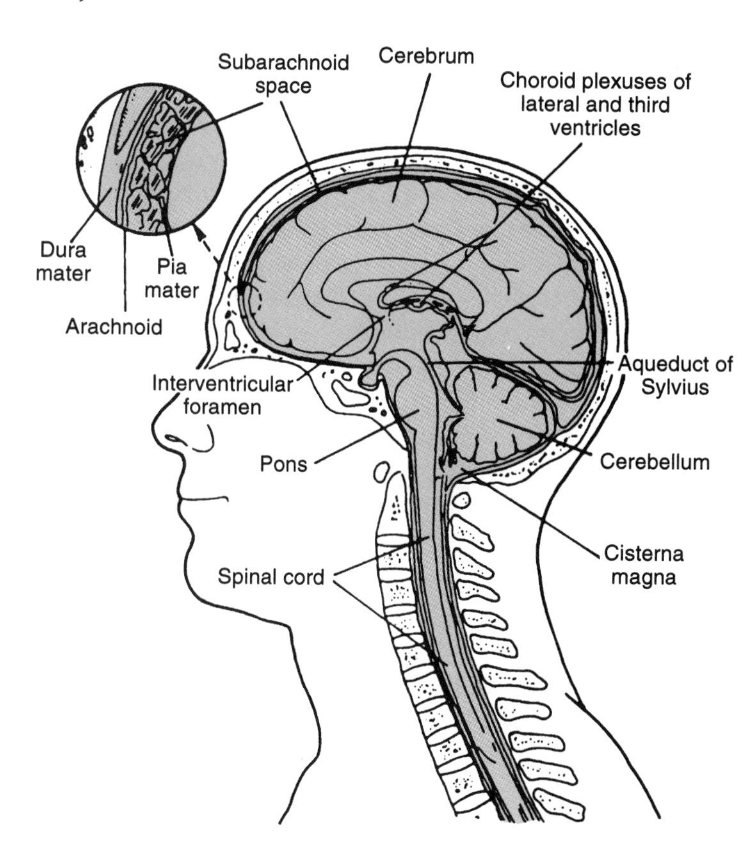

FIGURE 10 - 25
Central nervous system, illustrating the brain, the spinal cord, and the meninges. (Davis BG, Bishop ML: Clinical Laboratory Science: Strategies for Practice. Philadelphia, JB Lippincott, 1989)

laries make up the blood–brain barrier, supplying nutrients, but not allowing larger particles, such as microorganisms and most antibiotics, to pass from the blood into the brain.

There are no indigenous microflora of the nervous system. Microbes must gain access to the CNS through trauma (fracture or medical procedure), via the blood or lymph to the CSF, or along the peripheral nerves.

An infection of the meninges is called meningitis. Encephalitis is an infection of the brain itself, and *myelitis* is an infection of the spinal cord. Several viruses may produce viral or aseptic meningitis and encephalitis. The most common are enteroviruses, coxsackieviruses, echoviruses, and mumps viruses. Others include arboviruses, polioviruses, adenoviruses, measles, herpes, and varicella viruses. The arboviruses (arthropodborne viruses) are introduced by mosquito vectors and cause several forms of viral encephalitis. Overall, the three major causes of bacterial meningitis are *H. influenzae* (the primary cause in children), *N. meningitidis* (the primary cause in adolescents), and *S. pneumoniae* (the primary cause in the elderly) (see Fig. 7 – 4). The major causes of bacterial meningitis in neonates are *Streptococcus agalactiae* (group B, β-hemolytic streptococci), *E. coli* and other members of the family *Enterobacteriaceae,* and *Listeria monocytogenes.* Less common causes of bacterial meningitis are *S. aureus, P. aeruginosa, Salmonella,* and *Klebsiella.* Free-living amebae that may cause meningoencephalitis (an infection of both the brain and the meninges) are in the genera *Naegleria* and *Acanthamoeba.* Other protozoa that may invade the meninges are *Toxoplasma* and *Trypanosoma.* Occasionally fungal pathogens, especially *Cryptococcus neoformans* (an encapsulated yeast), cause meningitis.

Early symptoms of meningitis are similar to those of colds. Later, fever, severe headache, pain, and stiffness of the neck and back develop. Then neurological symptoms of dizziness, convulsions, minor paralysis, and coma occur; death may result within a few hours.

Diagnosis is usually made by a combination of patient symptoms, physical examination, and culture of the patient's cerebrospinal fluid. Treatment of viral meningitis is supportive. The use of specific antibacterial or antifungal agents will depend on the particular bacterial or fungal pathogen involved and the results of antimicrobial susceptibility testing.

Several CNS diseases are caused by toxins. Some examples of bacterial neurotoxins are botulinal toxin, the exotoxin of botulism; neurotoxins of *S. aureus* in toxic shock syndrome; and the *tetanospasmin* (a type of neurotoxin) of tetanus. Diseases caused by fungal toxins include ergot from grain molds and mushroom poisoning. *Gonyaulax,* an alga found in algal "blooms," produces neurotoxins, which may concentrate in bivalve shellfish and cause paralytic symptoms following ingestion of the contaminated shellfish.

Viral Nervous System Infections

POLIOMYELITIS, INFANTILE PARALYSIS

Characteristics An acute viral infection of the medulla oblongata, spinal cord, and nerves, with fever, headache, nausea, vomiting, sore throat, muscle pain and spasms, neck and back stiffness, with or without paralysis. Usually asymptomatic, or mild disease similar to influenza. Most often infects young children; occasionally, further damage occurs many years later.

Pathogen Poliovirus types 1, 2, and 3; small RNA enteroviruses (Fig. 10 – 26). Type 1 (wild type) most often causes paralysis. Types 2 and 3 are frequently vaccine-associated. Identified by tissue culture, cytopathic effects, and neutralizing antibody tests.

Reservoir Human gastrointestinal tract.

Transmission By direct contact or fecal-oral route in areas with poor sanitation. Virus is inhaled or ingested, carried by lymph fluid to lymph nodes, to blood, to CNS, damaging motor nerve impulses to muscles. Viruses are trans-

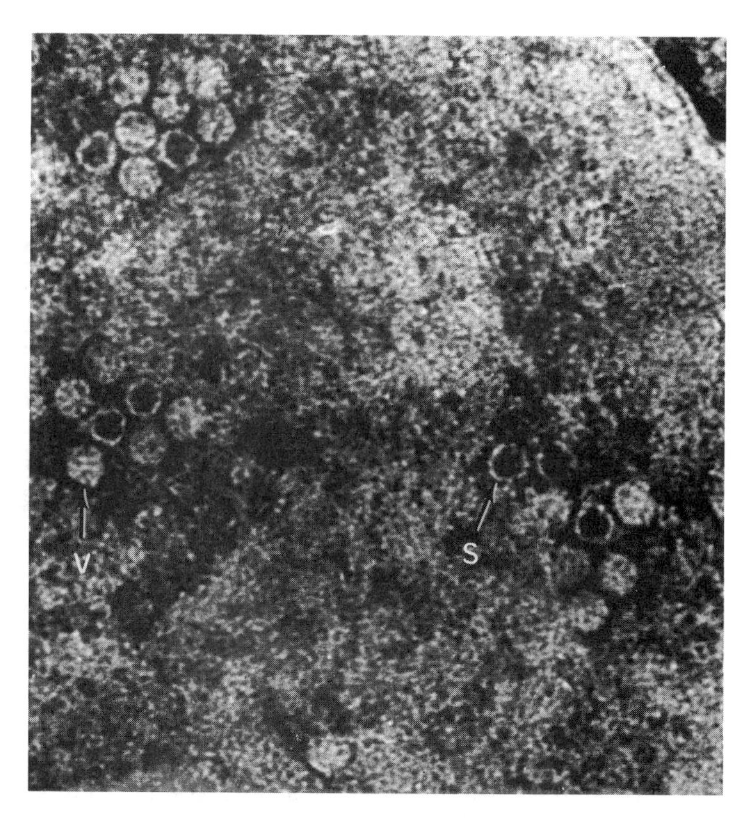

FIGURE 10 – 26
Development of poliovirus particles in pieces of cytoplasmic matrix of artificially disrupted cells. Particles in various stages of assembly from empty shells (*s*) to complete virions (*v*) can be seen (original magnification, ×200,000). (Horne RW, Nagington J: J Mol Biol 1:333, 1959. Copyright by Academic Press, Inc., Ltd.)

mitted by pharyngeal secretions and fecal material during infection and after vaccination with attenuated live vaccine.

Incubation period 3–35 days, usually 7–14 days.

Epidemiology In 1994, only one U.S. case was reported to the CDC. The World Health Organization (WHO) is attempting to eradicate polio worldwide by the year 2000.

Control Immunize all infants and children with a series of Salk inactivated polio vaccine (IPV) or Sabin oral attenuated poliovirus vaccine (OPV). In hospitalized patients, isolate with enteric precautions. Adequate sewage and water treatment precautions. Adequate sewage and water treatment.

Usual treatment None. Provide therapy and assistance for paralytic patient.

RABIES

Characteristics A fatal, acute viral encephalomyelitis (brain and spinal cord infection) of mammals, with depression, headache, fever, malaise, paralysis,

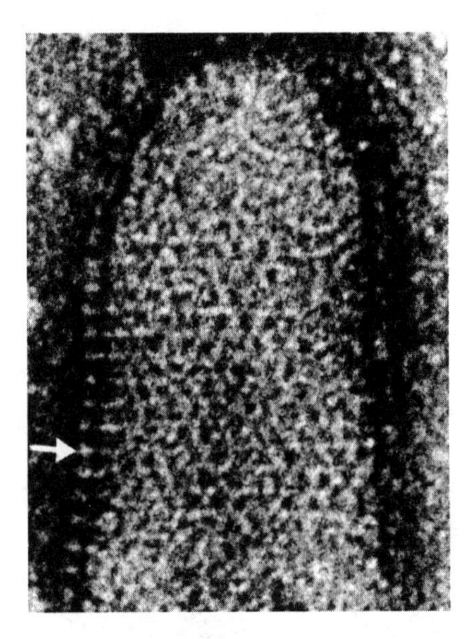

FIGURE 10 – 27
Intact rabies virus particle negatively stained with phosphotungstate. On the left (arrow) are well-resolved surface projections, 6–7 nm long (original magnification, ×400,000). (Hummeler K, et al: J Virol 1:152, 1967)

hydrophobia (fear of water), salivation, spasms of throat muscles, convulsions, and death caused by respiratory failure.

Pathogen Rabies virus (Fig. 10 – 27), a rhabdovirus; a large, complex, enveloped RNA virus. Identified by virus isolation and IFA tests to confirm direct observation of Negri inclusion bodies in animal brain tissue.

Reservoir Many wild and domestic mammals: dogs, foxes, coyotes, cats, skunks, raccoons, and bats.

Transmission Rabid animal bites introducing virus-laden saliva. Airborne transmission from bats in caves. Person-to-person by saliva.

Incubation period 2–8 weeks or longer.

Epidemiology In 1994, five human cases and 7,347 animal cases were reported to the CDC. Due to a high incidence of rabies during 1994, Texas imposed a quarantine on dogs entering the state.

Control Vaccinate all pets. Avoid all sick, aggressive, or friendly wild animals. Prophylactic immunization of high-risk persons (*e.g.*, veterinarians and veterinary technicians) with human diploid cell rabies vaccine (HDCV). Isolation of hospitalized patients.

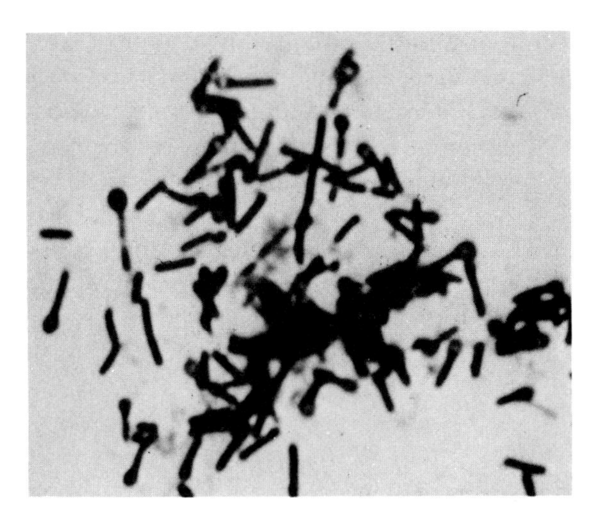

FIGURE 10 – 28
Cells of *Clostridium tetani* after 24 hours on a cooked meat–glucose medium. Note spherical, terminal endospores (original magnification, ×4,500). (Volk WA, et al: Essentials of Medical Microbiology, 4th ed. Philadelphia, JB Lippincott, 1991)

Usual treatment Prompt treatment of bite wounds with soap and water and local infusion of rabies immune globulin (RIG) antiserum. Vaccinate patient with HDCV.

Bacterial Nervous System Infections

TETANUS

Characteristics An acute infectious neuromuscular disease induced by the exotoxin, tetanospasmin, with convulsions and intermittent spasms of the voluntary muscles, especially masseter (the muscle that closes the jaw) and neck muscles; exhaustion, respiratory failure, and death may result.

Pathogen *C. tetani* (Fig. 10 – 28), a motile, gram-positive, anaerobic, spore-forming bacillus that produces a potent neurotoxin called tetanospasmin.

Reservoir Soil contaminated with feces; intestines of humans and animals.

Transmission Spores of *C. tetani* are introduced into a puncture wound, burn, or needle prick by contamination with soil, dust, or feces. Under anaerobic conditions in the wound, spores germinate into vegetative *C. tetani* cells that produce the exotoxin.

Incubation period 3–21 days or longer.

Epidemiology In 1994, 29 U.S. cases were reported to the CDC.

Control Active immunization of all children with DTP, a combined diphtheria-tetanus-pertussis vaccine, containing tetanus toxoid. Boosters of DT should be given at 10-year intervals and after a wound or accident with no booster for 5 years. Nonimmunized persons may be given human tetanus immune globulin (TIG) following a puncture wound.

Usual treatment Dirty wounds should be cleaned and left open, when feasible, to inhibit the anaerobic growth. Administer penicillin to destroy wound organisms, antitoxin (TIG) to neutralize the exotoxin, and vaccinate with DT.

SUMMARY

The major infectious diseases of skin, wounds, eyes, ears, mouth, respiratory system, gastrointestinal tract, urogenital areas, and cardiovascular and nervous systems have been described, mostly in outline format. In this way, the reader can discern the characteristics, pathogens, mechanisms of pathogenicity, reservoirs, modes of transmission, incubation period, means of control, and recommended treatment for each disease presented. A brief survey of human anatomy was shown by drawings. Indigenous microflora of each area were discussed because the source of the etiologic agent may be from opportunistic resident microflora in a compromised host whose defenses are impaired.

Many of these infections may move from one area of the body to another, involving several body systems. They were discussed relative to either the site of entry or where they usually cause the most damage.

PROBLEMS AND QUESTIONS

1. Name six factors that affect the number of microbes on the skin. How may opportunists invade through the skin?
2. Which bacteria cause most burn, wound, and bite infections? What types of toxins do they produce?
3. How do oral bacteria interact to cause dental caries? How can you prevent dental caries?
4. By what pathways may microbes invade to cause ear infections? Which bacteria are usually found in ear infections?
5. Why is pneumonia called a nonspecific disease? Name several pathogens that might cause pneumonia.
6. Differentiate between foodborne infections and foodborne intoxications. Cite some examples of each.
7. Which protozoa cause gastrointestinal infections? How do they get there?
8. How are urinary tract infections transmitted? Why might *E. coli* be found more often in female cystitis than in male cystitis?
9. How may sexually transmitted infections be prevented? List genital infections that may cause congenital and neonatal infections. How could transmission to fetus and newborn be prevented?
10. Differentiate between typhus fever and typhoid fever. How are they transmitted and controlled?
11. By what means might children be exposed to plague, tularemia, toxoplasmosis, and rabies? After exposure, how should children be treated to prevent or reduce the symptoms of these diseases?
12. List several anatomical sites, systems, or organs where there are no indigenous microflora. Why might this be so?
13. What are the usual symptoms of infectious diseases of the nervous system? Define meningitis, encephalitis, myelitis, meningoencephalitis, and encephalomyelitis.
14. List several diseases caused by neurotoxins. Cite several nonspecific diseases discussed in this chapter.

Self Test

After you have read Chapter 10, examined the objectives, reviewed the chapter outline, studied the new terms, and answered the problems and questions above, complete the following self test.

Matching Exercises

Match each of the diseases under Column I with the appropriate item in Column II. An answer from Column II may be used once or more than once.

DISEASES—TYPES OF ETIOLOGIC AGENTS

Column I

___ **1.** Gonorrhea
___ **2.** Trachoma
___ **3.** Tuberculosis
___ **4.** Histoplasmosis
___ **5.** Amebiasis
___ **6.** Candidiasis
___ **7.** Syphilis
___ **8.** AIDS
___ **9.** Common cold
___ **10.** "Strep" throat
___ **11.** Legionellosis
___ **12.** Rocky Mountain spotted fever
___ **13.** Warts
___ **14.** Athlete's foot
___ **15.** Botulism
___ **16.** Whooping cough
___ **17.** Measles
___ **18.** Mumps
___ **19.** Trichomoniasis
___ **20.** Hepatitis

Column II

a. Virus
b. Bacterium
c. Fungus
d. Protozoan

DISEASES—SYNONYMS

Column I

___ **1.** Measles
___ **2.** German measles
___ **3.** Chickenpox

Column II

a. Variola
b. Rubeola
c. Varicella
d. Rubella

DISEASES—ETIOLOGIC AGENTS

Column I

___ **1.** Primary atypical pneumonia
___ **2.** Pneumococcal pneumonia
___ **3.** Legionellosis
___ **4.** Histoplasmosis
___ **5.** Whooping cough
___ **6.** Diphtheria
___ **7.** "Strep" throat
___ **8.** Acute rhinitis
___ **9.** Otitis media
___ **10.** Thrush
___ **11.** Boils, carbuncles
___ **12.** Colds

Column II

a. *Bordetella pertussis*
b. *Streptococcus pneumoniae*
c. Influenza virus
d. *Mycoplasma pneumoniae*
e. *Legionella pneumophila*
f. *Histoplasma capsulatum*
g. *Coccidioides immitis*
h. *Candida albicans*
i. *Staphylococcus aureus*
j. Rhinovirus
k. *Corynebacterium diphtheriae*
l. *Streptococcus pyogenes*
m. *Haemophilus influenzae*
n. *Staphylococcus epidermidis*
o. *Streptococcus lactis*
p. Mumps virus

True and False (T or F)

___ **1.** Chickenpox can be diagnosed by finding Guarnieri bodies as cytoplasmic inclusions in infected cells.
___ **2.** Reye's syndrome is a severe encephalomyelitis and severe liver involvement in adults following a viral infection treated with aspirin.
___ **3.** Measles is usually a mild childhood disease easily controlled by vaccination.
___ **4.** Koplik's spots are small bluish-yellow spots that occur in the mouth on the buccal mucosa of rubella patients.
___ **5.** Infection of a fetus with rubella during the first trimester can cause severe birth defects.
___ **6.** Patients most susceptible to *P. aeruginosa* infections are hospitalized burn patients.
___ **7.** Trachoma is the leading cause of blindness worldwide and is especially prevalent where poor hygienic practices exist.
___ **8.** Lymphogranuloma venereum (LGV) is a syndrome with initial symptoms of conjunctivitis.
___ **9.** The causative agent of the most common form of bacterial pneumonia is *S. pneumoniae.*
___ **10.** The majority of respiratory tract infections are not bacterial but viral.

___ **11.** A vaccine against colds is possible because of the immunological similarity of the group of viruses that causes colds.

___ **12.** *Corynebacterium diphtheriae* excretes a powerful endotoxin that results in the formation of a pseudomembrane that can block respiratory passages.

___ **13.** Control of diphtheria is based entirely on the mass immunization of children.

___ **14.** Complications following some streptococcal infections include rheumatic fever and acute glomerulonephritis.

___ **15.** A positive tuberculin skin test is interpreted as indicating an active infection.

___ **16.** Typhoid fever is the most serious type of salmonellosis.

___ **17.** Urinary tract infections are almost always caused by strict pathogens.

___ **18.** Catheterization is not performed routinely for the collection of urine samples for microbiological examinations.

___ **19.** *Trichomonas vaginalis* appears to be spread as a sexually transmitted disease.

___ **20.** AIDS is caused by a virus that specifically destroys certain lymphocytes.

___ **21.** Mumps virus infects the sublingual salivary glands.

___ **22.** Rocky Mountain spotted fever (RMSF) is caused by a rickettsial organism, transmitted by body lice.

___ **23.** Pneumonic plague is almost always fatal and is very contagious.

___ **24.** An infection or inflammation of the membranes enclosing the brain and spinal cord is called meningitis.

___ **25.** Antibiotics are useful in the treatment of rabies.

Multiple Choice

1. Reye's syndrome is sometimes associated with
- **a.** influenza complications
- **b.** chickenpox
- **c.** using aspirin to treat fever during influenza
- **d.** all of the above

2. The eye is generally protected from damage by
- **a.** sebum lubricating the eyeball
- **b.** tears washing the eye
- **c.** mucous secretions that moisten the eye
- **d.** all of the above

3. *Chlamydia trachomatis* causes
- **a.** skin infections
- **b.** secondary infections of burns
- **c.** cold sores
- **d.** eye disease

4. *Bacteroides* is of concern in dental disease because it
- **a.** causes gum infections and loosens teeth
- **b.** is anaerobic
- **c.** survives in the gingival sulcus
- **d.** produces endotoxins
- **e.** all of the above

5. Saliva functions in oral health by
 a. rinsing teeth, buffering acids
 b. providing IgA to destroy oral flora
 c. providing proper anaerobic conditions
 d. all of the above

6. Otitis media is frequently caused by
 a. bacteria of the indigenous throat microflora
 b. *Streptococcus pneumoniae*
 c. *Haemophilus influenzae*
 d. all of the above

7. The most common pathogens of the upper respiratory tract are
 a. some bacterial species
 b. a variety of virus groups
 c. protozoa
 d. opportunistic fungi
 e. indigenous microflora

8. Serum hepatitis is caused by
 a. HAV
 b. HBV
 c. HCV
 d. rotavirus
 e. echovirus

9. Infections of the bladder are referred to as
 a. pyelonephritis
 b. glomerulonephritis
 c. cystitis
 d. bladder wall infection (BWI)

10. MMR refers to
 a. mortality and morbidity report
 b. a viral disease
 c. mumps-measles-rhinovirus
 d. mumps-measles-rubella vaccine

11. Enterotoxin of *Staphylococcus aureus*
 a. affects the nervous system reflexes
 b. produces gastrointestinal tract diarrhea
 c. induces bouts of repeated vomiting
 d. all of the above

12. Rabies immune globulin is used in which circumstance?
 a. to immunize pets
 b. to immunize humans
 c. treatment for rabies exposure
 d. all of the above

13. Dermatomycosis refers to
 a. a viral eye disease
 b. a bacterial skin infection
 c. a fungal skin infection
 d. a protozoan

14. Rhinitis refers to
 a. infections of the parotid salivary glands
 b. inflamed mucous membranes in the nose
 c. inflamed and infected tonsils
 d. inflamed and infected membranes of the throat
 e. none of the above

15. The effects of botulism are
 a. double vision, dizziness, respiratory paralysis
 b. diarrhea, fluid-electrolyte loss
 c. diarrhea, bloody stools, production of mucus
 d. liver cancer
 e. abortion in sheep

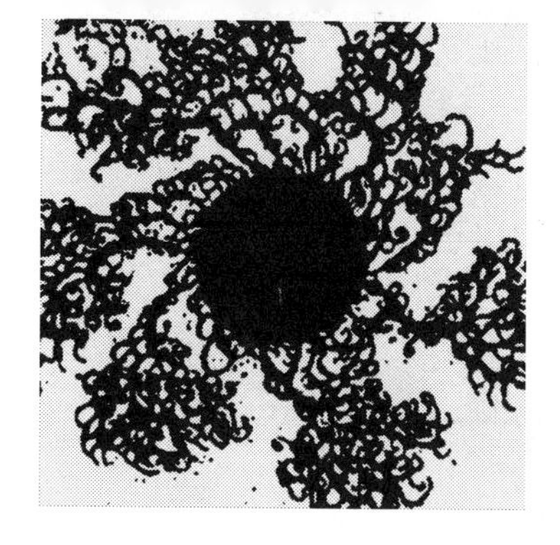

Appendices

APPENDIX A

CLASSIFICATION OF BACTERIA ACCORDING TO *BERGEY'S MANUAL OF SYSTEMATIC BACTERIOLOGY*

Kingdom Procaryotae

Divided into four divisions:

DIVISION I. GRACILICUTES
procaryotes with thin cell walls, implying a gram-negative type of cell wall.

DIVISION II. FIRMICUTES
procaryotes with thick and strong cell walls, indicating a gram-positive type of cell wall.

DIVISION III. TENERICUTES
procaryotes of a pliable and soft nature, indicating the lack of a rigid cell wall.

DIVISION IV. MENDOSICUTES
procaryotes with faulty cell walls, suggesting the lack of conventional peptidoglycan.

Volume 1

Section 1

THE SPIROCHETES

Order I. *Spirochaetales*
- Family I. *Spirochaetaceae*
  - Genus I. *Spirochaeta*
  - Genus II. *Cristispira*
  - Genus III. *Treponema*
  - Genus IV. *Borrelia*

Family II. *Leptospiraceae*
Genus I. *Leptospira*
Other organisms: Hindgut spirochetes of termites and *Cryptocercus punctulatus* (wood-eating cockroach).

Section 2

AEROBIC / MICROAEROPHILIC, MOTILE, HELICAL / VIBRIOID GRAM-NEGATIVE BACTERIA

Genus I. *Aquaspirillum*
Genus II. *Spirillum*
Genus III. *Azospirillum*
Genus IV. *Oceanospirillum*
Genus V. *Campylobacter*
Genus VI. *Bdellovibrio*
Genus VII. *Vampirovibrio*

Section 3

NONMOTILE (OR RARELY MOTILE), GRAM-NEGATIVE CURVED BACTERIA

Family I. *Spirosomaceae*
Genus I. *Spirosoma*
Genus II. *Runella*
Genus III. *Flectobacillus*
Other genera:
Genus *Microcyclus*
Genus *Meniscus*
Genus *Brachyarcus*
Genus *Pelosigma*

Section 4

GRAM-NEGATIVE AEROBIC RODS AND COCCI

Family I. *Pseudomonadaceae*
Genus I. *Pseudomonas*
Genus II. *Xanthomonas*
Genus III. *Frateuria*
Genus IV. *Zoogloea*
Family II. *Azotobacteraceae*
Genus I. *Azotobacter*
Genus II. *Azomonas*
Family III. *Rhizobiaceae*
Genus I. *Rhizobium*
Genus II. *Bradyrhizobium*
Genus III. *Agrobacterium*
Genus IV. *Phyllobacterium*
Family IV. *Methylococcaceae*
Genus I. *Methylococcus*
Genus II. *Methylomonas*
Family V. *Halobacteriaceae*
Genus I. *Halobacterium**
Genus II. *Halococcus**
Family VI. *Acetobacteraceae*
Genus I. *Acetobacter*
Genus II. *Gluconobacter*
Family VII. *Legionellaceae*
Genus I. *Legionella*
Family VIII. *Neisseriaceae*
Genus I. *Neisseria*
Genus II. *Moraxella*
Genus III. *Acinetobacter*
Genus IV. *Kingella*
Other genera:
Genus *Beijerinckia*
Genus *Derxia*
Genus *Xanthobacter*
Genus *Thermus*
Genus *Thermomicrobium*
Genus *Halomonas*
Genus *Alteromonas*
Genus *Flavobacterium*
Genus *Alcaligenes*
Genus *Serpens*
Genus *Janthinobacterium*
Genus *Brucella*
Genus *Bordetella*
Genus *Francisella*
Genus *Paracoccus*
Genus *Lampropedia*

*Genera that have been placed in more than one volume: *Gardnerella, Lachnospira,* and *Butyrivibrio* have thin, gram-positive walls but stain as gram-negatives. *Halobacterium, Halococcus,* and *Thermoplasma* are *Archaeobacteria* that are also included among the GRAM-NEGATIVE RODS AND COCCI or THE MYCOPLASMAS.

Section 5

FACULTATIVELY ANAEROBIC GRAM-NEGATIVE RODS

Family I. *Enterobacteriaceae*
- Genus I. *Escherichia*
- Genus II. *Shigella*
- Genus III. *Salmonella*
- Genus IV. *Citrobacter*
- Genus V. *Klebsiella*
- Genus VI. *Enterobacter*
- Genus VII. *Erwinia*
- Genus VIII. *Serratia*
- Genus IX. *Hafnia*
- Genus X. *Edwardsiella*
- Genus XI. *Proteus*
- Genus XII. *Providencia*
- Genus XIII. *Morganella*
- Genus XIX. *Yersinia*

Other genera of the family *Enterobacteriaceae:*
- Genus *Obesumbacterium*
- Genus *Xenorhabdus*
- Genus *Kluyvera*
- Genus *Rahnella*
- Genus *Cedecea*
- Genus *Tatumella*

Family II. *Vibrionaceae*
- Genus I. *Vibrio*
- Genus II. *Photobacterium*
- Genus III. *Aeromonas*
- Genus IV. *Plesiomonas*

Family III. *Pasteurellaceae*
- Genus I. *Pasteurella*
- Genus II. *Haemophilus*
- Genus III. *Actinobacillus*

Other genera:
- Genus *Zymomonas*
- Genus *Chromobacterium*
- Genus *Cardiobacterium*
- Genus *Calymmatobacterium*
- Genus *Gardnerella**
- Genus *Eikenella*
- Genus *Streptobacillus*

Section 6

ANAEROBIC GRAM-NEGATIVE STRAIGHT, CURVED, AND HELICAL RODS

Family I. *Bacteroidaceae*
- Genus I. *Bacteroides*
- Genus II. *Fusobacterium*
- Genus III. *Leptotrichia*
- Genus IV. *Butyrivibrio**
- Genus V. *Succinimonas*
- Genus VI. *Succinivibrio*
- Genus VII. *Anaerobiospirillum*
- Genus VIII. *Wolinella*
- Genus IX. *Selenomonas*
- Genus X. *Anaerovibrio*
- Genus XI. *Pectinatus*
- Genus XII. *Acetivibrio*
- Genus XIII. *Lachnospira**

Section 7

DISSIMILATORY SULFATE- OR SULFUR-REDUCING BACTERIA

- Genus *Desulfuromonas*
- Genus *Desulfovibrio*
- Genus *Desulfomonas*
- Genus *Desulfococcus*
- Genus *Desulfobacter*
- Genus *Desulfobulbus*
- Genus *Desulfosarcina*

Section 8

ANAEROBIC GRAM-NEGATIVE COCCI

FAMILY I. *Veillonellaceae*
- Genus I. *Veillonella*
- Genus II. *Acidaminococcus*
- Genus III. *Megasphaera*

Section 9

THE RICKETTSIAS AND CHLAMYDIAS

Order I. *Rickettsiales*
- Family I. *Rickettsiaceae*
  - Tribe I. *Rickettsieae*
    - Genus I. *Rickettsia*
    - Genus II. *Rochalimaea*
    - Genus III. *Coxiella*

Tribe II. *Ehrlichieae*
Genus IV. *Ehrlichia*
Genus V. *Cowdria*
Genus VI. *Neorickettsia*
Tribe III. *Wolbachieae*
Genus VII. *Wolbachia*
Genus VIII. *Rickettsiella*
Family II. *Bartonellaceae*
Genus I. *Bartonella*
Genus II. *Grahamella*
Family III. *Anaplasmataceae*
Genus I. *Anaplasma*
Genus II. *Aegyptianella*
Genus III. *Haemobartonella*
Genus IV. *Eperythrozoon*
Order II. *Chlamydiales*
Family I. *Chlamydiaceae*
Genus I. *Chlamydia*

Section 10

THE MYCOPLASMAS

Division Tenericutes
Class I. Mollicutes
Order I. *Mycoplasmatales*
Family I. *Mycoplasmataceae*
Genus I. *Mycoplasma*
Genus II. *Ureaplasma*
Family II. *Acholeplasmataceae*
Genus I. *Acholeplasma*
Family III. *Spiroplasmataceae*
Genus I. *Spiroplasma*
Other genera:
Genus *Anaeroplasma*
Genus *Thermoplasma**
Mycoplasmalike organisms of plants and invertebrates

Section 11

ENDOSYMBIONTS

Endosymbionts of protozoa
Genus I. *Holospora*
Genus II. *Caedibacter*
Genus III. *Pseudocaedibacter*
Genus IV. *Lyticum*
Genus V. *Tectibacter*
Endosymbionts of insects
Genus I. *Blattabacterium*
Endosymbionts of fungi and invertebrates other than arthropods

Volume 2

Section 12

GRAM-POSITIVE COCCI

Family I. *Micrococcaceae*
Genus I. *Micrococcus*
Genus II. *Stomatococcus*
Genus III. *Planococcus*
Genus IV. *Staphylococcus*
Family II. *Deinococcaceae*
Genus I. *Deinococcus*
Other organisms: Pyogenic hemolytic streptococci, Oral streptococci, Lactic acid streptococci, enterococci, and anaerobic streptococci
Genus *Leuconostoc*
Genus *Pediococcus*
Genus *Aerococcus*
Genus *Gemella*
Genus *Peptococcus*
Genus *Peptostreptococcus*
Genus *Ruminococcus*
Genus *Coprococcus*
Genus *Sarcina*

Section 13

ENDOSPORE-FORMING GRAM-POSITIVE RODS AND COCCI

Genus *Bacillus*
Genus *Sporolactobacillus*
Genus *Clostridium*
Genus *Desulfotomaculum*
Genus *Sporosarcina*
Genus *Oscillospira*

Section 14

REGULAR, NONSPORING, GRAM-POSITIVE RODS

Genus *Lactobacillus*

Genus *Listeria*
Genus *Erysipelothrix*
Genus *Brochothrix*
Genus *Renibacterium*
Genus *Kurthia*
Genus *Caryophanon*

Section 15

IRREGULAR, NONSPORING, GRAM-POSITIVE RODS

Genus *Corynebacterium*
Plant corynebacteria
Genus *Gardnerella**
Genus *Arcanobacterium*
Genus *Arthrobacter*
Genus *Brevibacterium*
Genus *Curtobacterium*
Genus *Caseobacter*
Genus *Microbacterium*
Genus *Aureobacterium*
Genus *Cellulomonas*
Genus *Agromyces*
Genus *Arachnia*
Genus *Rothia*
Genus *Propionibacterium*
Genus *Eubacterium*
Genus *Acetobacterium*
Genus *Lachnospira**
Genus *Butyrivibrio**
Genus *Thermoanaerobacter*
Genus *Actinomyces*
Genus *Bifidobacterium*

Section 16

MYCOBACTERIA

Family *Mycobacteriaceae*
Genus *Mycobacterium*

Section 17

NOCARDIOFORMS

Genus *Nocardia*
Genus *Rhodococcus*
Genus *Nocardioides*
Genus *Pseudonocardia*
Genus *Oerskovia*
Genus *Saccharopolyspora*
Genus *Micropolyspora*
Genus *Promicromonospora*
Genus *Intrasporangium*

Volume 3

Section 18

GLIDING, NONFRUITING BACTERIA

Order I. *Cytophagales*
Family I. *Cytophagaceae*
Genus I. *Cytophaga*
Genus II. *Sporocytophaga*
Genus III. *Capnocytophaga*
Genus IV. *Flexithrix*
Genus V. *Flexibacter*
Genus VI. *Microscilla*
Genus VII. *Saprospira*
Genus VIII. *Herpetosiphon*
Order II. *Lysobacterales*
Family I. *Lysobacteraceae*
Genus *Lysobacter*
Order III. *Beggiatoales*
Family I. *Beggiatoaceae*
Genus I. *Beggiatoa*
Genus II. *Thioploca*
Genus III. *Thiospirillopsis*
Genus IV. *Thiothrix*
Genus V. *Achromatium*
Family II. *Simonsiellaceae*
Genus I. *Simonsiella*
Genus II. *Alysiella*
Family III. *Leucotrichaceae*
Genus *Leucothrix*
Families and genera *incertae sedis* (status questionable):
Genus *Toxothrix*
Genus *Vitreoscilla*
Genus *Chitinophagen*
Genus *Desulfonema*
Family IV. *Pelonemataceae*
Genus *Pelonema*
Genus *Achroonema*

Genus *Peloploca*
Genus *Desmanthus*

Section 19

ANOXYGENIC PHOTOTROPHIC BACTERIA

Purple bacteria

Family I. *Chromatiaceae*
- Genus I. *Chromatium*
- Genus II. *Thiocystis*
- Genus III. *Thiospirillum*
- Genus IV. *Thiocapsa*
- Genus V. *Amoebobacter*
- Genus VI. *Lamprobacter*
- Genus VII. *Lamprocystis*
- Genus VIII. *Thiodictyon*
- Genus IX. *Thiopedia*

Family II. *Ectothiorhodospiraceae*
- Genus *Ectothiorhodospira*

Purple nonsulfur bacteria
- Genus *Rhodospirillum*
- Genus *Rhodopseudomonas*
- Genus *Rhodobacter*
- Genus *Rhodomicrobium*
- Genus *Rhodopila*
- Genus *Rhodocyclus*

Green Bacteria

Green sulfur bacteria
- Genus *Chlorobium*
- Genus *Prosthecochloris*
- Genus *Anacalochloris*
- Genus *Pelodictyon*
- Genus *Chloroherpeton*

Multicellular filamentous green bacteria
- Genus *Chloroflexus*
- Genus *Heliothrix*
- Genus *Oscillochloris*

Genera *incertae sedis*
- Genus *Heliobacterium*
- Genus *Erythrobacter*

Section 20

BUDDING AND/OR APPENDAGED BACTERIA

PROSTHECATE BACTERIA

Budding bacteria
- Genus *Hyphomicrobium*
- Genus *Hyphomonas*
- Genus *Pedomicrobium*
- Genus *"Filomicrobium"*
- Genus *"Dicotomicrobium"*
- Genus *"Tetramicrobium"*
- Genus *Stella*
- Genus *Ancalomicrobium*
- Genus *Prosthecomicrobium*

Nonbudding bacteria
- Genus *Caulobacter*
- Genus *Asticcacaulis*
- Genus *Prosthecobacter*
- Genus *Thiodendron*

Nonprosthecate Bacteria

Budding bacteria
- Genus *Planctomyces*
- Genus *Pasteuria*
- Genus *Blastobacter*
- Genus *Angulomicrobium*
- Genus *Gemmiger*
- Genus *Ensifer*
- Genus *Isophaera*

Nonbudding stalked bacteria
- Genus *Gallionella*
- Genus *Nevskia*

Morphologically unusual budding bacteria involved in iron and manganese deposition
- Genus *Seliberia*
- Genus *Metallogenium*
- Genus *Caulococcus*
- Genus *Kuznezovia*

Others: Spinate bacteria

Section 21

ARCHAEOBACTERIA METHANOGENIC BACTERIA
- Genus *Methanobacterium*
- Genus *Methanobrevibacter*
- Genus *Methanococcus*
- Genus *Methanomicrobium*
- Genus *Methanospirillum*

Genus *Methanosarcina*
Genus *Methanococcoides*
Genus *Methanothermus*
Genus *Methanolobus*
Genus *Methanoplanus*
Genus *Methanogenium*
Genus *Methanothrix*

Extreme Halophilic Bacteria

Genus *Halobacterium**
Genus *Halococcus**

Extreme Thermophilic Bacteria

Genus *Thermoplasma**
Genus *Sulfolobus*
Genus *Thermoproteus*
Genus *Thermofilum*
Genus *Thermococcus*
Genus *Desulfurococcus*
Genus *Thermodiscus*
Genus *Pyrodictium*

Section 22

SHEATHED BACTERIA

Genus *Sphaerotilus*
Genus *Leptothrix*
Genus *Haliscominobacter*
Genus *Lieskeella*
Genus *Phragmidiothrix*
Genus *Crenothrix*
Genus *Clonothrix*

Section 23

GLIDING, FRUITING BACTERIA

Order I. *Myxobacterales*
Family I. *Myxococcaceae*
Genus *Myxococcus*
Family II. *Archangiaceae*
Genus *Archangium*
Family III. *Cystobacteraceae*
Genus I. *Cystobacter*
Genus II. *Melittangium*
Genus III. *Stigmatella*
Family IV. *Polyangiaceae*
Genus I. *Polyangium*
Genus II. *Nannocystis*
Genus III. *Chondromyces*
Genus *incerta sedis*
Genus *Angiococcus*

Section 24

CHEMOLITHOTROPHIC BACTERIA NITRIFIERS

Family I. *Nitrobacteraceae*
Genus I. *Nitrobacter*
Genus II. *Nitrospina*
Genus III. *Nitrococcus*
Genus IV. *Nitrosomonas*
Genus V. *Nitrospira*
Genus VI. *Nitrosococcus*
Genus VII. *Nitrosolobus*

Sulfur Oxidizers

Genus *Thiobacillus*
Genus *Thiomicrospira*
Genus *Thiobacterium*
Genus *Thiospira*
Genus *Macromonas*

Obligate Hydrogen Oxidizers

Genus *Hydrogenbacter*

Metal Oxidizers and Depositors

Family I. *Siderocapsaceae*
Genus I. *Siderocapsa*
Genus II. *Naumaniella*
Genus III. *Ochrobium*
Genus IV. *Siderococcus*

Other Magnetotactic Bacteria

Section 25

CYANOBACTERIA

OTHERS

Order I. *Prochlorales*
Family I. *Prochloraceae*
Genus *Prochloron*

Volume 4

Section 26

Actinomycetes *that Divide in More Than One Plane*

Genus *Geodermatophilus*
Genus *Dermatophilus*
Genus *Frankia*
Genus *Tonsilophilus*

Section 27

SPORANGIATE *Actinomycetes*

Genus *Actinoplanes* (including *Amorphosporangium*)
Genus *Streptosporangium*
Genus *Ampullariella*
Genus *Spirillospora*
Genus *Pilimelia*
Genus *Dactylosporangium*
Genus *Planomonospora*
Genus *Planobispora*

Section 28

Streptomycetes AND THEIR ALLIES

Genus *Streptomyces*
Genus *Streptovertillium*
Genus *Actinopycnidium*
Genus *Actinosporangium*
Genus *Chainia*
Genus *Elytrosporangium*
Genus *Microellobosporia*
(The last five genera may be merged with *Streptomyces*)

Section 29

OTHER CONIDIATE GENERA

Genus *Actinopolyspora*
Genus *Actinosynnema*
Genus *Kineospora*
Genus *Kitasatosporia*
Genus *Microbispora*
Genus *Micromonospora*
Genus *Microtetrospora*
Genus *Saccharomonospora*
Genus *Sporichthya*
Genus *Streptoalloteichus*
Genus *Thermomonospora*
Genus *Actinomadura*
Genus *Nocardiopsis*
Genus *Excellospora*
Genus *Thermoactinomyces*

APPENDIX B

COMPENDIUM OF IMPORTANT BACTERIAL PATHOGENS OF HUMANS

Bordetella pertussis (Bor-duh-tel′-uh per-tus′-sis). A fastidious, gram-negative coccobacillus; the etiologic agent of whooping cough, which is also called pertussis

Campylobacter jejuni (Kam′-pih-low-bak′-ter juh-ju′-nee). A curved, gram-negative bacillus, having a characteristic corkscrew-like motility; a common cause of gastroenteritis with malaise, myalgia, arthralgia, headache, and cramping abdominal pain

Clostridium botulinum (Klos-trid′-ee-um bot-yu-ly′-num). An anaerobic, spore-forming, gram-positive bacillus; common in soil; produces a neurotoxin called botulinal toxin, which causes botulism, a very serious and sometimes fatal type of food poisoning

Clostridium perfringens (Klos-trid'-ee-um purr-frin-'jens). An anaerobic, spore-forming, gram-positive bacillus; common in feces and soil; the most common cause of gas gangrene (myonecrosis); produces an enterotoxin that causes relatively mild food poisoning

Clostridium tetani (Klos-trid'-ee-um tet'-an-eye). An anaerobic, spore-forming, gram-positive bacillus; common in soil; produces a neurotoxin called tetanospasmin, which causes tetanus

Corynebacterium diphtheriae (Kuh'-ry-nee-bak-teer'-ee-um dif-thee'-ree-ee). A pleomorphic, gram-positive bacillus; virulent strains are lysogenic; they cause diphtheria and produce a powerful exotoxin that causes tissue degeneration

Escherichia coli (Esh-er-ick'-ee-uh koh'-ly). A member of the family *Enterobacteriaceae*; a gram-negative bacillus; a facultative anaerobe; a very common member of the indigenous microflora of the colon; an opportunistic pathogen; the most common cause of urinary tract and nosocomial infections; some serotypes (the enterovirulent *E. coli*) are always pathogens

Haemophilus influenzae (He-mof'-uh-lus in-flu-en'-zee). A fastidious, gram-negative bacillus; a facultative anaerobe; encapsulated; found in low numbers as indigenous microflora of the upper respiratory tract; an opportunistic pathogen; one of the three most common causes of bacterial meningitis; causes about one-third of ear infections; causes respiratory infections, but is *not* the cause of influenza (caused by influenza viruses); some strains are ampicillin-resistant

Klebsiella pneumoniae (Kleb-see-el'-uh new-moh'-nee-ee). A member of the family *Enterobacteriaceae*; a gram-negative bacillus; a facultative anaerobe; a common member of the indigenous microflora of the colon; an opportunistic pathogen; a fairly common cause of pneumonia

Mycobacterium tuberculosis (My'-koh-bak-teer'-ee-um tu-ber'-kyu-loh'-sis). An acid-fast, gram-positive bacillus; causes tuberculosis; many strains are multiply drug-resistant

Neisseria gonorrhoeae (Ny-see'-ree-uh gon-or-ree'-ee). Also known as gonococci; a fastidious, gram-negative diplococcus; microaerophilic and capnophilic; always a pathogen; causes gonorrhea; many strains are penicillin-resistant

Neisseria meningitidis (Ny-see'-ree-uh men-in-jih'-tid-is). Also known as meningococci; an aerobic, gram-negative diplococcus; found as indigenous microflora of the upper respiratory tract of some people (carriers); one of the three most common causes of bacterial meningitis; also causes respiratory infections

***Proteus* species** (Pro'-tee-us). Members of the family *Enterobacteriaceae*; gram-negative bacilli; facultative anaerobes; common members of the indigenous microflora of the colon; opportunistic pathogens; a fairly common cause of cystitis

Pseudomonas aeruginosa (Su-doh-moh'-nas air-uj-in-oh'-suh). An aerobic, gram-negative bacillus; produces a characteristic blue-green pigment (pyocyanin); has a characteristic fruity odor; causes burn, wound, ear, urinary tract, and respiratory infections; one of the major causes of nosocomial infections; most strains are multiply drug-resistant and resistant to some disinfectants

***Salmonella* species** (Sal'-moh-nel'-uh). Members of the family *Enterobacteriaceae;* gram-negative bacilli; facultative anaerobes; a fairly common cause of food poisoning, especially cases caused by contaminated poultry; *Salmonella typhi* is the etiologic agent of typhoid fever

***Shigella* species** (She-gel'-uh). Members of the family *Enterobacteriaceae;* gram-negative bacilli; facultative anaerobes; a major cause of gastroenteritis and childhood mortality in the developing nations of the world

Staphylococcus aureus (Staf'-ih-low-kok'-us aw'-ree-us). Frequently referred to as "staph", as in staph infection; a gram-positive coccus in clusters; a facultative anaerobe; found in low numbers on skin; the nasal passages of some people (carriers) are colonized with *S. aureus;* an opportunistic pathogen; causes skin and wound infections; a very common cause of nosocomial infections; some strains produce a toxin that causes toxic shock syndrome; some strains (those that produce enterotoxin) cause food poisoning; some strains (the MRSA) are multiply drug-resistant

Streptococcus pneumoniae (Strep-toh-kok'-us new-moh'-nee-ee). Also known as pneumococci; a gram-positive diplococcus; a facultative anaerobe; found in low numbers as indigenous microflora of the upper respiratory tract; an opportunistic pathogen; the most common cause of bacterial pneumonia; one of the three most common causes of bacterial meningitis; causes about one-third of ear infections; some strains are penicillin-resistant

Streptococcus pyogenes (Strep-toh-kok'-us py-oj'-uh-nees). Also known as group A, beta-hemolytic streptococcus; a gram-positive coccus in chains; a facultative anaerobe; infrequently found in low numbers as indigenous microflora of the upper respiratory tract; an opportunistic pathogen; the cause of "strep" throat; causes skin and wound infections; some strains (those that produce erythrogenic toxin) are capable of causing scarlet fever; some strains are referred to as "flesh-eating bacteria"

Treponema pallidum (Trep-oh-nee'-muh pal'-luh-dum). A very thin, tightly coiled spirochete; the etiologic agent of syphilis

Vibrio cholerae (Vib'-ree-oh khol'-er-ee). An aerobic, curved (comma-shaped), gram-negative bacillus; halophilic; lives in salt water; causes cholera

Yersia pestis (Yer-sin'-ee-uh pes'-tis). A gram-negative bacillus; the etiologic agent of plague in humans, rodents, and other mammals; transmitted from rat-to-rat and rat-to-human by the rat flea

APPENDIX C

HELMINTHS

The word *helminth* means parasitic worm. Although worms are not microorganisms, a discussion of helminths is often included in microbiology courses and is always included in parasitology courses. Helminths infect humans, other animals, and plants, but we will limit our discussion to helminth infections of humans.

Helminths are multicellular, eucaryotic organisms in the Kingdom Animalia. The two major divisions of helminths are roundworms (*Nematoda* or *nematodes*) and flatworms (*Platyhelminthes*). The flatworms are further divided into tapeworms (*cestodes*) and flukes (*trematodes*). Table A – 1 lists some of the more common helminths that cause human infection.

Parasites are organisms that live *on* or *in* other living organisms (called *hosts*). Parasites that live on the outside of the host's body are called *ectoparasites,* whereas those that live inside are called *endoparasites.* Helminths are always endoparasites.

The typical helminth life cycle includes three major stages—the *egg,* the *larva,* and the *adult.* Adults produce eggs, from which larvae emerge, and the larvae mature into adult worms. The host that harbors the larval stage is called the *intermediate host,* whereas the host that harbors the adult worm is called the *definitive host.* Sometimes helminths have more than one intermediate host or more than one definitive host. The fish tapeworm, for example, has two intermediate hosts. Dogs, cats, or humans can serve as definitive hosts for the dog tapeworm. Table A – 2 contains information concerning intermediate and definitive hosts of various helminths.

Helminth infections are primarily acquired by ingesting the larval stage, although some larvae are injected into the body via the bite of infected insects, and others enter the body by penetrating skin. Table A – 3 explains how various helminth infections are acquired and how they are diagnosed in the clinical parasitology laboratory.

Many helminth diseases are endemic in the United States, although the United States does not have the variety or magnitude of helminth infections that occur in other parts of the world. Table A – 4 provides estimates of the total number of people, worldwide, who are infected with the most common helminths; Table A – 5 lists the most common helminth infections in the United States.

A 1990 World Health Organization listing of estimated annual deaths ranked six parasitic diseases, including three helminths, among the top 21 fatal diseases worldwide. Those parasitic diseases and the estimated number of annual deaths are shown in Table A – 6.

TABLE A – 1.
Helminths that Commonly Infect Humans

General Category	Scientific Name	Common Name	Disease
Intestinal nematodes	*Ascaris lumbricoides*	Large intestinal roundworm of humans	Ascariasis or *Ascaris* infection (adult worms in small intestine, although they can migrate to other parts of the body)
	Enterobius vermicularis	Pinworm	Pinworm infection or enterobiasis (adult worms in cecum)
	Necator americanus	New world hookworm	Hookworm infection (adult worms in small intestine)
	Strongyloides stercoralis	Threadworm	Strongyloidiasis (adult worms in small intestine)
	Trichuris trichiura	Whipworm	Trichuriasis or whipworm infection (adult worms in colon)
Blood and tissue nematodes	*Dracunculus medinensis*	Guinea worm	Dracunculiasis or guinea worm infection (adult worms in subcutaneous tissue)
	Trichinella spiralis	NA	Trichinosis (adult worms in lining of small intestine)
Filarial nematodes	*Brugia malayi* and *Wuchereria bancrofti*	NA	Filariasis (advanced stages are called elephantiasis) (adult worms in lymph nodes; prelarval stages called microfilariae are present in blood)
	Loa loa	Eye worm	Loiasis (adult worms in subcutaneous tissue; can migrate beneath conjunctiva of eye; microfilariae are present in blood)
	Onchocerca volvulus	Blinding worm	Onchocerciasis or "river blindness" (adult worms live in subcutaneous nodules; microfilariae in skin and eyes)
Intestinal cestodes	*Diphyllobothrium latum*	Fish tapeworm	Fish tapeworm infection or diphyllobothriasis (adult tapeworm in small intestine)

(*continued*)

TABLE A – 1. (Continued)

General Category	Scientific Name	Common Name	Disease
	Dipylidium caninum	Dog tapeworm	Dog tapeworm infection (adult tapeworm in small intestine)
	Hymenolepis diminuta	Rat tapeworm	Rat tapeworm infection (adult tapeworm in small intestine)
	Hymenolepis nana	Dwarf tapeworm	Dwarf tapeworm infection (adult tapeworm in small intestine
	Taenia saginata	Beef tapeworm	Beef tapeworm infection (adult tapeworm in small intestine)
	Taenia solium	Pork tapeworm	Pork tapeworm infection (adult tapeworm in small intestine) and/or cysticercosis (larvae in various tissues)
Tissue cestode	*Echinococcus granulosus*	NA	Hydatidosis or hydatid disease (larvae, called hydatid cysts, in organs or tissues)
Intestinal trematode	*Fasciolopsis buski*	Intestinal fluke	Fasciolopsiasis (adult worms in intestine)
Liver trematodes	*Clonorchis sinensis*	Chinese or oriental liver fluke	Clonorchiasis (adult worms in bile ducts)
	Fascioloa hepatica	Liver fluke	Fascioliasis (adult worms in bile ducts)
Lung trematode	*Paragonimus westermani*	Lung fluke	Paragonimiasis (adult worms in lungs)
Blood trematodes	*Schistosoma haematobium*	Blood fluke	Schistosomiasis (adult worms in blood vessels, especially vessels that surround the urinary bladder)
	Schistosoma japonicum	Blood fluke	Schistosomiasis (adult worms in blood vessels, especially vessels that surround the small intestine)
	Schistosoma mansoni	Blood fluke	Schistosomiasis (adult worms in blood vessels, especially vessels that surround the small intestine)

TABLE A – 2.
Intermediate and Definitive Hosts of Common Helminths

Helminth	Intermediate Host(s)	Definitive Host(s)
Ascaris lumbricoides	None	Human
Enterobius vermicularis	None	Human
Necator americanus	None	Human
Strongyloides stercoralis	None	Human
Trichuris trichiura	None	Human
Dracunculus medinensis	*Cyclops* (a fresh water crustacean)	Human
Trichinella spiralis	Pig, bear, walrus, human (a dead-end host), etc. (contain both the adult and larval stages)	Pig, bear, walrus, human, etc. (contain both the adult and larval stages)
Brugia malayi and *Wuchereria bancrofti*	Various species of mosquitoes	Human
Loa loa	*Chrysops* (mango fly)	Human
Onchocerca volvulus	*Simulium* (black fly)	Human
Diphyllobothrium latum	*Cyclops* (first intermediate host) and fresh water fish (second intermediate host)	Human
Dipylidium caninum	Flea	Dog, cat, human
Hymenolepis diminuta	Beetle	Rat, mouse, human
Hymenolepis nana	None	Human
Taenia saginata	Cow	Human
Taenia solium	Pig	Human
Echinococcus granulosus	Sheep, human (dead-end host)	Dog
Fasciolopsis buski	Fresh water snails	Human, dog, pig, rabbit
Clonorchis sinensis	Fresh water snails (first intermediate hosts), fresh water fish (second intermediate hosts)	Human, dog, cat, other fish-eating mammals
Fasciola hepatica	Fresh water snails	Human, cow, sheep
Paragonimus westermani	Fresh water snails (first intermediate hosts), crabs or crayfish (second intermediate hosts)	Human, dog, cat, carnivores
Schistosoma spp.	Fresh water snails	Human

TABLE A – 3.
How Helminth Infections are Acquired and Diagnosed

Helminth	How Infection is Acquired	How Infection is Diagnosed
Ascaris lumbricoides	Ingestion of eggs	Observation of eggs in stool specimens
Enterobius vermicularis	Ingestion of eggs	Observation of eggs in "Scotch tape preps"
Necator americanus	Penetration of skin by infective larvae	Observation of eggs in stool specimens
Strongyloides stercoralis	Penetration of skin by infective larvae	Observation of larvae in duodenal aspirates or stool specimens
Trichuris trichiura	Ingestion of eggs	Observation of eggs in stool specimens
Dracunculus medinensis	Ingestion of infected *Cyclops* in fresh water	Observation of adult worm beneath the skin or emerging from a blister (usually on the ankle or foot)
Trichinella spiralis	Ingestion of pork or bear meat containing larvae	Usually not diagnosed or an incidental finding at autopsy
Brugia malayi and *Wuchereria bancrofti*	Injection of infective larvae by mosquito	Observation of microfilariae in stained blood specimens
Loa loa	Injection of infective larvae by *Chrysops* (mango fly)	Observation of adult worm beneath the skin or in the conjunctiva of the eye; less often, by observing microfilariae in stained blood specimens
Onchocerca volvulus	Injection of infective larvae by *Simulium* (black fly)	Observing microfilariae in "skin snips"
Diphyllobothrium latum	Ingestion of fresh water fish containing second-stage larvae	Observation of worm segments (proglottids) or eggs in stool specimens
Dipylidium caninum	Ingestion of infected flea	Observation of proglottids or egg packets in stool specimens
Hymenolepis diminuta	Ingestion of infected beetle	Observation of eggs in stool specimens
Hymenolepis nana	Person-to-person (fecal-oral) transmission	Observation of eggs or proglottids (rarely) in stool specimens
Taenia saginata	Ingestion of infected beef	Observation of proglottids or eggs in stool specimens
Taenia solium	Ingestion of infected pork	Observation of proglottids or eggs in stool specimens; cysticercosis may be diagnosed by CT scans, MRI techniques, x-ray, or immunodiagnostic procedures

(*continued*)

TABLE A – 3. (Continued)

Helminth	How Infection is Acquired	How Infection is Diagnosed
Echinococcus granulosus	Ingestion of eggs	Observation of cysts by CT scans, MRI techniques, or x-ray; immunodiagnostic procedures
Fasciolopsis buski	Ingestion of raw or undercooked plants (water caltrops, water chestnuts, water bamboo) on which metacercariae are encysted	Observation of eggs in stool specimens
Clonorchis sinensis	Ingestion of infected freshwater fish	Observation of eggs in stool specimens
Fasciola hepatica	Ingestion of raw or undercooked aquatic vegetation (*e.g.*, watercress) on which metacercariae are encysted	Observation of eggs in stool specimens
Paragonimus westermani	Ingestion of infected crabs or crayfish	Observation of eggs in sputum or stool specimens
Schistosoma spp.	Penetration of skin by cercariae present in fresh water	Observation of eggs in urine (*S. haematobium*) or stool specimens (*S. japonicum* and *S. mansoni*)

TABLE A – 4.
Estimated Worldwide Prevalence for the Major Helminth Infections

Helminth Infection	Worldwide Prevalence
Ascariasis	1,000,000,000
Hookworm infection	900,000,000
Trichuriasis	750,000,000
Enterobiasis	400,000,000
Schistosomiasis	200,000,000
Filariasis	90,000,000
Strongyloidiasis	80,000,000
Taeniasis	70,000,000
Clonorchiasis/Opisthorchiasis	18,000,000
Fascioliasis	17,000,000
Trichinosis	11,000,000
Diphyllobothriasis	9,000,000
Paragonimiasis	6,000,000
Dracunculiasis	3,000,000

Reference: Hopkins DR: Homing in on helminths. Am J Trop Med Hyg 46:626–634, 1992.

TABLE A – 5.
The Most Common Helminth Infections in the United States.

Helminth	Percentage of Stools That Were Positive
Nematodes	
Hookworm	1.5
Trichuris trichiura	1.2
Ascaris lumbricoides	0.8
Strongyloides stercoralis	0.4
Enterobius vermicularis	0.4 (Also identified in 1094 of 9597 "Scotch tape preps;" the most common helminth infection in the U.S.)
Cestodes	
Hymenolepis nana	0.4
Taenia sp.	<0.1
Taenia saginata	<0.1
Hymenolepis diminuta	<0.1
Diphyllobothrium latum	<0.1
Dipylidium caninum	<0.1
Trematodes	
Clonorchis/Opisthorchis	0.6
Schistosoma mansoni	<0.1
Fasciola hepatica	<0.1
Paragonimus sp.	<0.1

Reference: Kappus KD, et al.: Intestinal parasitism in the United States: update on a continuing problem. Am J Trop Med Hyg 50:705–713, 1994

Note: Parasites were found in 20% of the 216,275 stool specimens examined by state diagnostic laboratories in the United States in 1987 (the last year such a survey was performed). Not all parasites listed are endemic in the United States; some infections were imported. In 1987, hookworms were the most frequently identified helminths in stool specimens. In all likelihood, most of these infections were acquired outside the United States. Thus, physicians should be aware of the possibility of hookworm infections and infections with other exotic helminths (*e.g., Clonorchis, Opisthorchis,* and *Schistosoma*) in the increasing numbers of immigrants and travelers from countries where these parasites are highly endemic.

TABLE A – 6.
Parasitic Diseases Ranked Among the Top 21 Fatal Diseases Worldwide by the World Health Organization (1990)

Parasitic Disease	Annual Deaths
Malaria	1–2 million
Schistosomiasis	200,000
Amebiasis	40,000–110,000
Hookworm infection	50,000–60,000
African trypanosomiasis	20,000
Ascariasis	20,000

PROTOZOA

TABLE A — 7
Protozoan Parasites that Infect Humans

Phylum/ Subphylum	Parasite	Disease	How Acquired	How Diagnosed
Sarcodina	*Entamoeba histolytica*	amebiasis; amebic dysentery; extra-intestinal amebic abscesses	usually, ingestion of cysts in contaminated food or water	observation of cysts and/or trophozoites (amebae) in fecal specimens
	Naegleria fowleri	primary amebic meningoencephalitis (PAM)	usually, diving into contaminated pond water (the "old swimming hole")	observation of trophozoites in CSF or cysts in autopsy tissue
Mastigophora	*Giardia lamblia*	giardiasis (a diarrheal disease)	usually ingestion of cysts in contaminated water or food	observation of cysts and/or trophozoites in fecal specimens
	Trichomonas vaginalis	trichomoniasis; causes about one-third of cases of vaginitis	direct contact; trichomoniasis is a sexually transmitted disease (STD)	observation of trophozoites in saline wet mounts of vaginal or urethral discharge material or prostatic secretions

(continued)

TABLE A — 7 (Continued)

Phylum/ Subphylum	Parasite	Disease	How Acquired	How Diagnosed
	Leishmania spp.	leishmaniasis	injection of the parasite when a *Phlebotomus* sand fly takes a blood meal	observation of parasite in aspirates or biopsy specimens
	subspecies of *Trypanosoma brucei*	African trypanosomiasis (African sleeping sickness)	injection of the parasite when a Tsetse fly takes a blood meal	observation of trypomastigotes in blood or CSF specimens or lymph node aspirates
	Trypanosoma cruzi	American trypanosomiasis (Chagas' disease)	parasites in the feces of a Reduviid bug get rubbed into bug bite wound	observation of trypomastigotes in blood or amastigotes in biopsy specimens
Ciliata (Ciliophora)	*Balantidium coli*	balantidiasis (a dysenteric disease)	ingestion of cysts in contaminated water	observation of cysts and/or trophozoites in fecal specimens
Sporozoa	*Cryptosporidium parvum*	cryptosporidiosis (a diarrheal disease)	ingestion of oocysts in contaminated water	observation of oocysts in fecal specimens
	Plasmodium spp.	malaria	injection of sporozoites when a female *Anopheles* mosquito takes a blood meal	observation of trophozoites, schizonts, and/or gametocytes in blood specimens
	Toxoplasma gondii	toxoplasmosis	ingestion of oocysts from cat feces or cysts in contaminated meat	immunodiagnostic procedures

SUGGESTED READING

Atlas RM: Microorganisms in our World. St. Louis, Mosby, 1995

Davis BE, et al: Microbiology, 4th ed. Philadelphia, JB Lippincott, 1990

Engelkirk PG: Introduction to Microbiology and the Clinical Microbiology Laboratory. Winston-Salem, Hunter Textbooks, 1995

Howard BJ, et al: Clinical and Pathogenic Microbiology, 2nd ed. St. Louis, Mosby, 1994

Koneman EW: Color Atlas and Textbook of Diagnostic Microbiology, 4th ed. Philadelphia, JB Lippincott, 1992

Koneman EW: Introduction to Diagnostic Microbiology. Philadelphia, JB Lippincott, 1994

Mahon CR, Manuselia G, Jr.: Textbook of Diagnostic Microbiology. Philadelphia, WB Saunders, 1995

McKane L, Kandel J: Microbiology: Essentials and Applications, 2nd ed. New York, McGraw-Hill, 1996.

Murray PR, et al. (eds): Manual of Clinical Microbiology, 6th ed. Washington, DC, American Society for Microbiology, 1995

Pelozar MJ, Jr., Chan ECS, Krieg NR: Microbiology Concepts and Applications. New York, McGraw-Hill, 1993

Talaro K, Talaro A: Foundations in Microbiology. Dubuque, Wm. C. Brown, 1993

Tortora GJ, Funke BR, Case CL: Microbiology: An Introduction, 5th ed. Redwood City, Benjamin/Cummings, 1995

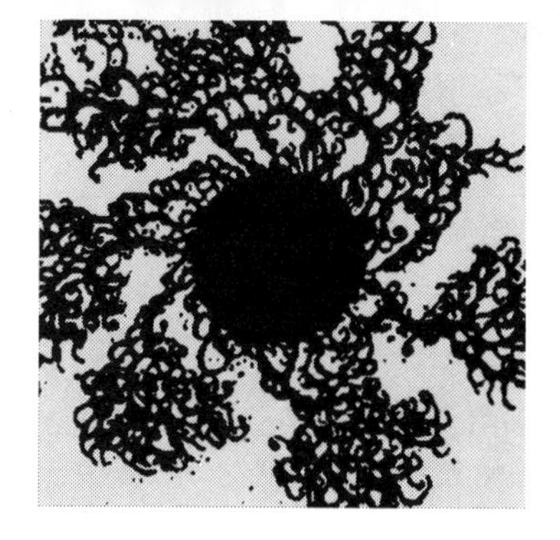

Glossary

Abiogenesis (ab′-ee-oh-jen-uh-sis). The theory that life may develop from nonliving matter

Acid (as′-id). A compound that yields a hydrogen ion in a polar solvent such as water; a solution that has a pH lower than 7.0 is said to be *acidic*

Acidophile (uh-sid′-oh-file). An organism that prefers acidic environments; such an organism is said to be *acidophilic*

Acquired immunity. Immunity or resistance acquired at some point in an individual's lifetime

Active acquired immunity. Immunity or resistance acquired as a result of the active production of antibodies

Active carrier. A person who has recovered from an infectious disease, but continues to harbor and transmit the pathogen

Acute disease. A disease having a sudden onset and short duration

Adenosine triphosphate (uh-den′-oh-seen try-fos′-fate). The major energy-carrying (energy-storing) molecule in a cell

Aerotolerant anaerobe (air-oh-tol′-er-ant an′-air-obe). An organism that can live in the presence of oxygen, but grows best in an anaerobic environment (one that contains no oxygen)

Agammaglobulinemia (ay-gam′-uh-glob′-yu-luh-nee′-me-uh). Absence of, or extremely low levels of, the gamma fraction of serum globulin; sometimes used to denote the absence of immunoglobulins

Agglutination (uh-glue-tuh-nay′-shun). The clumping of particles (including cells and latex beads) in solution

Agglutination tests. Laboratory procedures that result in agglutination, usually following reaction with antibodies and antigenic determinants on particles

AIDS. Acquired immunodeficiency syndrome

Algae (al′-gee), *sing.* alga. Eucaryotic, photosynthetic organisms that range in size from unicellular to multicellular; includes many seaweeds

Alkaliphile (al′-kuh-luh-file). An organism that prefers alkaline (basic) environments; such an organism is said to be *alkaliphilic*

Allergen (al′-ur-jin). An antigen to which one may become allergic

Allergy (al′-ur-jee). A disease resulting from acquired or induced sensitivity to an allergen

Ameba (uh-me′-bah), *pl.* amebae. A type of protozoan that moves by means of pseudopodia; in the phylum Sarcodina (a subphylum in some classification schemes)

Amino (uh-me′-no) **acids.** The basic units or building blocks of proteins

Amphitrichous (am-fit′-ri-kus) **bacteria.** Bacteria possessing one flagellum at each end (pole) of the cell

Anabolism (uh-nab′-oh-lizm). That part of metabolism concerned with the building of large compounds from smaller compounds; involves the creation of chemical bonds; requires energy; such chemical reactions are called anabolic or biosynthetic reactions

Anaerobe (an′-air-obe). An organism that does not require oxygen for survival; can exist in the absence of oxygen

Anamnestic (an-am-nes′-tick) **response.** An immune response following exposure to an antigen that the individual is already sensitized to; also known as a *secondary response* or *memory response*

Anaphylactic (an-uh-fuh-lak′-tick) **shock.** Shock following anaphylaxis; may lead to death

Anaphylaxis (an-uh-fuh-lak′-sis). An immediate, severe, sometimes fatal, systemic allergic reaction

Angstrom (ang′-strom). A unit of length, equivalent to 0.1 nanometer; roughly the diameter of an atom

Anion (an′-eye-on). An ion that carries a negative charge

Antagonism (an-tag′-ohn-izm). As used in this book, the killing, injury, or inhibition of one microorganism by products of another

Antibiosis (an′-tee-by-oh′-sis). An association of two organisms which is detrimental to one of them; an antagonistic relationship

Antibiotic (an′-tee-by-ot′-tik). A substance produced by a microorganism that inhibits or destroys other microorganisms

Antibody (an′-tee-bod-ee). A glycoprotein produced by lymphocytes in response to an antigen; often protective

Anticodon (an-tee-ko′-don). The trinucleotide sequence that is complementary to a codon; found on a transfer RNA molecule

Antigen (an′-tuh-jen). A substance, usually foreign, that stimulates the production of antibodies; an *anti*body *gen*erating substance; sometimes called an *immunogen*

Antigenic (an-tuh-jen′-ick) **determinant.** The smallest part of an antigen capable of stimulating the production of antibodies

Antimicrobial (an′-tee-my-kro′-be-ul) **agent.** A drug, disinfectant, or other substance that kills microorganisms or suppresses their growth

Antisepsis (an-tee-sep′-sis). Prevention of infection by inhibiting the growth of pathogens

Antiseptic (an-tee-sep′-tick). An agent or substance capable of effecting antisepsis; usually refers to a chemical disinfectant that is safe to use on living tissues

Antiseptic surgery. Surgery performed in a manner to prevent infection by inhibiting the growth of pathogens

Antiseptic technique. Procedures taken to effect antisepsis

Antiserum. Serum containing a particular antibody or antibodies; also called *immune serum.*

Antitoxin (an-tee-tok′-sin). An antibody produced in response to a toxin; often capable of neutralizing the toxin that stimulated its production

Arbovirus (are′-boh-vy′-rus). A virus that is transmitted by an arthropod; an arthropodborne virus

Archaebacteria (ark′-ee-back-tier′-ee-uh). Ancient bacteria, thought by some scientists to be the earliest types of bacteria

Arthropod (are′-throw-pod). An animal in the phylum Arthropoda; includes insects such as flies, mosquitoes, fleas, and lice, and arachnids such as mites and ticks

Asepsis (a-sep′-sis). A condition in which living pathogens are absent; a state of sterility

Aseptic (ay-sep′-tick) **techniques.** Measures taken to ensure that living pathogens are absent

Asymptomatic (ay′-simp-toh-mat′-ick) **disease.** A disease having no symptoms

Asymptomatic infection. The presence of a pathogen in or on the body, without any symptoms of disease

Atom (at′-um). The smallest particle of matter possessing the properties of an element; composed of protons, neutrons, and electrons

Atomic number. The number of negatively charged electrons in an uncharged atom or the number of protons in its nucleus; the atomic number of an element indicates the element's position in a periodic chart of the elements

Atomic weight. The weight or mass of an atom in relation to the mass of an atom of carbon-12 (^{12}C), which is set equal to 12.000

Atopic (ay-tope′-ick) **person.** Allergic person; one who suffers from allergies

Attenuated (uh-ten′-yu-ay-ted). An adjective meaning weakened, less pathogenic, used to describe certain microorganisms

Attenuation (uh-ten-yu-ay′-shun). The process by which microorganisms are attenuated

Autoclave (aw′-toe-klav). An apparatus used for sterilization by steam under pressure

Autogenous (aw-toj′-uh-nus) **vaccine.** A vaccine made from microorganisms or cells obtained from the person's own body

Autoimmune (aw-toh-im-myun′) **disease.** A disease in which the body produces antibodies directed against its own tissues

Autotroph (awe′-toe-trofe). An organism that uses carbon dioxide as its sole carbon source

Avirulent (ay-veer′-yu-lent). Not virulent

Axial (ak′-see-ul) **filament.** An organelle of motility possessed by spirochetes

B-cell. B lymphocyte; a type of leukocyte that plays an important role in the immune system

Bacillus (bah-sil′-us), *pl.* bacilli. A rod-shaped bacterium; there is a genus called *Bacillus,* made up of aerobic, gram-positive, spore-forming rods

Bacteremia (bak-ter-ee′-me-uh). The presence of bacteria in the bloodstream

Bacteria (back-tier′-ee-uh), *sing.* bacterium. Primitive, unicellular, procaryotic microorganisms

Bactericidal (bak-tear′-eh-sigh′-dull) **agent.** A chemical agent or drug that kills bacteria; a *bactericide*

Bacteriocins (bak-teer′-ee-oh-sinz). Proteins produced by certain bacteria (those possessing bacteriocinogenic plasmids) that can kill other bacteria

Bacteriologist (back′-tier-ee-ol′-oh-jist). One who studies or works with bacteria

Bacteriology (back′-tier-ee-ol′-oh-gee). That branch of science concerned with the study of bacteria

Bacteriophage (back-tier′-ee-oh-faj). A virus that infects a bacterium; also known simply as a phage

Bacteriostatic (bak-tear′-ee-oh-stat′-ick) **agent.** A chemical agent or drug that inhibits the growth of bacteria

Bacteriuria (bak-ter-ee′-yu′-ree-uh). The presence of bacteria in the urine

Barophile (bar′-oh-file). An organism that thrives under high environmental pressure; such an organism is said to be *barophilic*

Base. A compound that yields a hydroxyl ion in a polar solvent such as water; a solution that has a pH higher than 7.0 is said to be *basic*

Basophil (bay′-so-fil). A type of granulocyte found in blood; its granules contain acidic substances that attract basic dyes

Beta-lysin (β-lysin). An enzyme capable of destroying microorganisms in the bloodstream or within phagocytic cells

Binary (by′-nare-ee) **fission.** A method of reproduction whereby one cell divides to become two cells

Biochemistry (by-oh-kem′-is-tree). The chemistry of living organisms

Biogenesis (by-oh-gen′-uh-sis). The theory that life originates only from preexisting life and never from nonliving matter

Biological vector. A vector within which a pathogen either multiples or matures

Blocking antibodies. IgG antibodies that combine with allergens, but do not elicit allergic reactions; produced by the body in response to allergy shots

Botulinal (bot′-you-ly-nal) **toxin.** The neurotoxin produced by *Clostridium botulinum;* causes botulism

Candidiasis (kan-duh-dy′-uh-sis). Infection with, or disease caused by, a yeast in the genus *Candida*—usually *Candida albicans;* also known as moniliasis

Capnophile (cap′-no-file). An organism that grows best in the presence of increased concentrations of carbon dioxide

Capsid (kap′-syd). The external protein coat or covering of a virion

Capsomeres (kap′-so-meers). The protein units that make up the capsid of some virions

Capsule (kap′-sool). An organized layer of glycocalyx, firmly attached to the outer surface of the bacterial cell wall

Carbohydrates (kar-boh-high′-drates). Organic compounds containing carbon, hydrogen, and oxygen in a ratio of 1:2:1; also known as saccharides

Carrier (keh′-ree-er). An individual with an asymptomatic infection that can be transmitted to other susceptible individuals

Catabolism (kuh-tab′-oh-lizm). That part of metabolism concerned with breaking down large compounds into smaller compounds; involves the breaking of chemical bonds; energy is released; such chemical reactions are called catabolic or degradative reactions

Catalyst (kat′-uh-list). A substance (often an enzyme) that speeds up a chemical reaction, but is not itself consumed or permanently changed in the process

Catalyze (kat′-uh-lyz). To act as a catalyst; to speed up a reaction

Cation (kat′-eye-on). An ion that carries a positive charge

Cell (sell). The smallest unit of living structure capable of independent existence

Cell membrane (mem′-brain). The protoplasmic boundary of all cells; controls permeability and serves other important functions

Cell wall. The outermost layer of many cells (*e.g.*, algal, bacterial, fungal, and plant cells)

Cellulose (sell′-you-los). A polysaccharide found in the cell walls of algae and plants

Centrioles (sen′-tree-olz). Tubular structures thought to play a role in nuclear division (mitosis) in animal cells and the cells of lower plants

Cervicitis (sir-vuh-sigh′-tis). Inflammation of the neck of the uterus, the cervix uteri.

Chemoautotroph (keem′-oh-awe′-toe-trofe). An organism that uses chemicals as an energy source and carbon dioxide as a carbon source; a type of autotroph

Chemoheterotroph (keem′-oh-het′-er-oh-trofe). An organism that uses chemicals as a source of energy and organic molecules as a source of carbon; a type of heterotroph; sometimes referred to as a *chemoorganotroph*

Chemolithotroph (keem′-oh-lith′-oh-trofe). An organism that uses chemicals as a source of energy and inorganic molecules as a source of carbon; a type of lithotroph

Chemostat (keem′-oh-stat). A growth chamber designed to allow input of nutrients and output of cells on a controlled basis

Chemosynthesis (keem′-oh-syn′-thuh-sis). The process of obtaining energy and synthesizing organic compounds from simple inorganic reactions; carried out by some chemoautotrophic bacteria

Chemotaxis (keem-oh-tack′-sis). Movement of cells in response to a chemical (*e.g.*, the attraction of phagocytes to an area of injury)

Chemotherapeutic (keem′-oh-ther-uh-pyu′-tik) **agent.** Any chemical (drug) used to treat any disease

Chemotherapy (keem′-oh-ther′-uh-pee). The treatment of a disease (including an infectious disease) by means of chemical substances or drugs

Chemotroph (keem′-oh-trof). An organism that uses chemicals as a source of energy

Chitin (ky′-tin). A polysaccharide found in fungal cell walls, but not found in the cell walls of other microorganisms; also found in the exoskeleton of beetles and crabs

Chloroplast (klor′-oh-plast). A membrane-bound organelle found in the cytoplasm of algal and plant cells

Choleragin (kol′-er-uh-jen). The enterotoxin produced by *Vibrio cholerae*

Chromatin (kro′-muh-tin). The genetic material of the nucleus; consisting of DNA and associated proteins; during mitotic division, the chromatin condenses and is seen as chromosomes

Chromosome (kro′-mow-soam). A condensed form of chromatin; the location of genes; human diploid cells contain 46 chromosomes (23 pairs); bacterial cells usually contain only one chromosome, which divides to become two just prior to binary fission

Chronic disease. A disease of slow progress and long duration

Ciliates (sil′-ee-itz), *sing.* ciliate. Common name for ciliated protozoa

Ciliophora (sil′-ee-auf′-oh-rah). A phylum or protozoa that includes the ciliates

Cilia (sil′-ee-uh), *sing.* cilium. Thin hairlike organelles of motility

Cistron (sis′-tron). The smallest functional unit of heredity; a length of chromosomal DNA associated with a single biochemical function; a gene may consist of one or more cistrons; sometimes used synonymously with gene

Citric (sit′-rik) **acid cycle.** A series of chemical reactions that produces high-energy phosphate bonds; a biochemical pathway; also known as the tricarboxylic acid or Krebs cycle

Coagulase (ko-ag′-yu-lace). A bacterial enzyme that causes plasma to clot

Coccus (kok′-us), *pl.* cocci. A spherical bacterium

Codon (koh′-don). A sequence of three nucleotides in a strand of mRNA that provides the genetic information (code) for a certain amino acid to be incorporated into a growing protein chain

Coenzyme (koh′-en-zym). A substance that enhances or is necessary for the action of an enzyme; several vitamins are coenzymes; a type of cofactor

Cofactor (koh′-fak′-tor). An atom or molecule essential for the enzymatic action of certain proteins

Colicin. (kol′-uh-sin). A type of bacteriocin produced by *Escherichia coli* and other closely related bacteria

Collagen (kol′-luh-jen). The major protein in the white fibers of connective tissue, cartilage, and bone

Collagenase (kol′-uh-juh-nace). A bacterial enzyme that causes the breakdown of collagen

Commensalism (ko-men′-sul-izm). A symbiotic relationship in which one party derives benefit and the other party is unaffected in any way; most members of the indigenous microflora are commensals

Communicable (kuh-myun′-uh-kuh-bul) **disease.** A disease capable of being transmitted

Community-acquired infection. Any infection acquired outside of a hospital setting

Competence (kom′-puh-tense). As used in this book, the ability of a bacterial cell to take up free or naked DNA, which may lead to transformation; a bacterial cell capable of taking up free DNA is said to be competent

Complement (kom′-pluh-ment). A protein complex of 25 to 30 components (including C1 through C9) found in blood; involved in inflammation, chemotaxis, phagocytosis, and lysis of bacteria

Compound (kom′-pownd). A chemical substance formed by the covalent or electrostatic union of two or more elements

Conidium (ko-nid′-ee-um), *pl.* conidia. An asexual fungal spore

Conjugation (kon-ju-gay′-shun). As used in this book, the union of two bacterial cells, for the purpose of genetic transfer; *not* a reproductive process

Contagious (kon-tay′-jus). Capable of being transmitted from one person to another, as in contagious disease

Contagious (kon-tay′-jus) **disease.** A disease capable of person-to-person transmission; a type of communicable disease

Contamination (kon-tam-uh-nay′-shun). As used in this book, a condition indicating the presence of undesirable microorganisms (which would be referred to as *contaminants*)

Convalescent (kon-vuh-less′-ent) **carrier.** A person who no longer shows the signs of a particular infectious disease, but continues to harbor and transmit the pathogen during the convalescence period

Covalent (koh-vayl′-ent) **bond.** An atomic bond characterized by the sharing of two, four, or six electrons

Crenated (kree′-nay-ted). Wrinkled, shriveled (*e.g.*, the appearance of red blood cells placed into a hypertonic solution)

Crenation (kree-nay′-shun). The process of becoming, or state of being, crenated

Cyanobacteria (sigh-an-oh-bak-tier′-ee-uh). Photosynthetic bacteria

Cystitis (sis-ty′-tis). Inflammation or infection of the urinary bladder

Cytokines (sigh′-toe-kynz). Soluble protein mediators released by certain cells on contact with antigens; examples include *lymphokines* and *monokines*

Cytology (sigh-tol′-oh-gee). The study of cells

Cytoplasm (sigh-toe-plazm). That portion of a cell's protoplasm that lies outside the nucleus of the cell

Cytotoxic (sigh-toe-tok′-sik). Detrimental or destructive to cells

Cytotoxins (sigh′-tow-tok′-sinz). Toxic substances that inhibit or destroy cells

Death phase. That part of a bacterial growth curve during which no multiplication occurs and organisms are dying; the fourth and final phase in a bacterial growth curve

Dehydrogenation (dee-hy′-drah-jen-ay′-shun) **reactions.** Chemical reactions in which a pair of hydrogen atoms is removed from a compound, usually by the action of enzymes called dehydrogenases

Dehydrolysis (dee-hy-drol′-uh-sis). A chemical reaction in which two compounds are joined to form a larger compound and water is released in the process; also called dehydration synthesis

Deoxyribonucleic (dee-ox′-ee-ry′-bow-new-klay′-ick) **acid (DNA).** A macromolecule containing the genetic code in the form of genes

Dermatophytes (der-mah′-toh-fytes). Fungi that cause superficial mycoses of the skin, hair, and nails; the cause of tinea infections (ringworm)

Desiccation (des-uh-kay′-shun). The process of being desiccated (thoroughly dried)

Differential (dif-er-en′-shul) **media.** Culture media which enable microbiologists to readily differentiate one organism or group of organisms from another

Dipeptide (dy-pep′-tide). A protein consisting of two amino acids held together by a peptide bond

Diplococci (dip′-low-kok′-sigh). Cocci arranged in pairs

Disaccharide (die-sack′-uh-ride). A carbohydrate consisting of two monosaccharides; examples include sucrose (table sugar), lactose (milk sugar), and maltose (malt sugar)

Disinfect (dis-in-fekt′). To destroy pathogens in or on any substance or to inhibit their growth and vital activity

Disinfectant (dis-in-fek′-tent). A chemical agent used to destroy pathogens or inhibit their growth activity; usually refers to a chemical agent used on nonliving materials

Disinfection (dis-in-fek′-shun). The process of destroying pathogens and their toxins

DNA replication (rep-luh-kay′-shun). Production of two DNA molecules (called daughter molecules) from one parent molecule

DNA polymerase (poh-lim′-er-ace). The enzyme necessary for DNA replication

Ecology (ee-kol′-oh-jee). The branch of biology concerned with the total complex of interrelationships among living organisms; encompassing the relationships of organisms to each other, to the environment, and to the entire energy balance within a given ecosystem

Ecosystem (ee′-koh-sis-tem). An ecological system that includes all the organisms and the environment within which they occur naturally

Ectoparasite (ek′-toh-par′-uh-site). A parasite that lives on the external surface of its host

Edema (uh-dee′-muh). Swelling due to an accumulation of watery fluid in cells, tissues, or body cavities

Electrolyte (ee-lek′-troh-lite). A substance that decomposes into ions when placed into water

Electron (ee-lek′-tron) **microscope.** A type of microscope that uses electrons as a source of illumination

Electron transport system. A series of biochemical reactions by which energy is transferred stepwise; a major source of energy in some cells

Element (el′-uh-ment). A substance composed of atoms of only one kind

Encephalitis (en-sef-uh-ly′-tis). Inflammation or infection of the brain

Encephalomyelitis (en-sef-uh-low-my′-uh-ly′-tis). Inflammation or infection of the brain and spinal cord

Endemic (en-dem′-ick) **disease.** A disease that is always present in a particular community or region

Endoenzyme (en-doh-en′-zime). An enzyme produced by a cell that remains within the cell; an intracellular enzyme

Endoparasite (en-doh-par′-uh-site). A parasite that lives within the body of its host

Endoplasmic reticulum (end-oh-plaz′-mick re-tick′-you-lum). The network of cytoplasmic tubules and flattened sacs in a eucaryotic cell; rough or granular endoplasmic reticulum is endoplasmic reticulum (ER) with ribosomes attached

Endospore (en′-doh-spore). A resistant body formed within a bacterial cell

Endosymbiont (en′-doh-sym′-be-ont). The party in a symbiotic relationship that lives within the body of the other symbiont

Endotoxin (en-doh-tok′-sin). The lipid portion of the lipopolysaccharide found in the cell walls of gram-negative bacteria; intracellular toxin

Enriched media. Culture media which enable microbiologists to isolate fastidious organisms from samples or specimens and grow them in the laboratory

Enterotoxin (en-ter-oh-tok′-sin). A bacterial toxin specific for cells of the intestinal mucosa

Enzyme (en′-zime). A protein molecule that catalyzes (causes or speeds up) the occurrence of biochemical reactions; remains unchanged in the process; an organic catalyst

Eosinophil (ee-oh-sin′-oh-fil). A type of granulocyte found in blood; its granules contain basic substances that attract acidic dyes

Epidemic (ep-uh-dem′-ick) **disease.** A disease occurring in a higher than usual number of cases in a population during a given time interval

Epidemiology (ep-uh-dee-me-ol′-oh-jee). The study of relationships between the various factors that determine the frequency and distribution of diseases

Epididymitis (ep-uh-did-uh-my′-tis). Inflammation of the epididymis (a tubular structure within the testis)

Episome (ep′-eh-som). An extrachromosomal element (plasmid) that may either integrate into the host bacterium's chromosome or replicate and function stably when physically separated from the chromosome

Epitope (ep′-uh-tope). An antigenic determinant

Erythrocytes (ee-rith′-roh-sites). Red blood cells

Erythrogenic (ee-rith-roh-jen′-ick) **toxin.** A bacterial toxin that produces redness, usually in the form of a rash

Eucaryotic (you′-kar-ee-ah′-tick) **cell.** A cell containing a true nucleus; organisms having such cells are referred to as *eucaryotes*

Exfoliative (eks-foh′-lee-uh-tiv) **toxin.** A bacterial toxin that causes shedding or scaling

Exoenzyme (ek-soh-en′-zime). An enzyme produced by a cell that is released from the cell; an extracellular enzyme

Exotoxin (ek-soh-tok′-sin). A toxin that is released from the cell; an extracellular toxin

Exudate (eks′-yu-date). Any fluid (*e.g.*, pus) that exudes (oozes) from tissue, often as a result of injury, infection, or inflammation

Fastidious (fas-tid'-ee-us) **bacterium.** A bacterium that is difficult to isolate from samples or specimens and grow in the laboratory due to its complex nutritional requirements

Fatty acid. Any acid derived from fats by hydrolysis (see hydrolysis)

Fermentation (fer-men-tay'-shun). An anaerobic biochemical pathway in which substances are broken down and energy and reduced compounds are produced; oxygen does not participate in the process

Fibrinolysin (fy-brin-oh-ly'-sin). See *kinase*

Fibronectin (fi-bro-nek'-tin). A glycoprotein that acts as an adhesive and as a reticuloendothelial mediated host defense mechanism

Fimbriae (fim'-bree-ee), *sing.* fimbria. See pili

Flagellate (flaj'-eh-let). Common name for flagellated protozoa

Flagellin (flaj'-eh-lin). See Flagellum

Flagellum (fluh-jel'-um), *pl.* flagella. A whip-like organelle of motility; procaryotic and eucaryotic flagella differ in structure; procaryotic flagella are composed of a protein called *flagellin;* eucaryotic flagella are composed of nine double microtubules arranged around two single central microtubules

Fomite (foh'-mite). An inanimate object or substance capable of absorbing and transmitting a pathogen (*e.g.,* clothing, bed linens, towels, eating utensils)

Fungi (fun'-ji), *sing.* fungus. Eucaryotic, nonphotosynthetic microorganisms that are saprophytic or parasitic

Fungicidal (fun-juh-sigh'-dull) **agent.** A chemical agent or drug that kills fungi; a *fungicide* or *mycocide*

Gene (jeen). A functional unit of heredity that occupies a specific space (locus) on a chromosome; capable of directing the formation of an enzyme or other protein

Generalized infection. An infection that has spread throughout the body; also known as a *systemic infection*

Generation time. The time required for a cell to split into two cells; also called the *doubling time*

Genotype (jeen'-oh-type). The complete genetic constitution of an individual; *i.e.,* all of that individual's genes

Genus (jee'-nus), *pl.* genera. The first name in binomial nomenclature; contains closely related species

Germicidal (jer-muh-sigh'-dull) **agent.** A chemical agent or drug that kills microorganisms; a *germicide*

Gingivitis (jin-juh-vy'-tis). Inflammation or infection of the gingiva (gums)

Glucose (glue'-kohs). A biologically important six-carbon monosaccharide; a hexose; also called dextrose; the product of complete hydrolysis of polysaccharides such as cellulose, starch, and glycogen

Glycocalyx (gly-ko-kay′-licks). Extracellular material that may or may not be firmly attached to the outer surface of the cell wall; capsules and slime layers are examples

Glycogen (gly′-koh-jen). A polysaccharide stored by animal cells as a food reserve; composed of many glucose molecules

Glycolysis (gly-kol′-eh-sis). The anaerobic, energy-producing breakdown of glucose via a series of chemical reactions; a biochemical pathway; also called anaerobic glycolysis

Golgi (goal′-jee) **complex.** A membranous system located within the cytoplasm of a eucaryotic cell; associated with the transport and packaging of secretory proteins; also known as Golgi apparatus or Golgi body

Gram stain. A differential staining procedure named for its developer, Hans Christian Gram, a Danish bacteriologist; differentiates bacteria into those that stain purple (gram-positive) and those that stain red (gram-negative)

Granulocyte (gran′-yu-loh-site). A granular leukocyte; neutrophils, eosinophils, and basophils are examples

Growth curve. As used in this book, a graphic representation of the change in size of a bacterial population over a period of time; includes a lag phase, a log phase, a stationary phase, and a death phase

Haloduric (hail-oh-dur′-ick). Capable of surviving in a salty environment

Halophile (hail′-oh-file). An organism whose growth is enhanced by a high salt concentration; such an organism is said to be *halophilic*

Hapten (hap′-ten). A small, nonantigenic molecule that becomes antigenic when combined with a large molecule

HBV. Hepatitis B virus; the etiologic agent of serum hepatitis

Hemolysin (he-moll′-uh-sin). A bacterial enzyme capable of lysing erythrocytes, causing release of their hemoglobin

Hemolysis (he-moll′-uh-sis). Destruction of red blood cells (erythrocytes) in such a manner that hemoglobin is liberated into the surrounding environment

Heterotroph (het′-er-oh-trofe). An organism that uses organic chemicals as a source of carbon; sometimes called an *organotroph*

Histamine (his′-tuh-meen). Potent chemical released from cells (*e.g.,* basophils and mast cells) during some immune reactions; causes constriction of bronchial smooth muscles and vasodilation

Histiocyte (his′-tee-oh-site) or **histocyte** (his-toh-site). A fixed macrophage; *i.e.,* one that remains in tissue and does not wander

HIV. Human immunodeficiency virus; the etiologic agent of AIDS

Hospital-acquired infection. See *nosocomial infection*

Host. The organism on or in which a parasite lives

Hyaluronic (high′-uh-lu-ron′-ick) **acid.** A gelatinous, mucopolysaccharide that acts as an intracellular cement in body tissue

Hyaluronidase (high′-uh-lu-ron′-uh-dase). A bacterial enzyme that breaks down hyaluronic acid; sometimes called diffusing or spreading factor, because it enables bacteria to invade deeper into tissue

Hybridoma (high-brid-oh′-muh). A tumor produced in vitro by fusion of mouse tumor cells and specific-antibody producing cells; used in the production of monoclonal antibodies

Hydrocarbon (high-droh-kar′-bun). An organic compound consisting of only hydrogen and carbon atoms

Hydrogen (high′-droh-jen) **bond.** A bond arising from the sharing of a hydrogen atom; most commonly, where a hydrogen atom links a nitrogen atom to an oxygen atom or to another nitrogen atom

Hydrolysis (hi-drol′-eh-sis). A chemical process whereby a compound is cleaved into two or more simpler compounds with the uptake of the H and OH parts of a water molecule on either side of the chemical bond that is cleaved

Hypersensitivity (high′-per-sen-suh-tiv′-uh-tee). A condition in which there is an exaggerated immune response cell-mediated immunological reaction that causes tissue destruction or inflammation

Hypertonic (hi-per-tahn′-ick) **solution.** A solution having a greater osmotic pressure than cells placed into that solution; a higher concentration of solutes outside the cell

Hyphae (hy′-fee), *sing.* hypha. Long, branching, intertwining, cytoplasmic filaments of molds

Hypogammaglobulinemia (high′-poh-gam′-uh-glob-yu-luh-nee′-me-uh). Decreased quantity of the gamma fraction of serum globulin, including a decreased quantity of immunoglobulins

Hypotonic (hi-poh-tahn′-ick) **solution.** A solution having a lower osmotic pressure than cells placed into that solution; a lower concentration of solutes outside the cell

Iatrogenic (eye-at-roh-jen′-ick) **infection.** An infection induced by the treatment itself; literally, physician induced

Immune (im-yun′). Free from the possibility of acquiring a particular infectious disease

Immunity (im-u′-nuh-tee). Being immune or resistant

Immunocompetent (im′-u-noh-kom′-puh-tent). Able to produce a normal immune response

Immunodiagnostic (im′-u-noh-dy-ag-nos′-tick) **procedures.** Diagnostic test procedures that utilize the principles of immunology; used to detect either antigen or antibody in patients' specimens

Immunogen (im-u′-noh-jen). Another name for antigen

Immunoglobulin (im′-u-noh-glob′-yu-lin). A class of proteins, consisting of two light polypeptide chains and two heavy chains; all antibodies are immunoglobulins, but some immunoglobulins are not antibodies

Immunology (im-u-noll′-oh-gee). The science that deals with immunity from disease and the immune process

Immunosuppression (im′-u-no-sue-presh′-un). A condition in which a person is unable to mount a normal immune response due to suppression or depression of their immune system

In vitro (in vee′-trow). In an artificial environment, as in a laboratory setting; often used in reference to what occurs *outside* an organism

In vivo (in vee′-voh). In a living organism; used in reference to what occurs *within* a living organism

Inclusion bodies. Distinctive structures frequently formed in the nucleus and/or cytoplasm of cells infected with certain viruses

Incubatory (in′-kyu-buh-tor′-ee) **carrier.** A person capable of transmitting a pathogen during the incubation period of a particular infectious disease

Indigenous microflora (in-dij′-uh-nus my-crow-floor-uh). Microorganisms that live on and in the healthy body; also called *indigenous microbiota;* referred to in the past as normal flora

Infection (in-fek′-shun). The presence and multiplication of pathogens in or on the body, sometimes used as a synonym for infectious disease

Infectious disease (in-fek′-shus dee-zeez′). Any disease caused by a microorganism

Infestation (in-fes-tay′-shun). The presence of ectoparasites (*e.g.,* lice) on the body

Inflammation (in-fluh-may′-shun). A nonspecific pathologic process consisting of a dynamic complex of cytologic and histologic reactions that occur in response to an injury or abnormal stimulation by a physical, chemical, or biologic agent

Inorganic (in-or-gan′-ick) **chemistry.** The science concerned with compounds not involving covalent bonds

Inorganic compounds. Chemical compounds in which the atoms or radicals are held together by electrostatic forces rather than by covalent bonds

Interferon (in-ter-fear′-on). A class of small, antiviral glycoproteins, produced by cells infected with an animal virus; cell-specific and species-specific, but not virus-specific

Interleukins (in-ter-lu′-kinz). Lymphokines and polypeptide hormones; interleukin-1 is produced by monocytes; interleukin-2 is produced by lymphocytes

Ion (eye′-on). A positively or negatively charged atom or group of atoms

Ionic (eye-on′-ick) **bond.** An electrostatic bond

Isotonic (eye-soh-tahn′-ick) **solution.** A solution having the same osmotic pressure as cells placed into that solution; a concentration of solutes outside the cell equal to the concentration of solutes inside the cell

Isotopes (eye′-so-topes). Atoms that are chemically identical, but their nuclei contain different numbers of neutrons; for example, ^{12}C and ^{14}C are isotopes of carbon; ^{14}C has two more neutrons than ^{12}C

Killer cell. A type of cytotoxic T-cell involved in cell-mediated immune responses

Kinase (ky′-nace). As used in this book, a bacterial enzyme capable of dissolving clots; also known as *fibrinolysin*

Lag phase. That part of a bacterial growth curve during which multiplication of the organisms is very slow or scarcely appreciable; the first phase in a bacterial growth curve

Latent infection. An asymptomatic infection capable of manifesting symptoms under particular circumstances or if activated

Lecithin (less′-uh-thin). A name given to several types of phospholipids that are essential constituents of animal and plant cells

Lecithinase (less′-uh-thuh-nace). A bacterial enzyme capable of breaking down lecithin

Leukocidin (lu-koh-sigh′-din). A bacterial enzyme capable of destroying leukocytes

Leukocytes (lu′-koh-sites). White blood cells

Leukocytosis (lu′-koh-sigh-toe′-sis). An increased number of leukocytes in the blood

Leukopenia (lu-koh-pea′-nee-uh). A decreased number of leukocytes in the blood.

Light microscope. A type of microscope that uses visible light as a source of illumination; also called a brightfield microscope

Lipids (lip′-ids). Organic compounds containing carbon, hydrogen, and oxygen that are insoluble in water but soluble in so-called fat solvents such as diethyl ether and carbon tetrachloride

Lipopolysaccharide (lip′-oh-pol-ee-sack′-a-ride). A macromolecule of combined lipid and polysaccharide, found in the cell walls of gram-negative bacteria

Lithotroph (lith′-oh-trofe). An organism that uses inorganic molecules as a source of carbon

Local infection. An infection that remains localized; that does not spread; also known as a *focal infection*

Logarithmic (log′-uh-rith-mik) **growth phase.** That part of a bacterial growth phase during which maximal multiplication is occurring by geometrical progression; plotting the logarithm (log) of the number of organisms against time produces a straight upward-pointing line; the second phase in a bacterial growth curve; also known as the log phase or exponential growth phase

Logarithmic scale. A scale (as on graph paper) in which the values of a variable (*e.g.*, number of organisms at a particular point in time) are expressed as logarithms

Lophotrichous (low-fot′-ri-kus) **bacteria.** Bacteria possessing two or more flagella at one or both ends (poles) of the cell

Lymphadenitis (lim′-fad-uh-ny′-tis). Inflammation of a lymph node or lymph nodes

Lymphadenopathy (lim-fad-un-nop′-uh-thee). A disease process affecting a lymph node or lymph nodes

Lymphocytosis (lim′-foh-sigh-toe′-sis). An increased number of lymphocytes in the blood

Lymphokines (lim′-foh-kinz). Soluble protein mediators released by sensitized lymphocytes; examples include chemotactic factors and interleukins; lymphokines represent one category of *cytokines*

Lyophilization (ly-ahf′-eh-leh-zay′-shun). Freeze-drying; a method of preserving microorganisms and foods

Lysogenic (lye-so-jen′-ick) **bacteria.** Pertaining to bacteria in the state of lysogeny

Lysogenic conversion. Alteration of the genetic constitution of a bacterial cell due to the integration of viral genetic material into the host cell genome

Lysogenic cycle. When viral genetic material remains latent or inactive in an infected host cell

Lysogeny (lye-soj′-eh-nee). When viral genetic material is integrated into the genome of the host cell

Lysosomes (lye′-so-somz). Membrane-bound vesicles found in the cytoplasm of eucaryotic cells, containing a variety of digestive enzymes, including lysozyme

Lysozyme (lye′-so-zime). A digestive enzyme found in lysozymes, tears, and other body fluids; especially destructive to bacterial cell walls

Lytic cycle. When a virus takes over the metabolic machinery of the host cell, reproduces itself, and ruptures (lyses) the host cell to allow the newly assembled virions to escape

Macrophage (mak′-roh-faj). A large phagocytic cell that arises from a monocyte

Malaise (muh-laz′). A generalized feeling of discomfort or uneasiness

Mast cell. A tissue basophil

Mastigophora (mas′-ti-gof′-uh-rah). A subphylum of Protozoa in the phylum Sarcomastigophora; the flagellates; considered a phylum in some classification schemes

Mastitis (mass-ty′-tis). Inflammation of the breast

Mechanical vector. A vector that merely transports a pathogen to a susceptible host, but within with the pathogen neither multiplies nor matures

Medical asepsis (ay-sep′-sis). The absence of pathogens in a patient's environment

Medical aseptic (ay-sep′-tick) **technique.** Procedures followed and steps taken to ensure medical asepsis

Meninges (muh-nin′-jez), *sing.* meninx. As used in this book, the membranes that surround the brain and spinal cord

Meningitis (men-in-ji′-tis). Inflammation or infection of the meninges

Meningoencephalitis (muh-ning′-go-en-sef-uh-ly′-tis). Inflammation or infection of the brain and its surrounding membranes

Mesophile (meez′-oh-file). A microorganism having an optimum growth temperature between 25° C and 40° C; such an organism is said to be *mesophilic*

Mesosome (me′-so-zom). A procaryotic cell organelle (an infolding of the cytoplasmic membrane) possibly involved in cellular respiration

Messenger RNA (mRNA). The type of RNA that contains the exact same genetic information as a single gene on a DNA molecule; also called informational RNA

Metabolism (muh-tab′-oh-lizm). The sum of all the chemical reactions occurring in a cell; consists of *anabolism* and *catabolism*

Metabolite (muh-tab′-oh-lite). Any product of metabolism

Microbial (my-krow′-be-ul). Pertaining to microorganisms

Microbial antagonism. The killing, injury, or inhibition of one microbe by the substances produced by another

Microbicidal (my-krow′-buh-sigh′-dull) **agent.** A chemical or drug that kills microorganisms; a *microbicide*

Microbiologist (my′-crow-by-ol′-oh-jist). One who studies or works with microorganisms

Microbiology (my′-crow-by-ol′-oh-gee). That branch of science concerned with the study of microorganisms

Microbistatic (my-krow′-buh-stat′-ick) **agent.** A chemical agent or drug that inhibits the growth of microorganisms

Micrometer (my′-crow-me-ter). A unit of length, equal to one-millionth of a meter

Microorganisms (my′-crow-or′-gan-izms). Very small organisms; usually microscopic; usually single cells; also called microbes; includes algae, bacteria, fungi, protozoa, and viruses

Microscope (my′-crow-skope). An optical instrument that permits one to observe a small object by producing an enlarged image of the object

Microtubules (my-kro′-two-bules). Cylindrical, cytoplasmic tubules found in the cytoskeleton of eucaryotic cells; may be related to the movement of chromosomes during nuclear division

Mitochondria (my-toe-kon′-dree-uh), *sing.* mitochrondrion. Eucaryotic organelles involved in cellular respiration for the production of energy; energy factories of the cell

Mitosis (my-toe′-sis). A process of cell reproduction consisting of a sequence of modifications of the nucleus that result in the formation of two daughter cells with exactly the same chromosome and DNA content as that of the original cell

Molecule (mol′-ee-kyul). The smallest possible quantity of a di-, tri-, or polyatomic substance that retains the chemical properties of the substance

Monoclonal (mon-oh-klo′-nul) **antibodies.** Antibodies produced by a clone or genetically identical hybrid cells

Monocyte (mon′-oh-site). A relatively large mononuclear leukocyte

Monokines (mon′-oh-kinz). Soluble protein mediators released by sensitized monocytes and macrophages; monokines represent one category of *cytokines*

Monosaccharides (mon-oh-sak′-uh-rides). Carbohydrates that cannot be broken down into any simpler sugar by simple hydrolysis; simple sugars containing three to seven carbon atoms; the basic units of polysaccharides

Monotrichous (mah-not′-ri-kus) **bacteria.** Bacteria possessing only one flagellum

Motile (mow′-till). Possessing the ability to move

Mutagen (mew′-tah-jen). Any agent that can cause a mutation (*e.g.*, radioactive substances, x-rays, or certain chemicals); such an agent is said to be mutagenic

Mutant (mew′-tant). A phenotype in which a mutation is manifested

Mutation (mew-tay′-shun). An inheritable change in the character of a gene; a change in the sequence of base pairs in a DNA molecule

Mutualism (mew′-chew-ul-izm). A symbiotic relationship in which both parties derive benefit

Mycelium (my-see′-lee-um), *pl.* mycelia. A fungal colony; composed of a mass of intertwined hyphae

Mycologist (my-kol′-oh-jist). One who studies or works with fungi

Mycology (my-kol′-oh-gee). That branch of science concerned with the study of fungi

Mycosis (my-ko′-sis), *pl.* mycoses. A fungal disease

Myelitis (my-uh-ly′-tis). Inflammation or infection of the spinal cord

Myocarditis (my′-oh-kar-dy′-tis). Inflammation of the myocardium (the muscular walls of the heart)

Nanometer (nan′-oh-me-ter). A unit of length, equal to one-billionth of a meter

Natural killer cell. A type of cytotoxic human blood lymphocyte

Necrotoxin (nek′-roh-tok′-sin). An exotoxin that destroys cells

Negative stain. A staining procedure whereby unstained objects can be seen against a stained background

Nephritis (nef-ry′-tis). Inflammation of the kidneys

Neurotoxin (new′-roh-tok′-sin). A bacterial toxin that attacks the nervous system

Neutralism (new′-trul-izm). A symbiotic relationship in which organisms occupy the same niche but do not affect one another

Neutrophil (new′-tro-fil). A type of granulocyte found in blood; its granules contain neutral substances that attract neither acidic nor basic dyes; also called a *polymorphonuclear cell, "poly,"* or *PMN*

Nonendemic (non′-en-dem′-ic) **disease.** A disease that is not always present (not endemic) in a particular community or region

Nonpathogen (non′-path′-oh-jen). A microorganism that does not cause disease; such an organism is said to be *nonpathogenic*

Nosocomial (nose-oh-koh′-me-ul) **infection.** Any infection acquired while one is hospitalized; a hospital-acquired infection

Nuclear (new′-klee-er) **membrane.** The membrane that surrounds the chromosomes and nucleoplasm of a eucaryotic cell

Nucleic (new-klay′-ick) **acids.** Macromolecules consisting of linear chains of nucleotides; DNA, mRNA, tRNA, and rRNA are examples

Nucleolus (new-klee′-oh-lus). A dense portion of the nucleus, where ribosomal RNA (rRNA) is produced

Nucleoplasm (new′-klee-oh-plazm). That portion of a cell's protoplasm that lies within the nucleus

Nucleotides (new′-klee-oh-tides). The basic units or building blocks of nucleic acids, each consisting of a purine or pyrimidine combined with a pentose (ribose or deoxyribose) and a phosphate group

Nucleus (new′-klee-us), *pl.* nuclei. That portion of a eucaryotic cell that contains the nucleoplasm, chromosomes, and nucleoli

Obligate aerobe (air′-obe). An organism that requires 20–21% oxygen (the amount found in the air we breathe) to survive

Obligate anaerobe (an′-air-obe). An organism that cannot survive in oxygen

Oncogenic (ong-koh-jen′-ick). Capable of causing cancer

Oophoritis (oh-of-or-eye′-tis). Inflammation or infection on an ovary

Opportunist (op-poor-tune′-ist). A microbe with the potential to cause disease, but does not do so under ordinary circumstances; may cause disease in susceptible persons with lowered resistance; also called an opportunistic pathogen

Opsonin (op′-soh-nin). A substance (such as an antibody or complement component) that enhances phagocytosis

Opsonization (op′-suh-nuh-zay′-shun). The process by which bacteria are altered to be more readily and more efficiently engulfed by phagocytes; often involves coating the bacteria with antibodies and/or complement components

Orchitis (or-ky'-tis). Inflammation or infection of the testes

Organelles (or'-guh-nelz). General term for the various and diverse structures contained within a cell (*e.g.,* mitochondria, Golgi complex, nucleus, endoplasmic reticulum, and lysosomes)

Organic (or-gan'-ick) **chemistry.** The science concerned with covalently linked atoms, centering around carbon compounds of this type

Organic compounds. Chemical compounds composed of atoms held together by covalent bonds

Osmosis (oz-moh'-sis). The process by which a solvent (*e.g.,* water) moves through a semipermeable membrane from a solution having a lower concentration of solutes (dissolved substances) to a solution having a higher concentration of solutes

Osmotic (oz-maht'-ick) **pressure.** A measure of the tendency for water to move into a solution by osmosis; always a positive value

Oxidation (ok-seh-day'-shun). As used in this book, increasing the valence of an atom by the loss of one or more electrons, thus, making the atom more electropositive

Oxidation-reduction reactions. Any chemical reaction which must, *in toto,* comprise both oxidation and reduction; sometimes referred to as redox reactions

Pandemic (pan-dem'-ick) **disease.** A disease in epidemic proportions worldwide

Parasite (par'-uh-sight). An organism that lives on or in another living organism (called the host)

Parasitism (par'-uh-suh-tizm). A symbiotic relationship in which one party (the parasite) derives benefit at the expense of the other party (the host)

Parenteral (puh-ren'-ter-ul) **route.** Literally, by some route other than the gastrointestinal tract; usually refers to the introduction of substances into the body by injection

Parotitis (par-oh-ty'-tis) Inflammation of the parotid gland (a salivary gland located near the ear); also known as *parotiditis*

Passive acquired immunity. Immunity or resistance acquired as a result of receipt of antibodies produced by another person or by an animal

Passive carrier. A person who harbors a particular pathogen without ever having had the infectious disease it causes

Pasteurization (pas'-tour-i-zay'-shun). A heating process that kills pathogens in milk, wines, and other beverages

Pathogen (path'-oh-jen). Disease-causing microorganism; such an organism is said to be *pathogenic*

Pathogenicity (path'-oh-juh-nis'-uh-tee). The ability to cause disease; a microbe capable of causing infection is a said to be *pathogenic*

Peptidoglycan (pep'-ti-doe-gly'-kan). The rigid component of bacterial cell walls, consisting of polysaccharide chains linked together by peptide chains; cell walls of gram-positive bacteria contain a thicker layer of peptidoglycan than cell walls of gram-negative bacteria

Pericarditis (per'-ee-kar-dy'-tis). Inflammation of the pericardium (the membrane or sac around the heart)

Periodontitis (purr'-ee-oh-don-ty'-tis). Inflammation or infection of the *periodontium* (tissues that surround and support the teeth)

Peritrichous (peh-rit'-ri-kus) **bacteria.** Bacteria possessing flagella over their entire surface

Petri (pea'-tree) **dish.** A shallow, circular container made of thin glass or clear plastic, with a loosely fitting, overlapping cover; used in microbiology laboratories for cultivation of microorganisms on solid media

Phagocyte (fag'-oh-site). A cell capable of ingesting bacteria, yeasts, and other particulate matter by phagocytosis; amebae and certain white blood cells are examples of phagocytic cells

Phagocytosis (fag'-oh-sigh-toe'-sis). Ingestion of particulate matter involving the use of pseudopodia to surround the matter

Phagolysosome (fag-oh-ly'-soh-sohm). A membrane-bound vesicle formed by the fusion of a phagosome and a lysosome

Phagosome (fag'-oh-sohm). A membrane-bound vesicle containing an ingested particle (*e.g.*, a bacterium); found in phagocytic cells

Phenol (fee'-nol) **coefficient test.** A laboratory procedure used to determine the effectiveness of a disinfectant compared to the effectiveness of phenol

Phenotype (fee'-no-type). Manifestation of a genotype; all of the attributes or characteristics of an individual

Photoautotroph (foh'-toe-aw'-toe-trofe). An organism that uses light as an energy source and carbon dioxide as a carbon source; a type of autotroph

Photoheterotroph (foh'-toe-het'-er-oh-trofe). An organism that uses light as an energy source and organic compounds as a carbon source; a type of heterotroph; sometimes referred to as a *photoorganotroph*

Photolithotroph (foh'-toe-lith'-oh-trofe). An organism that uses light as an energy source and inorganic chemicals as a carbon source; a type of lithotroph

Photosynthesis (foe-toe-sin'-thuh-sis). Production of organic substances, using light energy; a cell that produces organic substances in this manner is said to be *photosynthetic*

Phototroph (foh'-toe-trofe). An organism that uses light as an energy source

Phycologist (fy-kol'-oh-jist). One who studies or works with algae

Phycology (fy-kol'-oh-gee). That branch of science concerned with the study of algae

Pili (py′-ly), *sing.* pilus. Hairlike surface projections possessed by some bacteria (called piliated bacteria); most are organelles of attachment; also called *fimbriae;* specialized pili, called *sex pili,* are described below

Pinocytosis (pin′-oh-sigh-toe′-sis). A process resembling phagocytosis, but used to engulf and ingest liquids rather than solid matter

Plasma (plaz′-muh). The liquid portion of circulating blood

Plasma cell. An antibody-secreting cell produced from a stimulated B-cell

Plasmid (plaz′-mid). An extrachromosomal genetic element; a molecule of DNA that can stably function and replicate while physically separate from the bacterial chromosome

Plasmolysis (plaz-moll′-uh-sis). Cell shrinkage due to a loss of water from the cell's cytoplasm

Pleomorphism (plee-oh-more′-fizm). Existing in more than one form; also known as polymorphism; an organism that exhibits pleomorphism is said to be *pleomorphic*

Polymer (pol′-uh-mer). A large molecule consisting of repeated units; nucleic acids, polypeptides, and polysaccharides are examples

Polypeptide (pol-ee-pep′-tide). A protein consisting of more than three amino acids held together by peptide bonds

Polyribosomes (pol-ee-ry′-boh-somz). Two or more ribosomes connected by a molecule of messenger RNA (mRNA)

Polysaccharide (pol-ee-sack′-uh-ride). Carbohydrate consisting of many sugar units; glycogen, cellulose, and starch are examples

Precipitate (pre-sip′-uh-tate). As a noun, a solid material that separates out of a solution or suspension

Precipitation (pre-sip-uh-tay′-shun) **tests.** Laboratory procedures involving formation of a precipitate

Precipitin (pre-sip′-uh-tin). A type of precipitate formed by the combination of antigen and specific antibody

Primary disease. The initial disease; may predispose the patient to secondary disease(s)

Prion (pry′-on). An infectious agent consisting of at least one protein, but no demonstrable nucleic acid

Procaryotic (pro′-kar-ee-ah′-tick) **cell.** A cell lacking a true nucleus; organisms having such cells are referred to as *procaryotes*

Proctitis (prok-ty-tis). Inflammation of the mucous membrane of the rectum

Properdin (pro-pare′-din). A normal serum gamma globulin that participates in the complement pathway

Prophage (pro′-faj). A temperate bacteriophage mutant whose genome does not contain all of the normal components and cannot become fully infective, yet can replicate indefinitely in the bacterial genome; also known as a *defective bacteriophage*

Prophylactic (pro′-fuh-lak′-tick) **agent.** A drug used to prevent a disease

Prophylaxis (pro-fuh-lak′-sis). Prevention of a disease or a process that can lead to a disease; *e.g.*, taking antimalarial medication in a malarious area

Prostaglandin (pros-tuh-glan′-din). Physiologically active tissue substances that cause many effects, including vasodilation, vasoconstriction, and stimulation of smooth muscle

Prostatitis (pros-tuh-ty′-tis). Inflammation or infection of the prostate

Prostration (pros-tray′-shun). Marked loss of strength; the patient is lying flat (prostrate)

Proteins (pro′-teens). Macromolecules consisting of two, three, or more amino acids

Protoplasm (pro′-toe-plazm). The semifluid matter within living cells; cytoplasm and nucleoplasm are examples

Protoplast (pro′-toe-plast). A bacterial cell which has lost its cell wall; the bacterium loses its characteristic shape and becomes round

Protozoa (pro-toe-zoe′-uh), *sing.* protozoan. Eucaryotic microorganisms frequently found in water and soil; some are pathogens; usually unicellular

Protozoologist (pro′-toe-zoe-ol′-oh-jist). One who studies or works with protozoa

Protozoology (pro′-toe-zoe-ol′-oh-gee). That branch of science concerned with the study of protozoa

Pseudopodium (sue-doe-poh′-dee-um), *pl.* pseudopodia. A temporary extension of protoplasm that is extended by an ameba or white blood cell for locomotion or the engulfment of particulate matter; also called a pseudopod

Psychroduric (sigh-krow-dur′-ick). Able to endure very cold temperatures

Psychrophile (sigh′-krow-file). An organism that grows best at a low temperature (0° to 32° C), with optimum growth occurring at 15° to 20° C; such an organism is said to be *psychrotrophic*

Purine (pure′-een). A molecule found in certain nucleotides and, therefore, in nucleic acids; adenine and guanine are purines found in both DNA and RNA

Pus. A fluid product of inflammation, containing leukocytes and dead cells

Pyelonephritis (py′-uh-low-nef-ry′-tis). Inflammation of certain areas of the kidneys, most often the result of bacterial infection

Pyogenic (pie-oh-jen′-ick). Pus-producing; causing the production of pus

Pyrimidine (pie-rim′-uh-deen). A molecule found in certain nucleotides and, therefore, in nucleic acids; thymine and cytosine are pyrimidines found in DNA; cytosine and uracil are pyrimidines found in RNA

Pyrogen (pie′-roh-jen). An agent that causes a rise in body temperature; such an agent is said to be *pyrogenic*

Reduction (ree-duk′-shun). As used in this book, decreasing the valence of an atom by the gain of one or more electrons, thus, making the atom more electronegative

Reservoirs (rez'-ev-wars) **of infection.** Living or nonliving material in or on which a pathogen multiples and/or develops

Resident microflora. Members of the indigenous microflora which are more or less permanent

Resolving power. The ability of the eye or an optical instrument to distinguish detail, such as the separation of closely adjacent objects; also called *resolution*

Reticuloendothelial (ree-tick'-yu-loh-en-doh-thee'-lee-ul) **system (RES).** A collection of phagocytic cells that includes macrophages and cells that line the sinusoids of the spleen, lymph nodes, and bone marrow

Reverse isolation. When a patient is isolated to protect him or her from infection

Ribonucleic (ry-boe-new-klay'-ick) **acid (RNA).** A macromolecule of which there are three main types: messenger RNA (mRNA), ribosomal RNA (rRNA), and transfer RNA, (tRNA); found in all cells, but only in certain viruses (RNA viruses)

Ribosomal (rye-boh-so'-mul) **RNA (rRNA).** The type of RNA molecule found in ribosomes

Ribosomes (ry'-boh-soams). Organelles which are the sites of protein synthesis in both procaryotic and eucaryotic cells

RNA polymerase (poh-lim'-er-ace). The enzyme necessary for transcription (see transcription)

Salpingitis (sal-pin-jy'-tis). Inflammation of the fallopian tube or eustachian tube

Salt. A compound formed by the interaction of an acid and a base; ionizable in water; sodium chloride (NaCl) is an example

Sanitization (san'-uh-tuh-zay'-shun). The process of making something sanitary (healthful)

Saprophyte (sap'-row-fight). An organism that lives on dead or decaying organic matter; such an organism is said to be *saprophytic*

Sarcodina (sar'-ko-dy'-nah). A subphylum of protozoa in the phylum Sarcomastigophora; includes the amebae; considered a phylum in some classification schemes

Sarcomastigophora (sar'-ko-mass-ti-gof'-oh-rah). A phylum of protozoa of the subkingdom Protozoa, characterized by flagella, pseudopodia, or both; contains the subphyla Sarcodina and Mastigophora

Sebum (see'-bum). The oily secretion produced by sebaceous glands of the skin

Secondary disease. A disease that follows the initial disease

Selective media. Culture media which allow a certain organism or group of organisms to grow while inhibiting growth of all other organisms

Selective permeability. An attribute of membranes whereby only certain substances are able to cross the membranes

Septicemia (sep-tuh-see'-me-uh). A disease consisting of chills, fever, prostration, and the presence of bacteria and/or their toxins in the blood

Serologic (ser-oh-loj'-ick) **procedures.** Immunodiagnostic test procedures performed using serum

Serology (suh-rol'-oh-jee). That branch of science concerned with serum and serologic procedures

Serum (seer'-um), *pl.* sera. The liquid portion of blood following coagulation (clotting)

Sex pilus. A specialized pilus through which one bacterial cell (the donor cell) transfers genetic material to another bacterial cell (the recipient cell), in a process called *conjugation*

Slime layer. A non-organized, non-attached layer of glycocalyx surrounding a bacterial cell

Solute (sol'-yute). The dissolved substance in a solution; for example, sucrose (table sugar) dissolved in water

Solution (soh-loo'-shun). A homogenous molecular mixture; generally, a substance dissolved in water (an aqueous solution)

Solvent (sol'-vent). A liquid in which another substance dissolves

Source isolation. When a patient is isolated to protect other persons from infection

Species (spe'-shez), *pl.* species. A specific member of a given genus (*e.g., Escherichia coli* is a species in the genus *Escherichia*); the name of a particular species consists of two parts—the generic name (the first name) and the specific epithet (the second name); singular species is abbreviated sp., while plural species is abbreviated spp.

Specific epithet. The second part (second name) in the name of a species

Spirochetes (spy'-roh-keets). Spiral-shaped bacteria; (*e.g., Treponema pallidum*), the etiologic agent of syphilis

Sporadic (spoh-rad'-ick) **disease.** A disease that occurs occasionally, usually affecting one person; neither endemic nor epidemic

Sporicidal (spor-uh-sigh'-dull) **agent.** A chemical agent that kills spores; a *sporicide*

Sporozoea (spor-oh-zoh'-ee-uh). A large class of protozoa, containing organisms that do not move by cilia, flagella, or pseudopodia; includes the malarial parasites; considered a phylum in some classification schemes

Sporulation (spor'-you-lay'-shun). Production of one or more spores

Staphylococci (staff'-eh-low-kok'-sigh). Pertaining to the genus *Staphylococcus*, cocci arranged in clusters

Staphylokinase (staf'-uh-low-ky'-nace). A kinase produced by *Staphylococcus aureus*

Stationary phase. That part of a bacterial growth phase during which organisms are dying at the same rate at which new organisms are being produced; the third phase in a bacterial growth curve

STD. Sexually transmitted disease
Sterile (stir′-ill). Free of all living microorganisms, including spores
Sterilization (stir′-uh-luh-zay′-shun). The destruction of *all* microorganisms in or on something (*e.g.*, in water or on surgical instruments)
Streptococci (strep′-toh-kok′-sigh). Pertaining to the genus *Streptococcus*, cocci arranged in chains of varying lengths
Streptokinase (strep′-toh-ky′-nace). A kinase produced by streptococci
Substrate (sub′-strayt). The substance that is acted upon or changed by an enzyme
Superinfection (sue′-per-in-fek′-shun). An overgrowth of one or more particular microorganisms; often microorganisms that are resistant to an antimicrobial agent that a patient is receiving
Surgical asepsis. The absence of microorganisms in a surgical environment (*e.g.*, an operating room)
Surgical aseptic technique. Procedures followed and steps taken to ensure surgical asepsis
Symbiont (sim′-bee-ont). One of the parties in a symbiotic relationship
Symbiosis (sim-bee-oh′-sis). The living together or close association of two dissimilar organisms
Synergism (sin′-er-jizm). As used in this book, the correlated action of two or more microorganisms so that the combined action is greater than that of each acting separately (*e.g.*, when two microbes accomplish more than either could do alone)
Systemic infection. See *generalized infection*

T-cell. T lymphocyte; a type of leukocyte that plays important roles in the immune system
Taxonomy (tak-sawn′-oh-me). The systematic classification of living things
Teichoic (tie-ko′-ick) **acids.** Polymers found in the cell walls of gram-positive bacteria
Temperate bacteriophage. A bacteriophage whose genome incorporates into and replicates with the genome of the host bacterium
Tetanospasmin (tet′-uh-noh-spaz′-min). The neurotoxin produced by *Clostridium tetani;* causes tetanus
Thermal death point (TDP). The temperature required to kill all microorganisms in a liquid culture in 10 minutes at pH 7
Thermal death time (TDT). The length of time required to kill all microorganisms in a liquid culture at a given temperature
Thermoduric (ther-mow-du′-rik). Able to survive high temperatures
Thermophile (ther′-mow-file). An organism that thrives at a temperature of 50° C or higher; such an organism is said to be *thermophilic*
Tinea (tin′-ee-uh) **infections.** Fungal infections of the skin, hair, and nails (ringworm); named for the part of the body that is affected (*e.g.*, tinea

capitis is a fungal infection of the scalp, tinea pedis is athlete's foot, tinea unguium is a fungal infection of the nails)

Toxemia (tok-see′-me-uh). The presence of toxins in the blood

Toxigenicity (tok′-suh-juh-nis′-uh-tee) or **toxinogenicity** (tok′-suh-no-juh-nis′-uh-tee). The capacity to produce toxin; a microorganism capable of producing a toxin is said to be *toxigenic* (or toxinogenic)

Toxin (tok′-sin). As used in this book, a poisonous substance produced by a microorganism

Toxoid (tok′-soyd). A toxin that has been altered in such a way as to destroy its toxicity but retain its antigenicity; toxoids are used as vaccines

Transcription (tran-skrip′-shun). Transfer of the genetic code from one type of nucleic acid to another; usually, the synthesis of an mRNA molecule from a DNA template

Transduction (trans-duk′-shun). Transfer of genetic material (and its phenotypic expression) from one bacterial cell to another via bacteriophages; in *general transduction,* the transducing bacteriophage is able to transfer any gene of the donor bacterium; in *specialized transduction,* the bacteriophage is able to transfer only one or some of the donor bacterium's genes

Transfer RNA (tRNA). The type of RNA molecule that is capable of combining with (and thus activating) a specific amino acid; involved in protein synthesis (translation); the anticodon on a tRNA molecule recognizes the codon on an mRNA molecule

Transformation (trans-for-may′-shun). In microbial genetics, transfer of genetic information between bacteria via uptake of naked DNA; bacteria capable of taking up naked DNA are said to be competent

Transient microflora. Temporary members of the indigenous microflora

Translation (trans-lay′-shun). The process by which mRNA, tRNA, and ribosomes effect the production of proteins from amino acids; protein synthesis

Tripeptide (try-pep′-tide). A protein consisting of three amino acids held together by peptide bonds

Tuberculocidal (too-bur′-kyu-low-sigh′-dull) **agent.** A chemical or drug that kills the bacteria that cause tuberculosis (*Mycobacterium tuberculosis*); a *tuberculocide*

Tyndallization (tin-dull-uh-zay′-shun). A process of boiling and cooling in which spores are allowed to germinate and then are killed by boiling again

Universal precautions. Safety precautions taken by health care workers to protect themselves from infection with HBV or HIV; the same precautions are taken for *all* patients and *all* patient specimens (body substances); also known as *universal body substance precautions*

Ureteritis (you-ree-ter-eye′-tis). Inflammation or infection of a ureter
Urethritis (you-ree-thry′-tis). Inflammation or infection of the urethra

Vacuoles (vak′-you-oles). Membrane-bound storage spaces in the cell
Variolation (var′-e-oh-lay′-shun). The obsolete process of inoculating a susceptible person with material obtained from a smallpox vesicle; also known as variolization
Vasoconstriction (vay′-so-kon-strik′-shun). Narrowing of blood vessels
Vasodilation (vay′-soh-die-lay′-shun). An increase in the diameter of blood vessels
Vector (vek′-tour). An invertebrate animal (*e.g.*, tick, mite, mosquito, flea) capable of transmitting pathogens among vertebrates
Virion (veer′-ee-on). A complete, infectious viral particle
Viroid (vi′-royd). An infectious agent of plants that consists of nucleic acid, but no protein; smaller than a virus
Virologist (vi-rol′-oh-jist). One who studies or works with viruses
Virology (vi-rol′-oh-gee). That branch of science concerned with the study of viruses
Virucidal (vi-ruh-sigh′-dull) **agent.** A chemical or drug that kills viruses; a virucide
Virulence (veer′-u-lenz). A measure of pathogenicity; *i.e.*, some pathogens are more *virulent* than others
Virulence factor. An attribute or property of a microorganism that contributes to its virulence or pathogenicity
Virulent (veer′-yu-lent). An adjective used to denote a degree of pathogenicity; *e.g.*, some pathogens are more or less virulent than others
Virulent bacteriophage. A bacteriophage that regularly causes lysis of the bacteria it infects
Viruses (vi′-rus-ez), *sing.* virus. Acellular microorganisms that are smaller than bacteria; intracellular parasites

Wandering macrophages. Macrophages that migrate in the bloodstream and tissues; sometimes called *free macrophages*

Zoonosis (zoh-oh-no′-sis), *pl.* zoonoses. An infectious disease or infestation transmissible from animals to humans

INDEX

Note: Page numbers followed by an *f* indicate figures; those followed by a *t* indicate tables. The abbreviation *pl.* is used for color plates.

H

I